Dynamical System and Stochastic Analysis

Dynamical System and Stochastic Analysis

Jun Huang
Yueyuan Zhang

MDPI

Basel • Beijing • Wuhan • Barcelona • Belgrade • Novi Sad • Cluj • Manchester

Editors

Jun Huang
School of Mechanical and
Electrical Engineering
Soochow University
Suzhou
China

Yueyuan Zhang
School of Mechanical and
Electrical Engineering
Soochow University
Suzhou
China

Editorial Office
MDPI AG
Grosspeteranlage 5
4052 Basel, Switzerland

This is a reprint of articles from the Special Issue published online in the open access journal *Mathematics* (ISSN 2227-7390) (available at: www.mdpi.com/journal/mathematics/special_issues/control_systemstoana).

For citation purposes, cite each article independently as indicated on the article page online and as indicated below:

Lastname, A.A.; Lastname, B.B. Article Title. *Journal Name* **Year**, *Volume Number*, Page Range.

ISBN 978-3-7258-2518-9 (Hbk)
ISBN 978-3-7258-2517-2 (PDF)
doi.org/10.3390/books978-3-7258-2517-2

Contents

Yassine Sabbar, Mohammad Izadi, Aeshah A. Raezah and Waleed Adel
Nonlinear Dynamics of a General Stochastic SIR Model with Behavioral and Physical Changes: Analysis and Application to Zoonotic Tuberculosis
Reprinted from: *Mathematics* **2024**, *12*, 1974, doi:10.3390/math12131974 . . . . **1**

Ghada AlNemer, Mohamed Hosny, Ramalingam Udhayakumar and Ahmed M. Elshenhab
Existence and Hyers–Ulam Stability of Stochastic Delay Systems Governed by the Rosenblatt Process
Reprinted from: *Mathematics* **2024**, *12*, 1729, doi:10.3390/math12111729 . . . . **18**

Junchao Zhang, Jun Huang and Changjie Li
Distributed Interval Observers with Switching Topology Design for Cyber-Physical Systems
Reprinted from: *Mathematics* **2024**, *12*, 163, doi:10.3390/math12010163 . . . . **33**

Lanai Huang, Xudong Zhao, Fengyu Lin and Junfeng Zhang
Combination of Functional and Disturbance Observer for Positive Systems with Disturbances
Reprinted from: *Mathematics* **2022**, *11*, 200, doi:10.3390/math11010200 . . . . **47**

Dongya Li, Xiaoping Zhang, Shuang Wang and Fengxiang You
Robust Synchronization of Fractional-Order Chaotic System Subject to Disturbances
Reprinted from: *Mathematics* **2022**, *10*, 4639, doi:10.3390/math10244639 . . . . **60**

Jing Xu and Jun Huang
An Overview of Recent Advances in the Event-Triggered Consensus of Multi-Agent Systems with Actuator Saturations
Reprinted from: *Mathematics* **2022**, *10*, 3879, doi:10.3390/math10203879 . . . . **75**

Sorin Lugojan, Loredana Ciurdariu and Eugenia Grecu
Another Case of Degenerated Discrete Chenciner Dynamic System and Economics
Reprinted from: *Mathematics* **2022**, *10*, 3782, doi:10.3390/math10203782 . . . . **97**

Xiaoxue Li, Xiaorong Hou, Jing Yang and Min Luo
Stability and Stabilization of 2D Linear Discrete Systems with Fractional Orders Based on the Discrimination System of Polynomials
Reprinted from: *Mathematics* **2022**, *10*, 1862, doi:10.3390/math10111862 . . . . **116**

Sorin Lugojan, Loredana Ciurdariu and Eugenia Grecu
Chenciner Bifurcation Presenting a Further Degree of Degeneration
Reprinted from: *Mathematics* **2022**, *10*, 1603, doi:10.3390/math10091603 . . . . **130**

Shitao Zhang, Peng Lin and Junfeng Zhang
Event-Triggered Asynchronous Filter of Nonlinear Switched Positive Systems with Output Quantization
Reprinted from: *Mathematics* **2022**, *10*, 599, doi:10.3390/math10040599 . . . . **147**

mathematics

Article

Nonlinear Dynamics of a General Stochastic SIR Model with Behavioral and Physical Changes: Analysis and Application to Zoonotic Tuberculosis

Yassine Sabbar [1], Mohammad Izadi [2,*], Aeshah A. Raezah [3] and Waleed Adel [4,5]

1 MAIS Laboratory, MAMCS Group, FST Errachidia, Moulay Ismail University of Meknes, P.O. Box 509, Errachidia 52000, Morocco; y.sabbar@umi.ac.ma
2 Department of Applied Mathematics, Faculty of Mathematics and Computer, Shahid Bahonar University of Kerman, Kerman 76169-14111, Iran
3 Department of Mathematics, Faculty of Science, King Khalid University, Abha 62529, Saudi Arabia; aalraezh@kku.edu.sa
4 Laboratoire Interdisciplinaire de l'Universite' Francaise d'Egypte (UFEID Lab), Universite' Francaise d'Egypte, Cairo 11837, Egypt; waleed.ouf@ufe.edu.eg
5 Department of Mathematics and Engineering Physics, Faculty of Engineering, Mansoura University, Mansoura 35511, Egypt
* Correspondence: izadi@uk.ac.ir

Abstract: This paper presents a comprehensive nonlinear analysis of an innovative stochastic epidemic model that accounts for both behavioral changes and physical discontinuities. Our research begins with the formulation of a perturbed model, integrating two general incidence functions and incorporating a Lévy measure to account for independent jump components. We start by confirming the well-posed nature of the model, ensuring its mathematical soundness and feasibility for further analysis. Following this, we establish a global threshold criterion that serves to distinguish between the eradication and the persistence of an epidemic. This threshold is crucial for understanding the long-term behavior of a disease within a population. To rigorously validate the accuracy of this threshold, we conducted extensive numerical simulations using estimated data on Zoonotic Tuberculosis in Morocco. These simulations provide practical insights and reinforce the theoretical findings of our study. A notable aspect of our approach is its significant advancement over previous works in the literature. Our model not only offers a more comprehensive framework but also identifies optimal conditions under which an epidemic can be controlled or eradicated.

Keywords: stochastic model; epidemic; behavioral change; jumps; tuberculosis

MSC: 34A12; 92D30; 37C10

Citation: Sabbar, Y.; Izadi, M.; Raezah, A.A.; Adel, W. Nonlinear Dynamics of a General Stochastic SIR Model with Behavioral and Physical Changes: Analysis and Application to Zoonotic Tuberculosis. *Mathematics* **2024**, *12*, 1974. https://doi.org/10.3390/math12131974

Academic Editor: Zhanybai T. Zhusubaliyev

Received: 4 June 2024
Revised: 21 June 2024
Accepted: 22 June 2024
Published: 26 June 2024

1. Introduction

During epidemics, the implementation of intervention measures is paramount in curbing the spread of infectious diseases and reducing their associated morbidity and mortality [1]. These measures encompass a spectrum of interventions ranging from pharmaceutical interventions like vaccination campaigns to non-pharmaceutical interventions such as social distancing measures and the promotion of hygienic practices. While the immediate goal of these interventions is to directly disrupt transmission chains, their indirect impact on the behavior of the susceptible population is equally significant [2]. Behavioral changes induced by intervention measures have emerged as a pivotal aspect in epidemic control strategies. The adoption of preventive behaviors, often prompted by the implementation of intervention measures, can significantly influence the trajectory of an outbreak [3]. For instance, the widespread adoption of mask-wearing during the COVID-19 pandemic and the consistent use of condoms to prevent sexually transmitted infections (STIs) exemplify the behavioral responses elicited by public health interventions.

Recently, in [4], the authors presented a new mathematical model aimed at simulating the dynamics of behavior changes within susceptible populations during epidemics. This innovative model distinguishes between two classes of non-infected individuals: the first comprises susceptible individuals who do not alter their behavior in response to intervention measures, while the second encompasses those who proactively change their behavior and adhere to the prescribed interventions. Central to this model is the recognition of the heterogeneous nature of human behavior in the face of epidemic threats. While some individuals may remain steadfast in their routines and habits, others may demonstrate a heightened awareness of the risks posed by the outbreak and willingly adopt preventive measures [5]. By categorizing susceptible individuals into distinct groups based on their behavioral responses, the model provides a nuanced understanding of how intervention measures influence population dynamics [6]. Key features include behavioral heterogeneity, incorporating fixed- and adaptive-behavior individuals and evaluating various intervention strategies such as public health campaigns, lockdowns, and vaccination drives to assess their impact on both groups [7]. The model dynamically simulates interactions between these groups and the infected population, considering the possibility of behavior change over time, and offers insights into how varying compliance levels with interventions affect epidemic control [8]. Real-world applications of this model include predicting outcomes for diseases like influenza and COVID-19, helping policymakers design targeted strategies to maximize compliance and effectiveness in mitigating infectious disease outbreaks [9].

In addition to incorporating heterogeneity in behavior changes, the authors of [4] also comprehensively addressed the impact of stochastic perturbations within their mathematical model. Stochastic perturbations refer to random fluctuations or disturbances that can influence the dynamics of epidemic transmission, introducing variability and unpredictability into the system. Therefore, the model under discussion is structured as follows:

$$\begin{cases} \mathrm{d}S_1(t) = \Big(r - \beta_1 S_1(t)I(t) - (u+c)S_1(t)\Big)\mathrm{d}t + \kappa_1 S_1(t)\mathrm{d}\rho_1(t), \\ \mathrm{d}S_2(t) = \Big(cS_1(t) - \beta_2 S_2(t)I(t) - uS_2(t)\Big)\mathrm{d}t + \kappa_2 S_2(t)\mathrm{d}\rho_2(t), \\ \mathrm{d}I(t) = \Big((\beta_1 S_1(t)I(t) + \beta_2 S_2(t)I(t) - (u+h_1+h_2)I(t)\Big)\mathrm{d}t + \kappa_3 I(t)\mathrm{d}\rho_3(t), \\ \mathrm{d}C(t) = \Big(h_1 I(t) - uC(t)\Big)\mathrm{d}t + \kappa_4 C(t)\mathrm{d}\rho_4(t). \end{cases} \tag{1}$$

According to model (1), the variables are defined as follows:

- S_1 denotes the susceptible individuals who maintain their behavior unchanged in response to the epidemic.
- S_2 represents the susceptible individuals who alter their behavior due to various interventions, such as media campaigns or governmental measures.
- I signifies the count of infected individuals within the population.
- C stands for the individuals who have recovered from the infection and acquired complete immunity.

In addition, the model parameters can be defined as follows:

- r represents the rate of population influx, encompassing births, immigration, or any other form of population input.
- β_1 and β_2 indicate the transmission rates of the epidemic, reflecting the likelihood of infection spread within the population.
- μ represents the natural death rate of the population.
- c signifies the rate at which susceptible individuals adjust their behavior and transition to the second class.
- h_1 denotes the recovery rate of infected individuals.
- h_2 represents the mortality rate attributed to infection.

In incorporating the stochastic component into the model, $(\rho_1(t), \rho_2(t), \rho_3(t), \rho_4(t))$ represents a four-dimensional Brownian motion with specific intensities κ_1, κ_2, κ_3, and

κ_4. A diagram depicting the flow dynamics among different classes is shown in Figure 1. In [4], the authors investigated the long-term characteristics of the solution by establishing adequate conditions for extinction and stationarity.

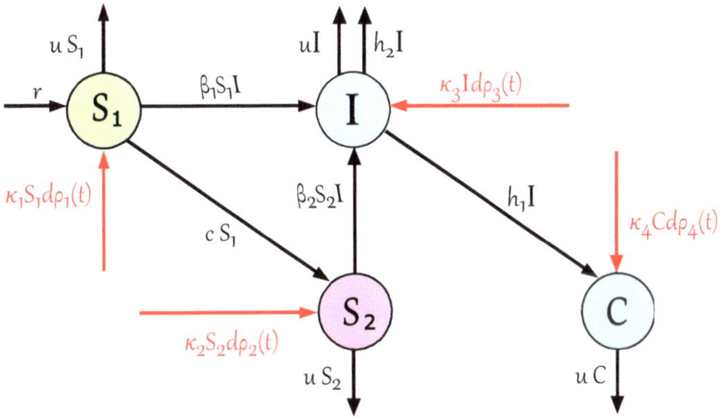

Figure 1. Schematic diagram illustrating the SIR model's behavioral change with white noises.

When discussing heterogeneity and variability, it is imperative to incorporate more realistic and generalized hypotheses. The study outlined in [4] has certain limitations as the authors employed a bilinear incidence rate and solely relied on white noise. However, in reality, interactions between individuals often deviate from the mass action principle, leading to nonlinear incidences that can assume various forms. In this study, we address this limitation by considering a broader spectrum of incidence rates that encompass diverse functions documented in the literature.

Stochastic epidemic models are crucial in biological mathematics because they capture the inherent randomness in disease transmission and progression, offering a realistic representation of epidemics. These models are particularly valuable in small populations where stochastic effects can significantly alter outcomes, explaining the variability and uncertainty in outbreak dynamics [10,11]. They incorporate complex biological processes and rare events, enhancing our understanding of disease behavior and the impact of interventions. In quantifying uncertainty and allowing for real-time data integration, stochastic models facilitate adaptive and robust epidemic management strategies. Additionally, they provide theoretical insights into disease dynamics and advance mathematical techniques, enriching the broader field of biological mathematics.

Generally, white noise fails to adequately simulate reality, particularly when interventions are implemented. Interventions can trigger abrupt jumps in the population, significantly altering the model dynamics [12]. Therefore, in this study, we extended our analysis to incorporate multidimensional jump processes, providing a more nuanced understanding of how interventions impact epidemic dynamics. By considering these factors, we aimed to enhance the realism and applicability of the model, enabling a more comprehensive exploration of epidemic dynamics in real-world scenarios [13,14]. Consequently, the system under consideration evolves through an interconnected perturbed formulation, presenting a holistic and sophisticated framework. The new formulation captures the complexities of sudden changes in population behaviors and disease transmission rates due to interventions such as vaccinations, quarantines, and public health policies. By integrating multidimensional jump processes, the model can account for both small fluctuations and significant shocks to the system, reflecting the stochastic nature of real-world epidemic events more accurately [15,16]. In particular, the inclusion of jump processes allows for the representation of non-continuous changes in state variables, which are critical for modeling events like sudden outbreaks or the rapid implementation of control measures. This approach provides a more flexible and detailed depiction of epidemic dynamics, accommodating the unpredictable and often abrupt nature of real-world events.

Technically, our newly introduced stochastic system diverges from conventional models by eschewing clearly delineated endemic or disease-free states [17]. Furthermore, our model exhibits independence in its stochastic component, adding a distinct dimension to the analysis. This independence characteristic enhances the depth and complexity of the analysis, providing unique insights into the system dynamics. Consequently, the conventional method of assessing disease persistence or extinction through the analysis of asymptotic behavior around these states falls short [18]. This underscores the imperative for an innovative approach grounded in stochastic analysis, which is adept at capturing the dynamic interactions among variables and uncertainties inherent in the system. This paper addresses the aforementioned issues and focuses on specific long-term properties of auxiliary equations. These properties play a crucial role in establishing the global threshold of our model.

The structure of this paper is organized as follows. Section 2 introduces the model formulation, accompanied by a detailed compilation of notations and hypotheses. In Section 3, we present the theoretical results, beginning with the verification of the model's well-posedness and outlining the global threshold that differentiates between epidemic extinction and persistence. Section 4 is devoted to conducting numerical simulations to rigorously validate our theoretical findings. Finally, we conclude with a summary of our key results and provide insights into potential future research directions in Section 5.

2. Model Formulation

In this section, we present the stochastic version of the behavioral model (1) with general incidence rates. The incorporation of stochastic components into the model is conducted by adding noise terms proportionally to each equation, ensuring that these stochastic perturbations maintain the independence property. This stochastic framework allows us to capture the inherent randomness and unpredictability in individual behaviors and disease transmission dynamics that deterministic models may overlook. By introducing stochasticity, we can better understand the variability and potential fluctuations in epidemic outcomes under different scenarios. Under this setting, we consider the following stochastic model:

$$\begin{cases} \mathrm{d}S_1(t) = \Big(r - f_1(S_1(t), I(t))I(t) - (u+c)S_1(t)\Big)\mathrm{d}t + \kappa_1 S_1(t)\mathrm{d}\rho_1(t) + \int_{\mathbb{R}^4\setminus\{0\}} z_1(\xi)S_1(t^-)\widetilde{\phi}_1(\mathrm{d}t,\mathrm{d}\xi), \\ \mathrm{d}S_2(t) = \Big(cS_1(t) - f_2(S_2(t), I(t))I(t) - uS_2(t)\Big)\mathrm{d}t + \kappa_2 S_2(t)\mathrm{d}\rho_2(t) + \int_{\mathbb{R}^4\setminus\{0\}} z_2(\xi)S_2(t^-)\widetilde{\phi}_2(\mathrm{d}t,\mathrm{d}\xi), \\ \mathrm{d}I(t) = \Big(\big(f_1(S_1(t), I(t)) + f_2(S_2(t), I(t))\big)I(t) - (u+h_1+h_2)I(t)\Big)\mathrm{d}t \\ \qquad + \kappa_3 I(t)\mathrm{d}\rho_3(t) + \int_{\mathbb{R}^4\setminus\{0\}} z_3(\xi)I(t^-)\widetilde{\phi}_3(\mathrm{d}t,\mathrm{d}\xi), \\ \mathrm{d}C(t) = \Big(h_1 I(t) - uC(t)\Big)\mathrm{d}t + \kappa_4 C(t)\mathrm{d}\rho_4(t) + \int_{\mathbb{R}^4\setminus\{0\}} z_4(\xi)C(t^-)\widetilde{\phi}_4(\mathrm{d}t,\mathrm{d}\xi). \end{cases} \tag{2}$$

where, f_1 and f_2 represent two general incidence rates that satisfy the following hypotheses:

- $(A0)$: The functions f_1 and f_2 are uniformly continuous and $\exists\ m,\ m_0 > 0$ such that

$$\begin{cases} m \geqslant \dfrac{\partial f_1}{\partial S_1}(S_1, I) \geqslant 0 \geqslant \dfrac{\partial f_1}{\partial I}(S_1, I) \geqslant -m_0, \\ m \geqslant \dfrac{\partial f_2}{\partial S_2}(S_2, I) \geqslant 0 \geqslant \dfrac{\partial f_2}{\partial I}(S_2, I) \geqslant -m_0. \end{cases} \tag{3}$$

The functions f_1 and f_2 represent two general incidence rates within the model, encapsulating the rates at which susceptible individuals become infected under varying conditions. The function f_1 corresponds to the incidence rate for the group of susceptible individuals who do not alter their behavior in response to intervention measures. This rate reflects the direct transmission dynamics between the non-compliant susceptible individuals and the infected population. Conversely, f_2 pertains to the incidence rate for the group of susceptible individuals who proactively change their behavior and adhere to intervention measures. This rate incorporates the effects of preventive actions, such as social distancing, mask-wearing, and other protective behaviors that reduce the probability of infection. In model (2), ρ_k $(k = 1, \cdots, 4)$ denote four mutually independent Brownian motions (BMs) of strengths $\kappa_k > 0$ $(k = 1, \cdots, 4)$ respectively. All these BMs are essentially defined on a filtered probability triple (stochastic basis) $\left(\Omega, \mathcal{F}_\Omega, (\mathcal{F}_{\{\Omega,t\}})_{t\geqslant 0}, \mathbb{P}\right)$ equipped with a filtration satisfying the usual criteria. ϕ_k $(k = 1, \cdots, 4)$ are four independent Poisson counting processes associated with four finite characteristic Lévy measures G_k $(k = 1, \cdots, 4)$. $\widetilde{\phi}_k$ $(k = 1, \cdots, 4)$ are four different compensated random measures such that $\widetilde{\phi}_k(\mathrm{d}t, \mathrm{d}\xi) = \phi_k(\mathrm{d}t, \mathrm{d}\xi) - G_k(\mathrm{d}\xi)\mathrm{d}t$. Finally, $z_k : \mathbb{R}^4 \setminus \{0\} \to \mathbb{R}$ are the jump size functions, which are postulated to be continuous on $\mathbb{R}^4 \setminus \{0\}$.

Prior to exploring the theoretical framework concerning our perturbed model represented by system (2), it is essential to introduce the following notations:

- $\alpha_1 := \max\limits_{k\in\{1,\cdots,4\}} \left\{ \int_{\mathbb{R}^4\setminus\{0\}} z_k^2(\xi) G_k(\mathrm{d}\xi) \right\}$.
- $\alpha_2 := \max\limits_{k\in\{1,\cdots,4\}} \left\{ \int_{\mathbb{R}^4\setminus\{0\}} \left\{ z_k(\xi) - \ln(1 + z_k(\xi)) \right\} G_k(\mathrm{d}\xi) \right\}$.
- $\alpha_3 := \max\limits_{k\in\{1,\cdots,4\}} \{\kappa_k^2\}$.
- $\alpha_4(\xi) := \max\limits_{k\in\{1,\cdots,4\}} \{z_k(\xi)\} = z_{\epsilon^*}(\xi)$, where ϵ^* indicates the index at which the maximum value is reached.
- $\alpha_5(\xi) := \min\limits_{k\in\{1,\cdots,4\}} \{z_k(\xi)\} = z_{\bar{\epsilon}}(\xi)$, where $\bar{\epsilon}$ indicates the index at which the minimum value is reached.
- $\alpha_6(\xi) := (1 + \alpha_4(\xi))^s - s \times \alpha_4(\xi) - 1$.
- $\alpha_7(\xi) := (1 + \alpha_5(\xi))^s - s \times \alpha_5(\xi) - 1$.
- $\alpha_8(\xi) := \max\{\alpha_6(\xi), \alpha_7(\xi)\}$.
- $\alpha_9 := \int_{\mathbb{R}^4\setminus\{0\}} \alpha_8(\xi) \mathbb{1}_{\{\alpha_6(\xi)\geq\alpha_7(\xi)\}} G_{\epsilon^*}(\mathrm{d}\xi) + \int_{\mathbb{R}^4\setminus\{0\}} \alpha_8(\xi) \mathbb{1}_{\{\alpha_7(\xi)>\alpha_6(\xi)\}} G_{\bar{\epsilon}}(\mathrm{d}\xi)$.

In addition, to maintain a meticulous balance between mathematical precision and biological relevance within the envisioned model, we present the following technical assumptions:

- *(A1)*: $z_k(\xi) + 1$ are positive $\forall k \in \{1, \cdots, 4\}$, and $\max\limits_{i\in\{1,2\}} \alpha_i < \infty$.
- *(A2)*: $\exists\, s > 2$ such that $\alpha_{10} = u - 0.5(s-1)\alpha_3 - s^{-1}\alpha_9 > 0$.

Remark 1. *Incidence rates account for the frequency and type of contacts between individuals, which are critical for disease transmission. Different diseases spread through different types of contact (e.g., respiratory droplets, direct physical contact), and general incidence rates help model these specific pathways accurately.*

Remark 2. *To illuminate the significance of (A2), it is important to reference the insights provided in [14] (Lemma 2.5). In this study, the authors elaborate on key perspectives that underpin the core findings of our research. By building upon the framework established in this lemma, we establish a robust link between the foundational concepts outlined therein and the overarching conclusions drawn from our investigation.*

Remark 3. *Differing from the results detailed in [13], our research introduces more sophisticated hypotheses and a nuanced framework. In particular, incorporating (A1) and (A2) enriches the accuracy and complexity of our analysis. Within this section, we operate under the hypothesis that all of the above assumptions are valid.*

3. Results

3.1. Evaluating the Mathematical Sufficiency of Stochastic System (2)

The main concern when examining an epidemic model is to ascertain if it has a unique and positive global solution over time. In this subsection, we will clarify specific conditions that ensure the existence of such a solution. We can represent the solution associated with initial data $s(0) = \big(S_1(0), S_2(0), I(0), C(0)\big)$ as $s(t) = \big(S_1(t), S_2(t), I(t), C(t)\big)$.

Theorem 1. *Under (A0) and (A1), we assert that for any initial data $s(0)$, there exists a unique solution $s(t)$ to system (2) for $t \geqslant 0$. This solution remains non-negative with near certainty for all time instances $t \geqslant 0$.*

Proof. In system (2), the coefficients involved demonstrate continuous differentiability within their defined domains, meeting the local Lipschitz criterion. As a result, for any initial solution $s(0)$ within the positive real four-dimensional space ($\mathbb{R}^4_+$), there exists a single maximal local solution $s(t)$ defined for t within the interval $(0, \gamma_e)$, where γ_e signifies the explosion time [19]. At this point, our aim is to ascertain the global characteristic of this solution, particularly to prove that $\gamma_e = \infty$ with near certainty. To achieve this, let us take into account a suitably large natural number $\beta_0 \in \mathbb{N}$ such that $s(0) \in [\beta_0^{-1}, \beta_0]$. For every integer $\beta \geqslant \beta_0$, we define the stopping time γ_k as follows:

$$\gamma_k = \inf\Big\{t \in (0, \gamma_e) \mid \min\big(S_1(t), S_2(t), I(t), C(t)\big) \leqslant \beta^{-1} \text{ or } \max\big(S_1(t), S_2(t), I(t), C(t)\big) \geqslant \beta\Big\}.$$

Let γ_∞ be defined as the limit of γ_β as β tends to infinity. It is clear that the sequence $(\gamma_\beta)_{\beta \geqslant \beta_0}$ is strictly increasing. Hence, the limit of γ_k as k tends to infinity equals the supremum of γ_β for $\beta \geqslant \beta_0$. According to the theory presented in [20], which states that the supremum of a sequence of stopping times is itself a stopping time, we conclude that γ_∞ is also a stopping time. Using the convention $\inf \emptyset = \infty$ throughout this paper, we can assert straightforwardly that $\gamma_\infty \leqslant \gamma_e$ almost surely. Therefore, establishing $\gamma_e = \infty$ almost surely directly depends on demonstrating that $\gamma_\infty = \infty$ almost surely. This precisely forms the objective we aim to achieve to conclude the proof. Now, let us assume that the assertion $\gamma_\infty = \infty$ almost surely is incorrect. This implies the existence of a positive value $D > 0$ such that $\mathbb{P}(\gamma_\infty \leqslant D) > 0$. Consequently, there exists a positive $x > 0$ such that

$$\mathbb{P}(\gamma_\beta \leqslant D) > x, \quad \forall \beta \geqslant \beta_0. \tag{4}$$

Examine the C^2 function F defined for $s(t) \in \mathbb{R}^4_+$ as follows:

$$F(s) = \Big(S_1 - q - q\ln\big(S_1 q^{-1}\big)\Big) + \Big(S_2 - q - q\ln\big(S_2 q^{-1}\big)\Big) + \big(I - 1 - \ln(I)\big).$$

Here, q denotes a positive constant to be selected meticulously at a subsequent stage. By employing Ito's multidimensional formula for $F(s(t))$, we obtain expressions applicable for all $\beta \geqslant \beta_0$ and $t \in (0, \gamma_\beta)$:

$$\begin{aligned} \mathrm{d}F(s(t)) = {} & \mathrm{L}F(s)\mathrm{d}t + \kappa_1(S_1(t) - q)\mathrm{d}\rho_1(t) + \kappa_2(S_2(t) - q)\mathrm{d}\rho_2(t) + \kappa_3(I(t) - 1)\mathrm{d}\rho_3(t) \\ & + \sum_{k=1}^{2} \int_{\mathbb{R}^4 \setminus \{0\}} \Big(z_k(\xi) S_k(t^-) - q\ln(1 + z_k(\xi))\Big)\widetilde{\phi}_k(\mathrm{d}t, \mathrm{d}\xi) \\ & + \int_{\mathbb{R}^4 \setminus \{0\}} \Big(z_3(\xi) I(t^-) - \ln(1 + z_3(\xi))\Big)\widetilde{\phi}_3(\mathrm{d}t, \mathrm{d}\xi), \end{aligned}$$

where

$$\begin{aligned}\mathrm{L}F(s(t)) = &\left(1-\frac{q}{S_1(t)}\right)\Big(r - f_1(S_1(t),I(t))I(t) - (u+c)S_1(t)\Big) \\ &+ \left(1-\frac{q}{S_2(t)}\right)\Big(cS_1(t) - f_2(S_2(t),I(t))I(t) - uS_2(t)\Big) \\ &+ \left(1-\frac{1}{I(t)}\right)\Big((f_1(S_1(t),I(t)) + f_2(S_2(t),I(t)))I(t) - (u+h_1+h_2)I(t)\Big) + 0.5\Big(q\kappa_1^2 + q\kappa_2^2 + \kappa_3^2\Big) \\ &+ q\sum_{k=1}^{2}\int_{\mathbb{R}^4\setminus\{0\}}\Big(z_k(\xi) - \ln(1+z_k(\xi))\Big)f_k(\mathrm{d}\xi) + \int_{\mathbb{R}^4\setminus\{0\}}\Big(z_3(\xi) - \ln(1+z_3(\xi))\Big)G_3(\mathrm{d}\xi).\end{aligned}$$

Then,

$$\begin{aligned}\mathrm{L}F(s(t)) \le &\Big(r - u(I(t)+S_2(t)+S_1(t)) - (h_1+h_2)I(t)\Big) - \left(\frac{qr}{S_1(t)} - q(u+c) - \frac{qf_1(S_1(t),I(t))}{S_1(t)}I(t)\right) \\ &- q\left(\frac{cS_1(t)}{S_2(t)} - u - \frac{f_2(S_2(t),I(t)))}{S_2(t)}I(t)\right) - \Big(f_1(S_1(t),I(t)) + f_2(S_2(t),I(t)) - (u+h_1+h_2)\Big) + \bar{C}.\end{aligned}$$

Here, $\bar{C} = 0.5(q\kappa_1^2 + q\kappa_2^2 + \kappa_3^2) + (2q+1)\alpha_2$. Then,

$$\begin{aligned}\mathrm{L}F(s(t)) &\leqslant (r + q(u+c) + qu + (u+h_1+h_2) + \bar{C}) \\ &\quad + \left(q\left(\frac{f_1(S_1(t),I(t))}{S_1(t)} + \frac{f_2(S_2(t),I(t))}{S_2(t)}\right) - (u+h_1+h_2)\right) \times I(t) \\ &\leqslant \underbrace{(\Lambda + (2q+1)u + h + \bar{C})}_{\triangleq \bar{D}} + (\underbrace{2mq}_{\text{by (A0)}} - (u+h_1+h_2)) \times I(t).\end{aligned}$$

Let us choose m such that $2mq - (u+h_1+h_2) < 0$. Then, for each $\beta \geqslant \beta_0$ and $t \in (0, \gamma_\beta)$, we obtain

$$\begin{aligned}\mathrm{d}F(s(t)) \leqslant\ &\bar{D}\,\mathrm{d}t + \kappa_1(S_1(t)-q)\mathrm{d}\rho_1(t) + \kappa_2(S_2(t)-q)\mathrm{d}\rho_2(t) + \kappa_3(I(t){-}1)\mathrm{d}\rho_3(t) \\ &+ \sum_{k=1}^{2}\int_{\mathbb{R}^4\setminus\{0\}}\Big(z_k(\xi)S_k(t^-) - q\ln(1+z_k(\xi))\Big)\widetilde{\phi}_k(\mathrm{d}t,\mathrm{d}\xi) \\ &+ \int_{\mathbb{R}^4\setminus\{0\}}\Big(z_3(\xi)I(t^-) - \ln(1+z_3(\xi))\Big)\widetilde{\phi}_3(\mathrm{d}t,\mathrm{d}\xi).\end{aligned}$$

So,

$$\mathbb{E}(F(s(D\wedge\gamma_\beta)) \leqslant F(s(0)) + \bar{D}\mathbb{E}(\gamma_k \wedge T) \leqslant F(s(0)) + \bar{D}T. \tag{5}$$

Since $F(\epsilon) \ge 0$ is true for all $\epsilon > 0$, the following implication arises:

$$\begin{aligned}\mathbb{E}(F(s(D\wedge\gamma_\beta)) &= \mathbb{E}\Big(F(s(D\wedge\gamma_\beta) \times \mathbb{1}_{\{\gamma_\beta \leqslant D\}}\Big) + \mathbb{E}\Big(F(s(t\wedge\gamma_\beta) \times \mathbb{1}_{\{\gamma_\beta > D\}}\Big) \\ &\geqslant \mathbb{E}\Big(F(s(\gamma_\beta) \times \mathbb{1}_{\{\gamma_\beta \leqslant D\}}\Big).\end{aligned} \tag{6}$$

We Observe that for any $\omega \in \Omega$ that verifies $\gamma_\beta(\omega) \leqslant D$, there exists a component of $F(s(\gamma_\beta))$ equal to either β or β^{-1}; thus,

$$\begin{aligned}\mathbb{E}\Big(F(s(\gamma_\beta) \times \mathbb{1}_{\{\gamma_\beta \leqslant D\}}\Big) \geqslant\ &\mathbb{P}(\gamma_\beta \leqslant D)\Big(\beta - q - q\ln\Big(\beta q^{-1}\Big)\Big) \wedge \Big(\beta^{-1} - q - q\ln\Big(\beta^{-1}q^{-1}\Big)\Big) \wedge (\beta - 1 - \ln(\beta)) \\ &\wedge \Big(\beta^{-1} - 1 - \ln\Big(\beta^{-1}\Big)\Big).\end{aligned} \tag{7}$$

By amalgamating (5), (6), and (7) with (4), we deduce that

$$F(s(0)) + \bar{D}D \geqslant x\Big(\beta - q - q\ln\big(\beta q^{-1}\big)\Big) \wedge \Big(\beta^{-1} - q - q\ln\big(\beta^{-1}q^{-1}\big)\Big) \wedge (\beta - 1 - \ln(\beta)) \wedge \Big(\beta^{-1} - 1 - \ln\big(\beta^{-1}\big)\Big).$$

Letting β tend toward infinity leads to the contradiction $F(s(0)) + \bar{D}D = \infty$, thus concluding the proof. □

3.2. Threshold Analysis of Stochastic System (2)

When exploring a mathematical model depicting the spread of a particular epidemic, our main focus is to determine whether the outbreak will eventually diminish or endure indefinitely. In this subsection, our goal is to reveal conditions that are both sufficiently rigorous and almost indispensable, shedding light on the asymptotic behavior of the epidemic's progression.

Before introducing the central theorem of this subsection, it is wise to begin a discourse on pertinent lemmas concerning an auxiliary subsystem derived from the initial two equations of (2). This additional subsystem becomes relevant when the population of infectious individuals is conspicuously absent from the context. Let us examine the following two novel stochastic processes:

$$\begin{cases} d\bar{S}_1(t) &= \Big(r - (u+c)\bar{S}_1(t)\Big)dt + \kappa_1\bar{S}_1(t)d\rho_1(t) + \int_{\mathbb{R}^4\setminus\{0\}} z_1(\xi)\bar{S}_1(t^-)\widetilde{\phi}_1(dt, d\xi), \\ d\bar{S}_2(t) &= \Big(c\bar{S}_1(t) - u\bar{S}_2(t)\Big)dt + \kappa_2\bar{S}_2(t)d\rho_2(t) + \int_{\mathbb{R}^4\setminus\{0\}} z_2(\xi)\bar{S}_2(t^-)\widetilde{\phi}_2(dt, d\xi), \end{cases} \tag{8}$$

with positive started data $\bar{S}_1(0) = S_1(0)$ and $\bar{S}_2(0) = S_2(0)$.

Lemma 1. *Stochastic system* (8) *is well-posed; and if*

$$(u+c) - 0.5\kappa_1^2 - \int_{\mathbb{R}^4\setminus\{0\}} \Big(z_1(\xi) - \ln(1 + z_1(\xi))\Big) G_1(d\xi) > 0, \tag{9}$$

then for any two integrable functions ψ_1 and ψ_2, we have

$$\begin{cases} \mathbb{P}\left(\lim_{t\to\infty} \frac{1}{t}\int_0^t \psi_1(\bar{S}_1(\epsilon))d\epsilon = \int_{(0,\infty)} \psi_1(\epsilon)\pi_1(d\epsilon)\right) = 1, \\ \mathbb{P}\left(\lim_{t\to\infty} \frac{1}{t}\int_0^t \psi_2(\bar{S}_2(\epsilon))d\epsilon = \int_{(0,\infty)} \psi_2(\epsilon)\pi_2(d\epsilon)\right) = 1, \end{cases} \tag{10}$$

where π_1 and π_2 are the single stationary distributions of $\bar{S}_1$ and $\bar{S}_2$, respectively.

Remark 4. *For an in-depth understanding of the proof for the above result, refer to Lemma 2.2 in [21]. The importance of considering* (9) *is elucidated in Lemma 2.4 of [13].*

Lemma 2. *Assuming that condition* (1) *holds true, we deduce*

$$\begin{cases} \mathbb{P}\left(\int_{(0,\infty)} f_1(S_1, 0)\,\pi_1(dS_1) \leqslant m\int_{(0,\infty)} S_1\,\pi_1(dS_1) = \frac{r}{u} < \infty\right) = 1, \\ \mathbb{P}\left(\int_{(0,\infty)} f_2(S_2, 0)\,\pi_2(dS_2) \leqslant m\int_{(0,\infty)} S_2\,\pi_2(dS_2) = \frac{cr}{(u+c)u} < \infty\right) = 1. \end{cases}$$

Proof. Drawing from $(A0)$, (10), and the ergodicity of processes $\bar{S}_1$ and $\bar{S}_2$, we can readily deduce the outcome of this lemma. □

When it comes to managing infectious diseases, a critical factor is determining a stochastic threshold that effectively distinguishes between the precarious brink of extinction and the resilient state of persistence for the infection. This threshold plays a crucial role, illuminating the complex dynamics that dictate the fate of the pathogen within a specific population. In this regard, by setting

$$\lambda_1 = \int_{(0,\infty)} f_1(S_1,0)\ \pi_1(dS_1) + \int_{(0,\infty)} f_2(S_2,0)\ \pi_2(dS_2),$$
$$\lambda_2 = (u + h_1 + h_2) + 0.5\kappa_3^2 + \int_{\mathbb{R}^4\setminus\{0\}} \Big(z_3(\xi) - \ln(1 + z_3(\xi))\Big) G_3(d\xi),$$

we have the following main theorem.

Theorem 2. *Consider $s(t)$ as the solution to system* (2) *with initial data $s(0) \in \mathbb{R}^4_+$. Then, we encounter two distinct real epidemic scenarios:*

- *The exponential extinction of the epidemic if $\lambda_1 < \lambda_2$.*
- *The stochastic continuation of the epidemic if $\lambda_1 > \lambda_2$.*

Proof. Using Itô's formula, we obtain the following equation:

$$d\ln\big(I(t)\big) = \left(\sum_{k=1}^{2} f_k(S_k(t), I(t)) - \lambda_2\right) dt + \kappa_3 d\rho_3(t) + \int_{\mathbb{R}^4\setminus\{0\}} \ln(1 + z_3(\xi))\widetilde{\phi}_3(dt, d\xi).$$

From $(A0)$, system (8), and the stochastic comparison theorem, we obtain

$$\begin{aligned} d\ln\big(I(t)\big) &\leqslant \left(\sum_{k=1}^{2} f_k(S_k(t),0) - \lambda_2\right) dt + \kappa_3 d\rho_3(t) + \int_{\mathbb{R}^4\setminus\{0\}} \ln(1 + z_3(\xi))\widetilde{\phi}_3(dt, d\xi) \\ &\leqslant \left(\sum_{k=1}^{2} f_k(\bar{S}_k(t),0) - \lambda_2\right) dt + \kappa_3 d\rho_3(t) + \int_{\mathbb{R}^4\setminus\{0\}} \ln(1 + z_3(\xi))\widetilde{\phi}_3(dt, d\xi). \end{aligned}$$

After integrating and then dividing both sides of the last inequality by t, we obtain

$$\frac{1}{t}\big(\ln\big(I(t)\big) - \ln\big(I(0)\big)\big) \leqslant \frac{1}{t}\sum_{k=1}^{2}\int_0^t f_k(\bar{S}_k(\epsilon),0)\ d\epsilon - \lambda_2 + \mathcal{A}_1(t), \tag{11}$$

where $\mathcal{A}_1(t) = \frac{\kappa_3}{t}\rho_3(t) + \frac{1}{t}\int_0^t \int_{\mathbb{R}^4\setminus\{0\}} \Big(\ln(1 + z_3 u)\Big)\widetilde{\phi}_3(ds, d\xi)$. In applying Kunita's inequality [22] to the discontinuous stochastic processes $\mathcal{A}_1$ and leveraging Lemma 2.2 of [23], it is straightforward to derive that $\mathbb{P}\left(\lim_{t\to\infty} \mathcal{A}_1(t) = 0\right) = 1$. Consequently, from Lemma 1, we deduce that

$$\begin{aligned} \limsup_{t\to\infty} \frac{1}{t}\ln\big(I(t)\big) &\leqslant \lim_{t\to\infty}\frac{1}{t}\int_0^t f_1(\bar{S}_1(\epsilon),0)\ d\epsilon + \lim_{t\to\infty}\frac{1}{t}\int_0^t f_2(\bar{S}_2(\epsilon),0)\ d\epsilon - \lambda_2 \\ &= \int_{(0,\infty)} f_1(S_1,0)\ \pi_1(dS_1) + \int_{(0,\infty)} f_2(S_2,0)\ \pi_2(dS_2) - \lambda_2 \\ &= \lambda_1 - \lambda_2. \end{aligned}$$

In the concept of the exponential extinction of the epidemic [19], a crucial factor is the value of $\lambda_1 - \lambda_2$. When this quantity is negative, it indicates a critical condition that leads to the disappearance of the epidemic.

Now, we shift our focus to the second scenario. To do so, let us define $\tilde{s}(t) = (S_1, S_2, I, \bar{S}_1, \bar{S}_2)$ and introduce the following function:

$$F_1(\tilde{s}(t)) = \frac{m}{u}\left(\left(\sum_{k=1}^{2}(\bar{S}_k - S_k)\right) - I\right) - \ln(I).$$

Leveraging Itô's formula in conjunction with the dynamics described by systems (2) and (8), we obtain

$$\begin{aligned}
\mathrm{d}F_1(\tilde{s}(t)) &= \mathrm{L}F_1(\tilde{s}(t))\mathrm{d}t + \sum_{k=1}^{2}\frac{m\kappa_k}{u}(\bar{S}_k(t) - S_k(t))\mathrm{d}\rho_k(t) - \kappa_3\Big(1+\frac{m}{u}I(t)\Big)\mathrm{d}\rho_3(t) \\
&+ \frac{m}{u}\sum_{k=1}^{2}\int_{\mathbb{R}^4\setminus\{0\}} z_k(\xi)(\bar{S}_k(t^-) - S_k(t^-))\widetilde{\phi}_k(\mathrm{d}t,\mathrm{d}\xi) - \frac{m}{u}\int_{\mathbb{R}^4\setminus\{0\}} z_3 uI(t^-)\widetilde{\phi}_3(\mathrm{d}t,\mathrm{d}\xi) \\
&- \int_{\mathbb{R}^4\setminus\{0\}} \ln(1+z_3(\xi))\widetilde{\phi}_3(\mathrm{d}t,\mathrm{d}\xi),
\end{aligned} \tag{12}$$

where

$$\mathrm{L}F_1(\tilde{s}(t)) = \lambda_2 - \sum_{k=1}^{2} f_k(S_k(t), I(t)) - m\sum_{k=1}^{2}(\bar{S}_k(t) - S_k(t)) + \frac{m}{u}(u+h_1+h_2)I(t).$$

As a result, we have

$$\begin{aligned}
\mathrm{L}F_1(\tilde{s}(t)) &\leqslant \lambda_2 - \sum_{k=1}^{2} f_k(\bar{S}_k(t),0) + \sum_{k=1}^{2}\Big(f_k(S_k(t),0) - f_k(S_k(t),I(t))\Big) \\
&+ \sum_{k=1}^{2}\Big(f_k(\bar{S}_k(t),0) - f_k(S_k(t),0)\Big) - m\sum_{k=1}^{2}(\bar{S}_k(t) - S_k(t)) + \frac{m}{u}(u+h_1+h_2)I(t).
\end{aligned}$$

Using $(A0)$, we can easily obtain that

$$\begin{aligned}
\mathrm{L}F_1(\tilde{s}(t)) &\leqslant \lambda_2 - \lambda_1 + \sum_{k=1}^{2}\Big(\int_{(0,\infty)} f_k(S_k,0)\,\pi_k(\mathrm{d}S_k) - f_k(\bar{S}_k(t),0)\Big) \\
&+ \sum_{k=1}^{2}\Big(f_k(S_k(t),0) - f_k(S_k(t),I(t))\Big) + \frac{m}{u}(u+h_1+h_2)I(t).
\end{aligned}$$

From Equation (12), we have

$$\begin{aligned}
\mathrm{d}F_1(\tilde{s}(t)) &\leqslant \Bigg(\lambda_2 - \lambda_1 + \sum_{k=1}^{2}\Big(\int_{(0,\infty)} f_k(S_k,0)\,\pi_k(\mathrm{d}S_k) - f_k(\bar{S}_k(t),0)\Big) \\
&+ \sum_{k=1}^{2}\Big(f_k(S_k(t),0) - f_k(S_k(t),I(t))\Big) + \frac{m}{u}(u+h_1+h_2)I(t)\Bigg)\mathrm{d}t \\
&+ \sum_{k=1}^{2}\frac{m\kappa_k}{u}(\bar{S}_k(t) - S_k(t))\mathrm{d}\rho_k(t) - \kappa_3\Big(1+\frac{m}{u}I(t)\Big)\mathrm{d}\rho_3(t) \\
&+ \frac{m}{u}\sum_{k=1}^{2}\int_{\mathbb{R}^4\setminus\{0\}} z_k(\xi)(\bar{S}_k(t^-) - S_k(t^-))\widetilde{\phi}_k(\mathrm{d}t,\mathrm{d}\xi) \\
&- \frac{m}{u}\int_{\mathbb{R}^4\setminus\{0\}} z_3 uI(t^-)\widetilde{\phi}_3(\mathrm{d}t,\mathrm{d}\xi) - \int_{\mathbb{R}^4\setminus\{0\}} \ln(1+z_3(\xi))\widetilde{\phi}_3(\mathrm{d}t,\mathrm{d}\xi).
\end{aligned}$$

Upon integrating the last inequality and subsequently dividing both sides of the last inequality by t, we obtain

$$
\begin{aligned}
\frac{F_1(\tilde{s}(t))}{t} \leqslant\ & \lambda_2 - \lambda_1 + \sum_{k=1}^{2}\Big(\int_{(0,\infty)} f_k(S_k,0)\,\pi_k(\mathrm{d}S_k) - \frac{1}{t}\int_0^t f_k(\bar{S}_k(s),0)\Big)\mathrm{d}s \\
& + \sum_{k=1}^{2}\frac{1}{t}\int_0^t \Big(f_k(S_k(s),0) - f_k(S_k(s),I(s))\Big)\mathrm{d}s + \frac{m(u+h_1+h_2)}{ut}\int_0^t I(s)\mathrm{d}s \\
& + \sum_{k=1}^{2}\frac{m\kappa_k}{ut}\int_0^t (\bar{S}_k(s) - S_k(s))\mathrm{d}\rho_k(s) - \frac{\kappa_3}{t}\int_0^t \Big(1 + \frac{m}{u}I(s)\Big)\mathrm{d}\rho_3(s) \\
& + \frac{m}{ut}\sum_{k=1}^{2}\int_0^t\int_{\mathbb{R}^4\setminus\{0\}} z_k(\xi)\big(\bar{S}_k(s^-) - S_k(s^-)\big)\widetilde{\phi}_k(\mathrm{d}s,\mathrm{d}\xi) + \frac{F_1(\tilde{s}(0))}{t} \\
& - \frac{m}{ut}\int_0^t\int_{\mathbb{R}^4\setminus\{0\}} z_3 u I(s^-)\widetilde{\phi}_3(\mathrm{d}s,\mathrm{d}\xi) - \frac{1}{t}\int_0^t\int_{\mathbb{R}^4\setminus\{0\}} \ln(1+z_3(\xi))\widetilde{\phi}_3(\mathrm{d}s,\mathrm{d}\xi).
\end{aligned}
$$

Using the properties of logarithm function, we obtain

$$
\begin{aligned}
\frac{m(u+h_1+h_2)}{ut}\int_0^t I(s)\mathrm{d}s \geqslant\ & \lambda_1 - \lambda_2 + \frac{m}{ut}\left(\left(\sum_{k=1}^{2}(\bar{S}_k(t) - S_k(t))\right) - I(t)\right) - \frac{I(t)}{t} \\
& - \sum_{k=1}^{2}\Big(\int_{(0,\infty)} f_k(S_k,0)\,\pi_k(\mathrm{d}S_k) - \frac{1}{t}\int_0^t f_k(\bar{S}_k(s),0)\mathrm{d}s\Big) \\
& - \sum_{k=1}^{2}\frac{1}{t}\int_0^t \Big(f_k(S_k(s),0) - f_k(S_k(s),I(s))\Big)\mathrm{d}s \\
& - \sum_{k=1}^{2}\frac{m\kappa_k}{ut}\int_0^t (\bar{S}_k(s) - S_k(s))\mathrm{d}\rho_k(s) + \frac{\kappa_3}{t}\int_0^t \Big(1 + \frac{m}{u}I(s)\Big)\mathrm{d}\rho_3(s) \\
& - \frac{m}{ut}\sum_{k=1}^{2}\int_0^t\int_{\mathbb{R}^4\setminus\{0\}} z_k(\xi)\big(\bar{S}_k(s^-) - S_k(s^-)\big)\widetilde{\phi}_k(\mathrm{d}s,\mathrm{d}\xi) - \frac{F_1(\tilde{s}(0))}{t} \\
& + \frac{m}{ut}\int_0^t\int_{\mathbb{R}^4\setminus\{0\}} z_3 u I(s^-)\widetilde{\phi}_3(\mathrm{d}s,\mathrm{d}\xi) + \frac{1}{t}\int_0^t\int_{\mathbb{R}^4\setminus\{0\}} \ln(1+z_3(\xi))\widetilde{\phi}_3(\mathrm{d}s,\mathrm{d}\xi).
\end{aligned}
$$

From $(A0)$, we derive

$$
\begin{aligned}
\frac{m(u+h_1+h_2)}{ut}\int_0^t I(s)\mathrm{d}s \geqslant\ & \lambda_1 - \lambda_2 + \frac{m}{ut}\left(\left(\sum_{k=1}^{2}(\bar{S}_k(t) - S_k(t))\right) - I(t)\right) - \frac{I(t)}{t} \\
& - \sum_{k=1}^{2}\Big(\int_{(0,\infty)} f_k(S_k,0)\,\pi_k(\mathrm{d}S_k) - \frac{1}{t}\int_0^t f_k(\bar{S}_k(s),0)\mathrm{d}s\Big) - \frac{2m_0}{t}\int_0^t I(s)\mathrm{d}s \\
& - \sum_{k=1}^{2}\frac{m\kappa_k}{ut}\int_0^t (\bar{S}_k(s) - S_k(s))\mathrm{d}\rho_k(s) + \frac{\kappa_3}{t}\int_0^t \Big(1 + \frac{m}{u}I(s)\Big)\mathrm{d}\rho_3(s) \\
& - \frac{m}{ut}\sum_{k=1}^{2}\int_0^t\int_{\mathbb{R}^4\setminus\{0\}} z_k(\xi)\big(\bar{S}_k(s^-) - S_k(s^-)\big)\widetilde{\phi}_k(\mathrm{d}s,\mathrm{d}\xi) - \frac{F_1(\tilde{s}(0))}{t} \\
& + \frac{m}{ut}\int_0^t\int_{\mathbb{R}^4\setminus\{0\}} z_3 u I(s^-)\widetilde{\phi}_3(\mathrm{d}s,\mathrm{d}\xi) + \frac{1}{t}\int_0^t\int_{\mathbb{R}^4\setminus\{0\}} \ln(1+z_3(\xi))\widetilde{\phi}_3(\mathrm{d}s,\mathrm{d}\xi).
\end{aligned}
$$

Consequently,

$$
\frac{m(u+h_1+h_2) + 2(u+c)m_0}{ut}\int_0^t I(s)\mathrm{d}s \geqslant \lambda_1 - \lambda_2 + \mathcal{A}_2(t) + \mathcal{A}_3(t) + \mathcal{A}_4(t),
$$

where

$$\mathcal{A}_2(t) = \frac{m}{ut}\left(\left(\sum_{k=1}^{2}(\bar{S}_k(t) - S_k(t))\right) - I(t)\right) - \frac{I(t)}{t},$$

$$\mathcal{A}_3(t) = -\sum_{k=1}^{2}\left(\int_{(0,\infty)} f_k(S_k, 0)\ \pi_k(\mathrm{d}S_k) - \frac{1}{t}\int_0^t f_k(\bar{S}_k(s), 0)\mathrm{d}s\right),$$

$$\mathcal{A}_4(t) = -\sum_{k=1}^{2}\frac{m\kappa_k}{ut}\int_0^t (\bar{S}_k(s) - S_k(s))\mathrm{d}\rho_k(s) + \frac{\kappa_3}{t}\int_0^t \left(1 + \frac{m}{u}I(s)\right)\mathrm{d}\rho_3(s)$$
$$- \frac{m}{ut}\sum_{k=1}^{2}\int_0^t\int_{\mathbb{R}^4\setminus\{0\}} z_k(\xi)\left(\bar{S}_k(s^-) - S_k(s^-)\right)\widetilde{\phi}_k(\mathrm{d}s, \mathrm{d}\xi) - \frac{F_1(\tilde{s}(0))}{t}$$
$$+ \frac{m}{ut}\int_0^t\int_{\mathbb{R}^4\setminus\{0\}} z_3 u I(s^-)\widetilde{\phi}_3(\mathrm{d}s, \mathrm{d}\xi) + \frac{1}{t}\int_0^t\int_{\mathbb{R}^4\setminus\{0\}} \ln(1 + z_3(\xi))\widetilde{\phi}_3(\mathrm{d}s, \mathrm{d}\xi).$$

Considering $(A0)$ and leveraging Lemma 1, we can firmly establish that $\mathbb{P}\left(\lim_{t\to\infty}\mathcal{A}_2(t) = 0\right) = 1$. Furthermore, the application of Lemma 2.2 from [23] enables a straightforward deduction that $\mathbb{P}\left(\lim_{t\to\infty}\mathcal{A}_3(t) = 0\right) = 1$ and $\mathbb{P}\left(\lim_{t\to\infty}\mathcal{A}_4(t) = 0\right) = 1$. Then,

$$\mathbb{P}\left(\liminf_{t\to\infty}\frac{1}{t}\int_0^t I(s)\,\mathrm{d}s \geqslant \frac{u(\lambda_1 - \lambda_2)}{m(u + h_1 + h_2) + 2(u + c)m_0}\right) = 1.$$

In scenarios where λ_1 exceeds λ_2, the persistence of the epidemic in the future becomes more pronounced. This concludes the demonstration of the theorem. □

4. Numerical Application: Zoonotic Tuberculosis

In this section, we present a series of numerical demonstrations to corroborate the conclusions outlined in our research, employing authentic data on Zoonotic Tuberculosis as detailed in (Example 1, [24]). This dataset meticulously records reported instances within Morocco, offering a detailed insight into the current scenario. The dataset encompasses a range of parameters and initial conditions reflective of the epidemiological landscape of Zoonotic Tuberculosis in the region. To ensure the robustness of our findings, we draw upon deterministic parameters and initial data outlined in Table 1. These parameters were carefully selected based on the latest epidemiological studies and statistical analyses pertinent to the spread and control of Zoonotic Tuberculosis. Our approach involves integrating these parameters into our mathematical model to simulate various outbreak scenarios, thereby enabling us to examine the potential efficacy of different intervention strategies.

Table 1. The numerical values corresponding to the deterministic parameters and initial data governing the dynamics of system (2).

Parameters	Extinction Case	Persistence Case	Source
r	0.177	0.177	[24]
c	0.15	0.15	Estimated
β_1	0.249	3.3	[24]
β_2	1.3	2.3	[24]
u	0.167	0.167	[24]
h_1	0.2	0.2	[24]
h_2	0.01	0.01	[24]
m_1	0.1	0.1	Estimated
m_2	0.2	0.2	Estimated
m_3	0.1	0.1	Estimated
m_4	0.2	0.2	Estimated
$S_1(0)$	2	2	Assumed
$S_2(0)$	1	1	Assumed
$I(0)$	0.5	0.5	Assumed
$C(0)$	0.1	0.1	Assumed

About the dual incidence functions g_1 and g_2, we consider the following general nonlinear incidences:

$$f_1(S_1, I) = \frac{\beta_1 S_1}{1 + m_1 S_1 + m_2 I}, \qquad f_2(S_2, I) = \frac{\beta_2 S_2}{1 + m_3 S_2 + m_4 I},$$

where m_1, m_2, $m3$, and $m4$ are four saturated coefficients. We begin our check by illustrating the rigor of condition (9). We take $\kappa = \ell_2 = 0.3$ and $\kappa(\xi) = z_2(\xi) = 0.08$. In this case, we obtain

$$(u + c) - 0.5\kappa_1^2 - \int_{\mathbb{R}^4 \setminus \{0\}} \Big(z_1(\xi) - \ln(1 + z_1(\xi))\Big) G_1(\mathrm{d}\xi) = 0.017 > 0.$$

According to Figure 2, we visually depict the existence of an ergodic stationary distribution for the stochastic processes $(\bar{S}_1, \bar{S}_2)$. This graphical representation unmistakably highlights the enduring nature of the processes over time.

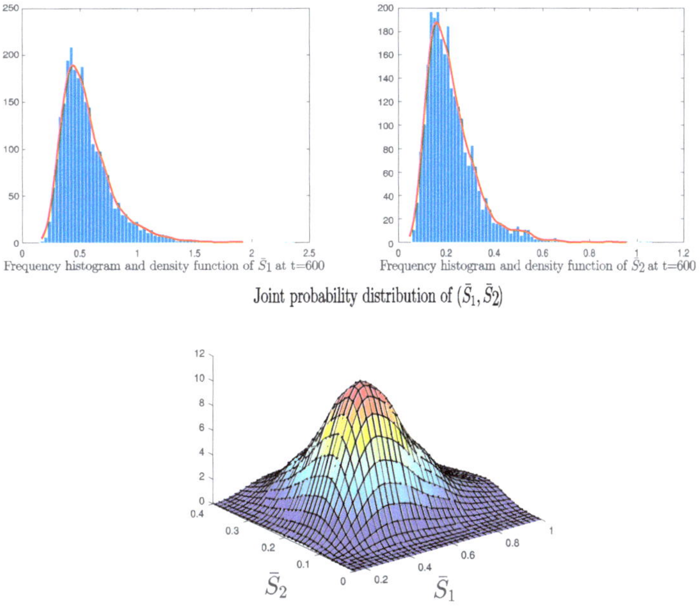

Figure 2. Numerical simulation of the existence of a stationary distribution: histograms and joint distribution under condition (9).

Firstly, we take specific values for the parameters κ_k ($k \in \{1, \cdots, 4\}$), setting them as 0.12, 0.11, 0.15, and 0.1, respectively. The jump intensities are defined using the function $z_k(\xi) = \frac{-u_k \xi}{0.5 + \xi^2}$, where $k \in 1, \cdots, 4$ and $\xi = 0.3$. The corresponding values for u_k ($k \in \{1, \cdots, 4\}$) are established as 0.02, 0.03, -0.0770, and 0.03, respectively. Here, we obtain $\lambda_1 = 0.0592$ and $\lambda_2 = 0.2061$. As a result, we have verified that the prerequisite outlined in Theorem 2 has been met and

$$\limsup_{t \to \infty} \frac{1}{t} \ln\left(I(t)\right) = \lambda_1 - \lambda_2 = -0.1469 < 0.$$

To solidify this discovery through numerical analysis, we depict two distinct trajectory types corresponding to system (2) in Figure 3. From this visualization, several observations can be drawn. Initially, we note the extinction of Tuberculosis in both trajectories, yet the solution involving jumps demonstrates a faster extinction compared to the solution with only white noise. Additionally, we observe that the incorporation of jumps helps in capturing abrupt changes, particularly in behavioral shifts. At time $t = 38$, a significant jump occurs within the educated class S_2, leading to complete extinction in the infected class.

Now, let us explore the scenario of persistent Tuberculosis. In this experimental setup, our focus is on κ_k $(k \in 1, \cdots, 4)$, with specific values assigned as 0.09, 0.09, 0.028, and 0.026, respectively. The jump intensities are determined by the function $z_k(\xi) = \frac{-u_k \xi}{0.5+\xi^2}$, where k spans from 1 to 4, and ξ is set at 0.24. The corresponding values for u_k $(k \in 1, \cdots, 7)$ are defined as 0.02, 0.03, 0.02, and 0.03, respectively. By utilizing the numerical values provided in [24] and adjusting β_1 to 3.3 and β_2 to 2.3, we can readily confirm the validity of our hypotheses, resulting in $\lambda_1 = 3.9876$ and $\lambda_2 = 1.231$. Consequently, in line with the assertions of Theorem 2, we can confidently affirm that our model exhibits persistence on average, a trend consistently reflected in the patterns illustrated in Figure 4. Notably, the endemic equilibrium characterizing the deterministic version no longer functions as the stable state for the stochastic model (2). Therefore, over an extended temporal span, the influence of noise intensity becomes a significant factor, shaping the extent to which the solution fluctuates around the deterministic equilibrium states.

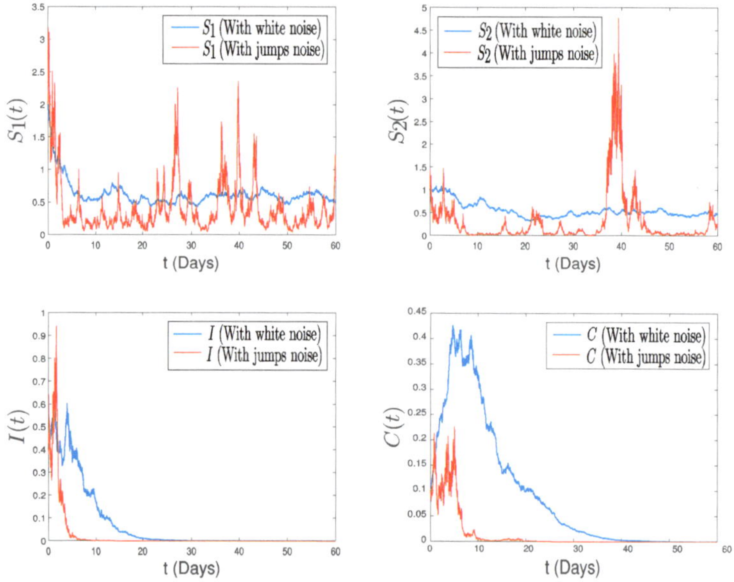

Figure 3. Numerical simulation of system (2) with two trajectories: one characterized by white noise and the other by jump diffusion.

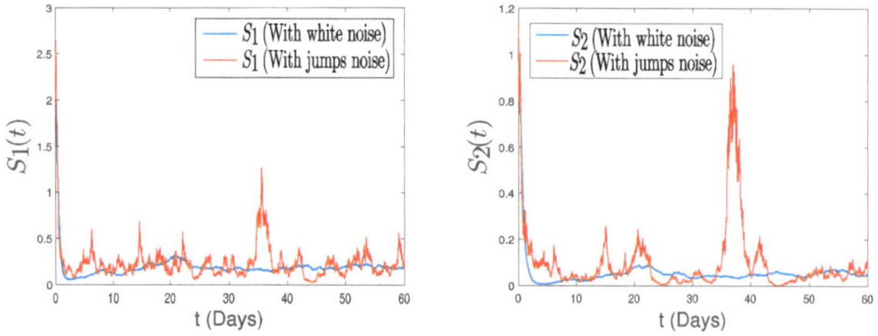

Figure 4. *Cont.*

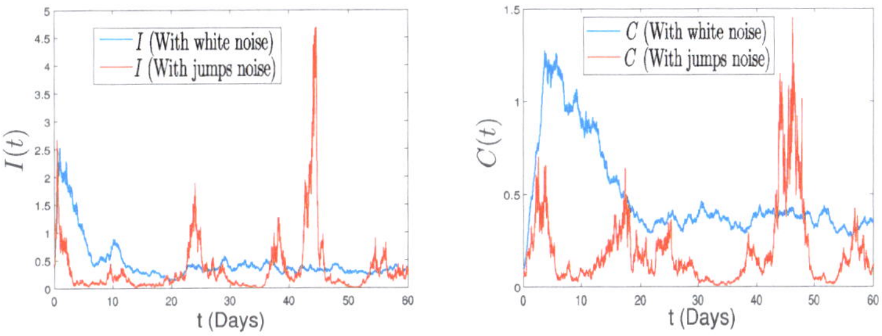

Figure 4. Numerical simulation of system (2) with two trajectories: one characterized by white noise and the other by jump diffusion.

5. Conclusions

The dynamic shift in behavior during an epidemic profoundly impacts infection probabilities. Building upon the model introduced in [4], we significantly extended the framework by incorporating general incidence functions and Lévy jumps. This extension ensures the well-posedness of the model and allows us to devise an auxiliary system that is pivotal in determining the global threshold distinguishing between extinction and persistence. Our practical application, validated within the context of Zoonotic Tuberculosis, underscores the pivotal role of Lévy jumps in accurately modeling epidemics, shedding light on their intricate dynamics. Unlike prior works, our paper's distinctiveness lies in the innovative mathematical techniques employed, introducing a more sophisticated set of hypotheses and a nuanced analytical framework. The inclusion of Assumptions 1 and 2 enhances the precision and depth of our analysis, emphasizing the importance of a meticulous approach. Furthermore, our expanded assumption framework accommodates a broader range of functions, using a non-standard analytical approach to delineate the threshold between the eradication and continuation of infection.

These theoretical advancements hold significant implications for public health, providing crucial insights that can inform and improve epidemic management and mitigation strategies. Understanding the impact of behavior shifts on infection probabilities allows public health officials to design more effective interventions. For instance, targeted public health campaigns can be developed to encourage behaviors that reduce transmission rates, such as promoting vaccination, social distancing, and the use of personal protective equipment. Additionally, the integration of Lévy jumps into epidemic models enables better prediction and rapid response to sudden changes in infection rates, facilitating quicker and more effective public health responses to emerging outbreaks. By identifying precise thresholds for extinction and persistence, health authorities can allocate resources more efficiently, prioritizing areas with the highest risk of sustained transmission. Our model also assists policymakers in developing evidence-based policies that account for behavioral dynamics and nonlinear transmission patterns, leading to more effective control measures.

Furthermore, the potential for broader applications of our theoretical framework is vast. Future research should explore extending our model to various infectious diseases, potentially creating a universal framework for epidemic modeling. This extension would significantly enhance our ability to manage a wide range of public health challenges, improving preparedness and response strategies. By integrating these findings into public health strategies, we can enhance epidemic management, protect vulnerable populations, and ultimately save lives. The implications of our work extend beyond the immediate findings, offering a robust foundation for future innovations in epidemic modeling and public health policy.

Author Contributions: Conceptualization, Y.S. and A.A.R.; methodology, Y.S. and A.A.R.; software, Y.S. and A.A.R.; validation, Y.S., M.I., A.A.R. and W.A.; formal analysis, Y.S. and A.A.R.; funding acquisition, M.I. and W.A.; investigation, Y.S., M.I., A.A.R. and W.A.; writing—original draft preparation, Y.S. and A.A.R.; writing—review and editing, Y.S., M.I., A.A.R. and W.A. All authors have read and agreed to the published version of the manuscript.

Funding: This research was funded by Deanship of Research and Graduate Studies at King Khalid University, Abha, Saudi Arabia, under grant number RGP2/174/45.

Institutional Review Board Statement: Not applicable.

Informed Consent Statement: Not applicable.

Data Availability Statement: The real data used were derived from [24].

Acknowledgments: The authors extend their appreciation to the Deanship of Research and Graduate Studies at King Khalid University for funding this work through Large Research Project under grant number RGP2/174/45.

Conflicts of Interest: The authors declare no conflicts of interest.

References

1. Capasso, V. *Mathematical Structures of Epidemic Systems*; Springer Science & Business Media: Berlin/Heidelberg, Germany, 2008; Volume 97.
2. Brauer, F.; Castillo-Chavez, C. *Mathematical Models in Population Biology and Epidemiology*; Springer Science & Business Media: Berlin/Heidelberg, Germany, 2013; Volume 40.
3. Kermack, W.O.; McKendrick, A.G. Contributions to the mathematical theory of epidemics–I. 1927. *Bull. Math. Biol.* **1991**, *53*, 33–55. [PubMed]
4. Nguyen, D.T.; Du, N.H.; Nguyen, S.L. Asymptotic behavior for a stochastic behavioral change SIR model. *J. Math. Anal. Appl.* **2024**, *538*, 128361. [CrossRef]
5. Frieswijk, K.; Zino, L.; Ye, M.; Rizzo, A.; Cao, M. A mean-field analysis of a network behavioral-epidemic model. *IEEE Control Syst. Lett.* **2022**, *6*, 2533–2538. [CrossRef]
6. Osi, A.; Ghaffarzadegan, N. Parameter estimation in behavioral epidemic models with endogenous societal risk-response. *PLoS Comput. Biol.* **2024**, *20*, e1011992. [CrossRef]
7. Marsudi, T.; Suryanto, A.; Darti, I. Global stability and optimal control of an HIV/AIDS epidemic model with behavioral change and treatment. *Eng. Lett.* **2021**, *29*, 575–591. [CrossRef]
8. Li, J.; Wang, X.; Lin, X. Impact of behavioral change on the epidemic characteristics of an epidemic model without vital dynamics. *Math. Biosci. Eng.* **2018**, *15*, 1425–1434. [CrossRef]
9. Ward, C.; Deardon, R.; Schmidt, A.M. Bayesian modeling of dynamic behavioral change during an epidemic. *Infect. Dis. Model.* **2023**, *8*, 947–963. [CrossRef]
10. Li, S. SIR epidemic model with general nonlinear incidence rate and Lévy jumps. *Mathematics* **2024**, *12*, 215. [CrossRef]
11. Wang, H.; Zhang, G.; Chen, T.; Li, Z. Threshold analysis of a stochastic SIRS epidemic model with logistic birth and nonlinear incidence. *Mathematics* **2023**, *11*, 1737. [CrossRef] [PubMed]
12. Zhao, Y.; Jiang, D. The behavior of an SVIR epidemic model with stochastic perturbation. *Abstr. Appl. Anal.* **2014**, *2014*, 742730. [CrossRef]
13. Sabbar, Y.; Khan, A.; Din, A.; Tilioua, M. New method to investigate the impact of independent quadratic alpha stable Poisson jumps on the dynamics of a disease under vaccination strategy. *Fractal Fract.* **2023**, *7*, 226. [CrossRef]
14. Kiouach, D.; Sabbar, Y.; El-idrissi, S.E.A. New results on the asymptotic behavior of an SIS epidemiological model with quarantine strategy, stochastic transmission, and Lévy disturbance. *Math. Meth. Appl. Sci.* **2021**, *44*, 13468–13492.
15. Rosinski, J. Tempering stable processes. *Stoch. Process. Their Appl.* **2007**, *117*, 677–707.
16. Koponen, I. Analytic approach to the problem of convergence of truncated Lévy flights towards the gaussian stochastic process. *Phys. Rev. E* **1995**, *52*, 1197–1199. [CrossRef]
17. Zhang, X.; Yang, Q. Threshold behavior in a stochastic SVIR model with general incidence rates. *Appl. Math. Lett.* **2021**, *121*, 107403.
18. Zhou, B.; Zhang, X.; Jiang, D. Dynamics and density function analysis of a stochastic SVI epidemic model with half saturated incidence rate. *Chaos Solit. Fract.* **2020**, *137*, 109865. [CrossRef]
19. Mao, X. *Stochastic Differential Equations and Applications*; Elsevier: Amsterdam, The Netherlands, 2007. [CrossRef]
20. Karatzas, I.; Shreve, S.E. *Brownian Motion and Stochastic Calculus*; Springer: Berlin/Heidelberg, Germany, 1998.
21. Sabbar, Y.; Kiouach, D.; Rajasekar, S.P.; El-Idrissi, S.E.A. The influence of quadratic Lévy noise on the dynamic of an SIC contagious illness model: New framework, critical comparison and an application to COVID-19 (SARS-CoV-2) case. *Chaos Solit. Fract.* **2022**, *159*, 112110.
22. Øksendal, B.K.; Sulem, A. *Applied Stochastic Control of Jump Diffusions*; Springer: Berlin/Heidelberg, Germany, 2007; Volume 498.

23. Privault, N.; Wang, L. Stochastic SIR Lévy jump model with heavy tailed increments. *J. Nonlinear Sci.* **2021**, *31*, 15.
24. El Attouga, S.; Bouggar, D.; El Fatini, M.; Hilbert, A.; Pettersson, R. Lévy noise with infinite activity and the impact on the dynamic of an SIRS epidemic model. *Physica A* **2023**, *618*, 128701.

mathematics

MDPI

Article

Existence and Hyers–Ulam Stability of Stochastic Delay Systems Governed by the Rosenblatt Process

Ghada AlNemer [1], Mohamed Hosny [2], Ramalingam Udhayakumar [3] and Ahmed M. Elshenhab [4,*]

[1] Department of Mathematical Sciences, College of Science, Princess Nourah bint Abdulrahman University, P.O. Box 84428, Riyadh 11671, Saudi Arabia; gnnemer@pnu.edu.sa
[2] Department of Electrical Engineering, Benha Faculty of Engineering, Benha University, Benha 13511, Egypt; mohamed.hosny@bhit.bu.edu.eg
[3] Department of Mathematics, School of Advanced Sciences, Vellore Institute of Technology, Vellore 632014, India; udhayakumar.r@vit.ac.in
[4] Department of Mathematics, Faculty of Science, Mansoura University, Mansoura 35516, Egypt
* Correspondence: ahmedelshenhab@mans.edu.eg

Abstract: Under the effect of the Rosenblatt process, time-delay systems of nonlinear stochastic delay differential equations are considered. Utilizing the delayed matrix functions and exact solutions for these systems, the existence and Hyers–Ulam stability results are derived. First, depending on the fixed point theory, the existence and uniqueness of solutions are proven. Next, sufficient criteria for the Hyers–Ulam stability are established. Ultimately, to illustrate the importance of the results, an example is provided.

Keywords: Hyers–Ulam stability; stochastic delay system; Rosenblatt process; delayed matrix function; Krasnoselskii's fixed point theorem

MSC: 37A50; 34K50; 34K20; 47H10

Citation: AlNemer, G.; Hosny, M.; Udhayakumar, R.; Elshenhab, A.M. Existence and Hyers–Ulam Stability of Stochastic Delay Systems Governed by the Rosenblatt Process. *Mathematics* **2024**, *12*, 1729. https://doi.org/10.3390/math12111729

Academic Editor: Jun Huang

Received: 22 April 2024
Revised: 16 May 2024
Accepted: 23 May 2024
Published: 2 June 2024

1. Introduction

Many researchers have paid significant attention to stochastic delay differential equations (SDDEs) and their applications because of their effective modeling in several scientific and engineering fields, such as physics, economics, biology, fluid dynamics, finance, medicine, and so forth (see, for instance, [1–9]). Recently, determining the exact solutions of differential systems has been attempted. Specifically, many new results regarding how to represent solutions for time-delay systems were obtained from the novel study [10,11], which were applied to stability analysis and control problems (see, [12–17] and the references therein).

The Wiener–Ito multiple integral of order q is defined as

$$Z_H^q(\ell) = a(H,q)\int_{R^q}\left(\int_0^{\ell}\prod_{j=1}^{q}(\varsigma-\Im_j)_+^{-\left(\frac{1}{2}+\frac{1-H}{q}\right)}d\varsigma\right)d\mathcal{G}(\Im_1)\ldots d\mathcal{G}(\Im_q), \tag{1}$$

in terms of the standard Wiener process, $(\mathcal{G}(\Im))_{\Im\in\mathbb{R}}$, where $\mathbf{E}\left(Z_H^q(1)\right)^2 = 1$ and $\Im_+ = \max(\Im,0)$ are the conditions under which $a(H,q)$ is a normalizing constant. The process $\left(Z_H^q(\ell)\right)_{\ell\geq 0}$, provided by (1), is called the Hermite process. The Hermite process is the fractional Brownian motion (fBm) with a Hurst parameter of $H\in\left(\frac{1}{2},1\right)$ for $q=1$, while it is not Gaussian for $q=2$. Additionally, the Hermite process, denoted by (1) for $q=2$, is referred to as the Rosenblatt process. Most of the studies [18–20] involved fBm because of its self-similarity, long-range dependence, and more straightforward calculus of the Gaussian. But, fBm fails in the concrete case of having non-Gaussianity smooth-tongued in the models. In that situation, the Rosenblatt process is applicable. Non-Gaussian processes like the Rosenblatt process have numerous intriguing characteristics such as stationarity of the increments, long-range dependence, and self-similarity (for more details, see [21–29]). Therefore, it seems interesting to study a new class of stochastic differential equations driven by the Rosenblatt process.

On the other hand, studying the stability of (SDDEs) solutions is essential, and Hyers–Ulam stability (HUS) is a crucial topic. In 1940, Ulam [30] created the first proposal that functional equations are stable, during a lecture at Wisconsin University. In 1941, Hyers [31] provided a solution to this problem, after which HUS was established. In addition to providing a solid theoretical foundation for the well-posedness and HUS for SDDEs, the study of HUS for SDDEs also provides a solid theoretical foundation for the approximate solution of SDDEs. When it is rather difficult to acquire a precise solution for the system with HUS, we may substitute an approximate solution for an accurate one, and the HUS can, to a certain extent, ensure the dependability of the estimated solution.

Recently, many researchers have examined the HUS of diverse kinds of stochastic differential equations (see, [32–35] and the references therein).

However, as far as we know, the standard literature has not dealt with the existence and HUS of second-order nonlinear SDDEs driven by the Rosenblatt process. Therefore, in this study, we try, for the first time, to analyze such a topic.

Our study focuses on determining the existence and HUS of the nonlinear SDDEs driven by the Rosenblatt process, taking into account the previous research.

$$\begin{aligned} &\aleph''(\ell) + \mathbb{D}\aleph(\ell - \zeta) = \hbar(\ell, \aleph(\ell)) + \Delta(\ell, \aleph(\ell))\frac{\mathrm{d}Z_H(\ell)}{\mathrm{d}\ell}, \ \ell \in \mp := [0, \varpi], \\ &\aleph(\ell) \equiv \psi(\ell), \ \aleph'(\ell) \equiv \psi'(\ell), \ \ell \in \mp_1 := [-\zeta, 0], \end{aligned} \tag{2}$$

where $\aleph(\ell) \in \mathbb{R}^n$ represents the state vector, $\zeta > 0$ denotes a delay, $\varpi > (m-1)\zeta$, $m = 1, 2, \ldots$, $\psi \in C([-\zeta, 0], \mathbb{R}^n)$, $\mathbb{D} \in \mathbb{R}^{n \times n}$ is any matrix, and $\hbar \in C(\mp \times \mathbb{R}^n, \mathbb{R}^n)$ is a provided function. In the separable Hilbert space $\mathbb{R}^n$, let $\aleph(\cdot)$ have value, and let the norm be $\|\cdot\|$ and the inner product be $\langle \cdot, \cdot \rangle$ with parameter $H \in \left(\frac{1}{2}, 1\right)$, $Z_H(\ell)$ is a Rosenblatt process on an another real separable Hilbert space $(\mathcal{A}, \|\cdot\|_{\mathcal{A}}, \langle \cdot, \cdot \rangle_{\mathcal{A}})$. Furthermore, consider $\Delta \in C(\mp \times \mathbb{R}^n, L_2^0)$, where $L_2^0 = L_2\left(Q^{\frac{1}{2}}\mathcal{A}, \mathbb{R}^n\right)$.

The remaining sections of this paper are structured as follows: In Section 2, we present some notations and necessary preliminaries. In Section 3, by utilizing Krasnoselskii's fixed point theorem, some sufficient conditions are established for the existence and uniqueness of solutions to the system (2). In Section 4, we prove the Hyers–Ulam stability of (2) via Grönwall's inequality lemma approach. Finally, we provide a numerical example to illustrate the effectiveness of the derived results.

2. Preliminaries

During the entire paper, consider $(\Sigma, \eth, \mathbb{P})$ to represent the complete probability space with a probability measure $\mathbb{P}$ on Σ and a filtration $\{\eth_\ell | \, \ell \in \mp\}$ produced by $\{Z_H(s) | \, s \in [0, \ell]\}$. For some $1 < \mu < \infty$, consider the Hilbert space $L^\mu(\Sigma, \eth_\varpi, \mathbb{R}^n)$ to express all $\eth_\varpi$-measurable μth-integrable variables having values in $\mathbb{R}^n$ with norm $\|\aleph\|_{L^\mu}^\mu = \mathbf{E}\|\aleph(\ell)\|^\mu$, where the expectation $\mathbf{E}$ is defined by $\mathbf{E}\aleph = \int_\Sigma \aleph \mathrm{d}\mathbb{P}$. Assume that $\mathcal{A}$ and $\mathcal{B}$ are two Banach spaces, $Q \in L_b(\mathcal{A}, \mathcal{A})$ indicates an operator on $\mathcal{A}$ that is self-adjoint trace class and non-negative, and $L_b(\mathcal{A}, \mathcal{B})$ is the space of the bounded linear operators from $\mathcal{A}$ to $\mathcal{B}$. Let $L_2^0 = L_2\left(Q^{\frac{1}{2}}\mathcal{A}, \mathcal{B}\right)$ be the space of all Q-Hilbert–Schmidt operators from $Q^{\frac{1}{2}}\mathcal{A}$ into $\mathcal{B}$, equipped with the norm

$$\|\Xi\|_{L_2^0}^2 = \left\|\Xi Q^{\frac{1}{2}}\right\|^2 = \mathrm{Tr}\left(\Xi Q \Xi^T\right).$$

Provided a norm $\|\Xi\|_{\mathcal{Q}} = \left(\sup_{\ell\in\mp} \mathbf{E}\|\Xi(\ell)\|^{\mu}\right)^{1/\mu}$, let $\mathcal{Q} := C([-\zeta,\varpi], L^{\mu}(\Sigma, \eth_{\varpi}, \mathbb{P}, \mathbb{R}^n))$ be the Banach space of all μth-integrable and $\eth_{\varpi}$-adapted processes Ξ. A norm $\|\cdot\|$ on $\mathbb{R}^n$ can be represented by the matrix norm

$$\|\mathbb{D}\| = \max\left\{\sum_{i=1}^{n}|d_{i1}|, \sum_{i=1}^{n}|d_{i2}|, \ldots, \sum_{i=1}^{n}|d_{in}|\right\},$$

where $\mathbb{D} : \mathbb{R}^n \longrightarrow \mathbb{R}^n$. Furthermore, consider

$$\begin{aligned} &C^1(\mp, L^{\mu}(\Sigma, \eth_{\varpi}, \mathbb{P}, \mathbb{R}^n)) \\ &= \left\{\aleph \in C(\mp, L^{\mu}(\Sigma, \eth_{\varpi}, \mathbb{P}, \mathbb{R}^n)) : \aleph' \in C(\mp, L^{\mu}(\Sigma, \eth_{\varpi}, \mathbb{P}, \mathbb{R}^n))\right\}. \end{aligned}$$

Finally, we assume the initial values

$$\|\psi\|_C^{\mu} = \sup_{s\in\mp_1} \mathbf{E}\|\psi(s)\|^{\mu} \text{ and } \|\psi'\|_C^{\mu} = \sup_{s\in\mp_1} \mathbf{E}\|\psi'(s)\|^{\mu}.$$

Some of the basic definitions and lemmas employed in this study are discussed.

Definition 1 ([13]). *Let the $n \times n$ identity matrix and null matrix be symbolized by $\mathbb{I}$ and Θ, respectively. Then, for $\iota = 0, 1, 2, \ldots$, the delayed matrix functions $\mathcal{H}_{\zeta}(\mathbb{D}\ell)$ and $\mathcal{M}_{\zeta}(\mathbb{D}\ell)$ are defined, respectively, by*

$$\mathcal{H}_{\zeta}(\mathbb{D}\ell) := \begin{cases} \Theta, & -\infty < \ell < -\zeta, \\ \mathbb{I}, & -\zeta \le \ell < 0, \\ \mathbb{I} - \mathbb{D}\frac{\ell^2}{2!}, & 0 \le \ell < \zeta, \\ \vdots & \vdots \\ \mathbb{I} - \mathbb{D}\frac{\ell^2}{2!} + \mathbb{D}^2\frac{(\ell-\zeta)^4}{4!} & \\ +\cdots + (-1)^{\iota}\mathbb{D}^{\iota}\frac{(\ell-(\iota-1)\zeta)^{2\iota}}{(2\iota)!}, & (\iota-1)\zeta \le \ell < \iota\zeta, \end{cases} \tag{3}$$

and

$$\mathcal{M}_{\zeta}(\mathbb{D}\ell) := \begin{cases} \Theta, & -\infty < \ell < -\zeta, \\ \mathbb{I}(\ell+\zeta), & -\zeta \le \ell < 0, \\ \mathbb{I}(\ell+\zeta) - \mathbb{D}\frac{\ell^3}{3!}, & 0 \le \ell < \zeta, \\ \vdots & \vdots \\ \mathbb{I}(\ell+\zeta) - \mathbb{D}\frac{\ell^3}{3!} + \mathbb{D}^2\frac{(\ell-\zeta)^5}{5!} & \\ \quad +\cdots + (-1)^{\iota}\mathbb{D}^{\iota}\frac{(\ell-(\iota-1)\zeta)^{2\iota+1}}{(2\iota+1)!}, & (\iota-1)\zeta \le \ell < \iota\zeta, \end{cases} \tag{4}$$

Lemma 1 ([13]). *The solution of (2) can be expressed in the following form:*

$$\begin{aligned} \aleph(\ell) = {} & \mathcal{H}_{\zeta}(\mathbb{D}(\ell-\zeta))\psi(0) + \mathcal{M}_{\zeta}(\mathbb{D}(\ell-\zeta))\psi'(0) \\ & -\mathbb{D}\int_{-\zeta}^{0} \mathcal{M}_{\zeta}(\mathbb{D}(\ell-2\zeta-\varsigma))\psi(\varsigma)d\varsigma \\ & +\int_0^{\ell} \mathcal{M}_{\zeta}(\mathbb{D}(\ell-\zeta-\varsigma))\hbar(\varsigma,\aleph(\varsigma))d\varsigma \\ & +\int_0^{\ell} \mathcal{M}_{\zeta}(\mathbb{D}(\ell-\zeta-\varsigma))\Delta(\varsigma,\aleph(\varsigma))dZ_H(\varsigma). \end{aligned}$$

Lemma 2 ([29]). *If $\sigma : \mp \longrightarrow L_2^0$ satisfies*

$$\int_0^{\varpi} \|\sigma(\varsigma)\|_{L_2^0}^2 d\varsigma < \infty,$$

then

$$\mathbf{E}\left\|\int_0^{\ell}\sigma(\varsigma)\mathrm{d}Z_H(\varsigma)\right\|^2 \leq 2H\ell^{2H-1}\int_0^{\ell}\|\sigma(\varsigma)\|^2_{L_2^0}\mathrm{d}\varsigma.$$

Lemma 3 ([36]). *For $\Lambda : \mp \longrightarrow L_2^0$, such that*

$$\int_0^{\ell}\|\Lambda(\varsigma)\|^{\mu}_{L_2^0}\mathrm{d}\varsigma < \infty,$$

and applying Hölder's inequality and the Kahane–Khintchine inequality, there is a constant τ_μ, such that

$$\begin{aligned}
\mathbf{E}\left\|\int_0^{\ell}\Lambda(\varsigma)\mathrm{d}Z_H(\varsigma)\right\|^{\mu} &\leq \tau_\mu\left\{\mathbf{E}\left\|\int_0^{\ell}\Lambda(\varsigma)\mathrm{d}Z_H(\varsigma)\right\|^2\right\}^{\mu/2} \\
&\leq \tau_\mu\left\{2H\ell^{2H-1}\int_0^{\ell}\|\Lambda(\varsigma)\|^2_{L_2^0}\mathrm{d}\varsigma\right\}^{\mu/2} \\
&\leq \tau_\mu\left(2H\ell^{2H-1}\right)^{\mu/2}\left(\int_0^{\ell}\mathrm{d}\varsigma\right)^{\mu/2-1}\int_0^{\ell}\left(\|\Lambda(\varsigma)\|^2_{L_2^0}\right)^{\mu/2}\mathrm{d}\varsigma \\
&= \tau_\mu(2H)^{\mu/2}\ell^{\mu H-1}\int_0^{\ell}\|\Lambda(\varsigma)\|^{\mu}_{L_2^0}\mathrm{d}\varsigma.
\end{aligned}$$

Definition 2 ([37]). *When considering a specific constant $\kappa > 0$, and a function $\Pi \in C(\mp, \mathbb{R}^n)$ fulfilling*

$$\mathbf{E}\left\|\Pi''(\ell) + \mathbb{D}\Pi(\ell-\zeta) - \hbar(\ell,\Pi(\ell)) - \Delta(\ell,\Pi(\ell))\mathrm{d}Z_H(\ell)\right\|^{\mu} \leq \kappa, \quad \ell \in [0,\varpi], \tag{5}$$

implies that there exist a solution $\aleph \in C(\mp, \mathbb{R}^n)$ of (2) and a number $W > 0$ such that

$$\mathbf{E}\|\Pi(\ell) - \aleph(\ell)\|^{\mu} \leq W\kappa, \quad \textit{for all } \ell \in [0,\varpi].$$

The system (2) is Hyers–Ulam stable on $[0,\varpi]$.

Remark 1 ([37]). *A function $\Pi \in C(\mp, \mathbb{R}^n)$ is a solution of the inequality (5) if and only if there exists a function $\mathcal{E} \in C(\mp, \mathbb{R}^n)$, such that*

(i) $\mathbf{E}\|\mathcal{E}(\ell)\|^{\mu} \leq \kappa$, $\ell \in \mp$.
(ii) $\Pi''(\ell) = -\mathbb{D}\Pi(\ell-\zeta) + \Delta(\ell,\Pi(\ell))\mathrm{d}Z_H(\ell) + \hbar(\ell,\Pi(\ell)) + \mathcal{E}(\ell)$, $\ell \in \mp$.

Definition 3 ([38]). *The Mittag–Leffler function, containing two parameters, is defined as*

$$\mathbb{E}_{\alpha,\epsilon}(\ell) = \sum_{\iota=0}^{\infty}\frac{\ell^{\iota}}{\Gamma(\alpha\iota+\epsilon)}, \quad \alpha,\ \epsilon > 0,\ \ell \in \mathbb{C}.$$

If $\epsilon = 1$, then

$$\mathbb{E}_{\alpha,1}(\ell) = \mathbb{E}_{\alpha}(\ell) = \sum_{\iota=0}^{\infty}\frac{\ell^{\iota}}{\Gamma(\alpha\iota+1)}, \quad \alpha > 0.$$

Lemma 4 ([15]). *For any $\ell \in [(\iota-1)\zeta, \iota\zeta]$, $\iota = 1,2,\ldots$, we obtain*

$$\|\mathcal{H}_\zeta(\mathbb{D}(\ell))\| \leq \mathbb{E}_2\left(\|\mathbb{D}\|\ell^2\right),$$

and

$$\|\mathcal{M}_\zeta(\mathbb{D}(\ell))\| \leq (\ell+\zeta)\mathbb{E}_{2,2}\left(\|\mathbb{D}\|(\ell+\zeta)^2\right).$$

Lemma 5. *(Grönwall's inequality, [39]). Let $\hbar(\ell)$ and $\wp(\ell)$ be nonnegative, continuous functions on $0 \le \ell \le T$, for which the inequality*

$$\hbar(\ell) \le \eta + \int_0^\ell \wp(s)\hbar(s)\mathrm{d}s, \quad \text{for } \ell \in [0,T],$$

holds, where $\eta \ge 0$ is a constant. Then,

$$\hbar(\ell) \le \eta \exp\left(\int_0^\ell \wp(s)\mathrm{d}s\right), \quad \text{for } \ell \in [0,T].$$

Lemma 6. *(Krasnoselskii's fixed point theorem, [40]). Assume that $\mathcal{J}$ is a closed, bounded, and non-empty convex subset of a Banach space $\mathcal{U}$. If O_1 and O_2 are mappings from $\mathcal{J}$ into $\mathcal{U}$, such that*
(i) $O_1\ell + O_2\aleph \in \mathcal{J}$ for every pair $\ell, \aleph \in \mathcal{J}$,
(ii) O_2 is a contraction mapping,
(iii) O_1 is continuous and compact,
then there is $\Im \in \mathcal{J}$, such that $\Im = O_1\Im + O_2\Im$.

3. Main Results

In this section, we present and prove the existence, uniqueness, and Hyers–Ulam stability results of (2). To prove our main results, the assumptions listed below are assumed:

(G1): There exist a function $\Delta : \mp \times \mathbb{R}^n \longrightarrow L_2^0$ that is continuous, and a constant $U_\Delta \in L^{r_2}(\mp, \mathbb{R}^+)$ and $r_2 > 1$, such that

$$\mathbf{E}\|\Delta(\ell,\aleph_1) - \Delta(\ell,\aleph_2)\|_{L_2^0}^\mu \le U_\Delta(\ell)\mathbf{E}\|\aleph_1 - \aleph_2\|^\mu, \quad \text{for all } \ell \in \mp, \aleph_1, \aleph_2 \in \mathbb{R}^n.$$

Let $\mu \in [2,\infty)$ and $\sup_{\ell\in\mp} \mathbf{E}\|\Delta(\ell,0)\|_{L_2^0}^\mu = W_\Delta < \infty$.

(G2): There exist a function $\hbar : \mp \times \mathbb{R}^n \longrightarrow L_2^0$ that is continuous, and a constant $U_\hbar \in L^{r_2}(\mp, \mathbb{R}^+)$ and $r_2 > 1$, such that

$$\mathbf{E}\|\hbar(\ell,\aleph_1) - \hbar(\ell,\aleph_2)\|^\mu \le U_\hbar(\ell)\mathbf{E}\|\aleph_1 - \aleph_2\|^\mu, \ \mathbf{E}\|\hbar(\ell,\aleph)\|^\mu \le U_\hbar(\ell)\left(1 + \mathbf{E}\|\aleph\|^\mu\right),$$

for all $\ell \in \mp, \aleph_1, \aleph_2 \in \mathbb{R}^n$.

Using Krasnoselskii's fixed point theorem, we now prove the existence and uniqueness results.

Theorem 1. *If* **(G1)**–**(G2)** *holds, then there exists a unique mild solution of the nonlinear stochastic system (2), provided that*

$$2^{\mu-1}W_2 + W_3 < 1, \tag{6}$$

where

$$W_2 := \frac{\tau_\mu (2H)^{\mu/2} \varpi^{\mu(H+1)-\frac{1}{r_2}}}{(\mu r_1 + 1)^{\frac{1}{r_1}}} \left(\mathbb{E}_{2,2}\left(\|\mathbb{D}\|\varpi^2\right)\right)^\mu \|U_\Delta\|_{L^{r_2}(\mp,\mathbb{R}^+)},$$

and

$$W_3 := \frac{\varpi^{\mu+\frac{1}{r_1}}}{(\mu r_1 + 1)^{\frac{1}{r_1}}} \left(\mathbb{E}_{2,2}\left(\|\mathbb{D}\|\varpi^2\right)\right)^\mu \|U_\hbar\|_{L^{r_2}(\mp,\mathbb{R}^+)},$$

for $\frac{1}{r_1} + \frac{1}{r_2} = 1$, $r_1, r_2 > 1$.

Proof. We deal with the set

$$\mathcal{T}_\varrho = \left\{\aleph \in \mathcal{Q} : \|\aleph\|_{\mathcal{Q}}^\mu = \sup_{\ell\in\mp} \mathbf{E}\|\aleph(\ell)\|^\mu \le \varrho\right\},$$

for each positive number ϱ. Let $\ell \in \mp$. Applying Lemma 1, we then transform problem (2) into a fixed point problem and define an operator $F : \mathcal{Q} \longrightarrow \mathcal{Q}$ by

$$\begin{aligned}(F\aleph)(\ell) = {} & \mathcal{H}_\zeta(\mathbb{D}(\ell-\zeta))\psi(0) + \mathcal{M}_\zeta(\mathbb{D}(\ell-\zeta))\psi'(0) \\ & -\mathbb{D}\int_{-\zeta}^{0} \mathcal{M}_\zeta(\mathbb{D}(\ell-2\zeta-\varsigma))\psi(\varsigma)d\varsigma \\ & +\int_0^\ell \mathcal{M}_\zeta(\mathbb{D}(\ell-\zeta-\varsigma))\hbar(\varsigma,\aleph(\varsigma))d\varsigma \\ & +\int_0^\ell \mathcal{M}_\zeta(\mathbb{D}(\ell-\zeta-\varsigma))\Delta(\varsigma,\aleph(\varsigma))dZ_H(\varsigma),\end{aligned}$$

for $\ell \in \mp$. Decomposing the operator F, the operators $\mathcal{L}_1$ and $\mathcal{L}_2$ can be described on $\mathcal{T}_\varrho$, as provided below:

$$\begin{aligned}(\mathcal{L}_1\aleph)(\ell) = {} & \mathcal{H}_\zeta(\mathbb{D}(\ell-\zeta))\psi(0) + \mathcal{M}_\zeta(\mathbb{D}(\ell-\zeta))\psi'(0) \\ & -\mathbb{D}\int_{-\zeta}^{0} \mathcal{M}_\zeta(\mathbb{D}(\ell-2\zeta-\varsigma))\psi(\varsigma)d\varsigma \\ & +\int_0^\ell \mathcal{M}_\zeta(\mathbb{D}(\ell-\zeta-\varsigma))\hbar(\varsigma,\aleph(\varsigma))d\varsigma,\end{aligned} \tag{7}$$

$$(\mathcal{L}_2\aleph)(\ell) = \int_0^\ell \mathcal{M}_\zeta(\mathbb{D}(\ell-\zeta-\varsigma))\Delta(\varsigma,\aleph(\varsigma))dZ_H(\varsigma). \tag{8}$$

At this point, we observe that $\mathcal{T}_\varrho$ is a convex set, closed and bounded of $\mathcal{Q}$. Consequently, our proof consists of three essential steps:

Step 1. We show the existence of $\varrho > 0$, such that $\mathcal{L}_1\aleph + \mathcal{L}_2\Im \in \mathcal{T}_\varrho$ for all $\aleph, \Im \in \mathcal{T}_\varrho$. For each $\ell \in \mp$ and $\aleph, \Im \in \mathcal{T}_\varrho$, and using (7) and (8), we obtain

$$\begin{aligned}& \|\mathcal{L}_1\aleph + \mathcal{L}_2\Im\|_{\mathcal{Q}}^\mu \\ & = \sup_{\ell\in\mp} \mathbf{E}\|(\mathcal{L}_1\aleph + \mathcal{L}_2\Im)(\ell)\|^\mu \\ & \leq 5^{\mu-1}\Big[\|\mathcal{H}_\zeta(\mathbb{D}(\ell-\zeta))\|^\mu \mathbf{E}\|\psi(0)\|^\mu + \|\mathcal{M}_\zeta(\mathbb{D}(\ell-\zeta))\|^\mu \mathbf{E}\|\psi'(0)\|^\mu \\ & \quad +\|\mathbb{D}\|^\mu \mathbf{E}\left\|\int_{-\zeta}^{0} \mathcal{M}_\zeta(\mathbb{D}(\ell-2\zeta-\varsigma))\psi(\varsigma)d\varsigma\right\|^\mu \\ & \quad +\mathbf{E}\left\|\int_0^\ell \mathcal{M}_\zeta(\mathbb{D}(\ell-\zeta-\varsigma))\hbar(\varsigma,\aleph(\varsigma))d\varsigma\right\|^\mu \\ & \quad +\mathbf{E}\left\|\int_0^\ell \mathcal{M}_\zeta(\mathbb{D}(\ell-\zeta-\varsigma))\Delta(\varsigma,\Im(\varsigma))dZ_H(\varsigma)\right\|^\mu\Big] \\ & = \textstyle\sum_{n=1}^{5}\mathbb{I}_n.\end{aligned} \tag{9}$$

From Lemma 4, we have

$$\begin{aligned}\mathbf{I}_1 & = 5^{\mu-1}\|\mathcal{H}_\zeta(\mathbb{D}(\ell-\zeta))\|^\mu \mathbf{E}\|\psi(0)\|^\mu \\ & \leq 5^{\mu-1}\Big(\mathbb{E}_2\Big(\|\mathbb{D}\|(\ell-\zeta)^2\Big)\Big)^\mu \mathbf{E}\|\psi\|_C^\mu,\end{aligned}$$

$$\begin{aligned}\mathbf{I}_2 & = 5^{\mu-1}\|\mathcal{M}_\zeta(\mathbb{D}(\ell-\zeta))\|^\mu \mathbf{E}\|\psi'(0)\|^\mu \\ & \leq 5^{\mu-1}\Big(\ell\mathbb{E}_{2,2}\Big(\|\mathbb{D}\|\ell^2\Big)\Big)^\mu \mathbf{E}\|\psi'\|_C^\mu,\end{aligned}$$

$$\begin{aligned}\mathbf{I}_3 &= 5^{\mu-1}\|\mathbb{D}\|^{\mu}\mathbf{E}\left\|\int_{-\zeta}^{0}\mathcal{M}_{\zeta}(\mathbb{D}(\ell-2\zeta-\varsigma))\psi(\varsigma)\mathrm{d}\varsigma\right\|^{\mu}\\&\leq 5^{\mu-1}\|\mathbb{D}\|^{\mu}\zeta^{\mu-1}\mathbf{E}\|\psi\|_C^{\mu}\int_{-\zeta}^{0}\|\mathcal{M}_{\zeta}(\mathbb{D}(\ell-2\zeta-\varsigma))\|^{\mu}\mathrm{d}\varsigma\\&\leq 5^{\mu-1}\|\mathbb{D}\|^{\mu}\zeta^{\mu}\left(\ell\mathbb{E}_{2,2}\left(\|\mathbb{D}\|\ell^2\right)\right)^{\mu}\mathbf{E}\|\psi\|_C^{\mu},\end{aligned}$$

$$\begin{aligned}\mathbf{I}_4 &= 5^{\mu-1}\mathbf{E}\left\|\int_{0}^{\ell}\mathcal{M}_{\zeta}(\mathbb{D}(\ell-\zeta-\varsigma))\Delta(\varsigma,\Im(\varsigma))\mathrm{d}Z_H(\varsigma)\right\|^{\mu}\\&= 5^{\mu-1}\mathbf{E}\left\{\left\|\int_{0}^{\ell}\mathcal{M}_{\zeta}(\mathbb{D}(\ell-\zeta-\varsigma))\Delta(\varsigma,\Im(\varsigma))\mathrm{d}Z_H(\varsigma)\right\|^{2}\right\}^{\mu/2}.\end{aligned}$$

Applying Lemmas 2 and 3, we obtain

$$\begin{aligned}\mathbf{I}_4 &\leq 5^{\mu-1}\tau_{\mu}\left\{\mathbf{E}\left\|\int_{0}^{\ell}\mathcal{M}_{\zeta}(\mathbb{D}(\ell-\zeta-\varsigma))\Delta(\varsigma,\Im(\varsigma))\mathrm{d}Z_H(\varsigma)\right\|^{2}\right\}^{\mu/2}\\&\leq 5^{\mu-1}\tau_{\mu}\left\{2H\ell^{2H-1}\int_{0}^{\ell}\mathbf{E}\|\mathcal{M}_{\zeta}(\mathbb{D}(\ell-\zeta-\varsigma))\Delta(\varsigma,\Im(\varsigma))\|_{L_2^0}^{2}\mathrm{d}\varsigma\right\}^{\mu/2}\\&\leq 5^{\mu-1}\tau_{\mu}\left(2H\ell^{2H-1}\right)^{\mu/2}\left\{\int_{0}^{\ell}\mathbf{E}\|\mathcal{M}_{\zeta}(\mathbb{D}(\ell-\zeta-\varsigma))\Delta(\varsigma,\Im(\varsigma))\|_{L_2^0}^{2}\mathrm{d}\varsigma\right\}^{\mu/2}\\&\leq 5^{\mu-1}\tau_{\mu}\left(2H\ell^{2H-1}\right)^{\mu/2}\\&\quad\times\left\{\left(\int_{0}^{\ell}\left(\mathbf{E}\|\mathcal{M}_{\zeta}(\mathbb{D}(\ell-\zeta-\varsigma))\Delta(\varsigma,\Im(\varsigma))\|_{L_2^0}^{2}\right)^{\mu/2}\mathrm{d}\varsigma\right)^{2/\mu}\left(\int_{0}^{\ell}\mathrm{d}\varsigma\right)^{\frac{\mu-2}{\mu}}\right\}^{\mu/2}\\&\leq 5^{\mu-1}\tau_{\mu}(2H)^{\mu/2}\varpi^{\mu H-1}\int_{0}^{\ell}\mathbf{E}\|\mathcal{M}_{\zeta}(\mathbb{D}(\ell-\zeta-\varsigma))\Delta(\varsigma,\Im(\varsigma))\|_{L_2^0}^{\mu}\mathrm{d}\varsigma.\end{aligned}$$

Using Lemma 4 and (**G1**), we obtain

$$\begin{aligned}\mathbf{I}_4 &\leq 5^{\mu-1}\tau_{\mu}(2H)^{\mu/2}\varpi^{\mu H-1}\int_{0}^{\ell}\left((\ell-\varsigma)\mathbb{E}_{2,2}\left(\|\mathbb{D}\|(\ell-\varsigma)^2\right)\right)^{\mu}\mathbf{E}\|\Delta(\varsigma,\Im(\varsigma))\|_{L_2^0}^{\mu}\mathrm{d}\varsigma\\&\leq 5^{\mu-1}\tau_{\mu}(2H)^{\mu/2}\varpi^{\mu H-1}\\&\quad\times 2^{\mu-1}\left\{\int_{0}^{\ell}\left((\ell-\varsigma)\mathbb{E}_{2,2}\left(\|\mathbb{D}\|(\ell-\varsigma)^2\right)\right)^{\mu}\mathbf{E}\|\Delta(\varsigma,\Im(\varsigma))-\Delta(\varsigma,0)\|_{L_2^0}^{\mu}\mathrm{d}\varsigma\right.\\&\qquad\left.+\int_{0}^{\ell}\left((\ell-\varsigma)\mathbb{E}_{2,2}\left(\|\mathbb{D}\|(\ell-\varsigma)^2\right)\right)^{\mu}\mathbf{E}\|\Delta(\varsigma,0)\|_{L_2^0}^{\mu}\mathrm{d}\varsigma\right\}\\&\leq (10)^{\mu-1}\tau_{\mu}(2H)^{\mu/2}\varpi^{\mu H-1}\\&\quad\times\left\{\int_{0}^{\ell}\left((\ell-\varsigma)\mathbb{E}_{2,2}\left(\|\mathbb{D}\|(\ell-\varsigma)^2\right)\right)^{\mu}U_{\Delta}(\varsigma)\mathbf{E}\|\Im(\varsigma)\|^{\mu}\mathrm{d}\varsigma\right.\\&\qquad\left.+W_{\Delta}\int_{0}^{\ell}\left((\ell-\varsigma)\mathbb{E}_{2,2}\left(\|\mathbb{D}\|(\ell-\varsigma)^2\right)\right)^{\mu}\mathrm{d}\varsigma\right\}\\&\leq (10)^{\mu-1}\tau_{\mu}(2H)^{\mu/2}\varpi^{\mu H-1}\left\{\|\Im\|_{\mathcal{Q}}^{\mu}\int_{0}^{\ell}\left((\ell-\varsigma)\mathbb{E}_{2,2}\left(\|\mathbb{D}\|(\ell-\varsigma)^2\right)\right)^{\mu}U_{\Delta}(\varsigma)\mathrm{d}\varsigma\right.\\&\qquad\left.+\frac{\varpi^{\mu+1}W_{\Delta}}{\mu+1}\left(\mathbb{E}_{2,2}\left(\|\mathbb{D}\|\varpi^2\right)\right)^{\mu}\right\}.\end{aligned}\tag{10}$$

Additionally, using Hölder inequality and (**G1**), we obtain

$$\begin{aligned}
&\int_0^\ell \Big((\ell-\varsigma)\mathbb{E}_{2,2}\Big(\|\mathbb{D}\|(\ell-\varsigma)^2\Big)\Big)^\mu U_\Delta(\varsigma)d\varsigma \\
&\leq \left(\int_0^\ell \Big((\ell-\varsigma)\mathbb{E}_{2,2}\Big(\|\mathbb{D}\|(\ell-\varsigma)^2\Big)\Big)^{\mu r_1} d\varsigma\right)^{\frac{1}{r_1}} \left(\int_0^\ell U_\Delta^{r_2}(\varsigma)d\varsigma\right)^{\frac{1}{r_2}} \\
&\leq \Big(\mathbb{E}_{2,2}\Big(\|\mathbb{D}\|\varpi^2\Big)\Big)^\mu \left(\int_0^\ell (\ell-\varsigma)^{\mu r_1} d\varsigma\right)^{\frac{1}{r_1}} \left(\int_0^\ell U_\Delta^{r_2}(\varsigma)d\varsigma\right)^{\frac{1}{r_2}} \\
&\leq \frac{\varpi^{\mu+\frac{1}{r_1}}}{(\mu r_1+1)^{\frac{1}{r_1}}} \Big(\mathbb{E}_{2,2}\Big(\|\mathbb{D}\|\varpi^2\Big)\Big)^\mu \|U_\Delta\|_{L^{r_2}(\mp,\mathbb{R}^+)}.
\end{aligned} \tag{11}$$

Substituting (11) into (10), we obtain

$$\begin{aligned}
\mathbf{I}_4 &\leq (10)^{\mu-1}\tau_\mu(2H)^{\mu/2}\varpi^{\mu H-1} \\
&\quad\times\left\{\frac{\varrho\varpi^{\mu+\frac{1}{r_1}}}{(\mu r_1+1)^{\frac{1}{r_1}}}\Big(\mathbb{E}_{2,2}\Big(\|\mathbb{D}\|\varpi^2\Big)\Big)^\mu \|U_\Delta\|_{L^{r_2}(\mp,\mathbb{R}^+)} + \frac{\varpi^{\mu+1}W_\Delta}{\mu+1}\Big(\mathbb{E}_{2,2}\Big(\|\mathbb{D}\|\varpi^2\Big)\Big)^\mu\right\} \\
&= (10)^{\mu-1}W_2\varrho + \frac{(10)^{\mu-1}\tau_\mu(2H)^{\mu/2}\varpi^{\mu(H+1)}W_\Delta}{\mu+1}\Big(\mathbb{E}_{2,2}\Big(\|\mathbb{D}\|\varpi^2\Big)\Big)^\mu.
\end{aligned}$$

Furthermore, using (11) and (**G2**), we obtain

$$\begin{aligned}
\mathbf{I}_5 &= 5^{\mu-1}\mathbf{E}\left\|\int_0^\ell \mathcal{M}_\zeta(\mathbb{D}(\ell-\zeta-\varsigma))\hbar(\varsigma,\aleph(\varsigma))d\varsigma\right\|^\mu \\
&\leq 5^{\mu-1}\int_0^\ell \Big((\ell-\varsigma)\mathbb{E}_{2,2}\Big(\|\mathbb{D}\|(\ell-\varsigma)^2\Big)\Big)^\mu \mathbf{E}\|\hbar(\varsigma,\aleph(\varsigma))\|^\mu d\varsigma \\
&\leq 5^{\mu-1}\int_0^\ell \Big((\ell-\varsigma)\mathbb{E}_{2,2}\Big(\|\mathbb{D}\|(\ell-\varsigma)^2\Big)\Big)^\mu U_\hbar(\varsigma)(1+\mathbf{E}\|\aleph\|^\mu)d\varsigma \\
&\leq \frac{5^{\mu-1}(1+\varrho)\varpi^{\mu+\frac{1}{r_1}}}{(\mu r_1+1)^{\frac{1}{r_1}}}\Big(\mathbb{E}_{2,2}\Big(\|\mathbb{D}\|\varpi^2\Big)\Big)^\mu \|U_\hbar\|_{L^{r_2}(\mp,\mathbb{R}^+)} \\
&= 5^{\mu-1}(1+\varrho)W_3.
\end{aligned}$$

From $\mathbf{I}_1$ to $\mathbf{I}_5$, (9) becomes

$$\begin{aligned}
&\|\mathcal{L}_1\aleph + \mathcal{L}_2\Im\|_Q^\mu \\
&\leq 5^{\mu-1}\Big\{\Big(\mathbb{E}_2\Big(\|\mathbb{D}\|(\ell-\zeta)^2\Big)\Big)^\mu \mathbf{E}\|\psi\|_C^\mu \\
&\quad+\Big(\ell\mathbb{E}_{2,2}\Big(\|\mathbb{D}\|\ell^2\Big)\Big)^\mu \mathbf{E}\|\psi'\|_C^\mu \\
&\quad+\|\mathbb{D}\|^\mu\zeta^\mu\Big(\ell\mathbb{E}_{2,2}\Big(\|\mathbb{D}\|\ell^2\Big)\Big)^\mu \mathbf{E}\|\psi\|_C^\mu \\
&\quad+2^{\mu-1}W_2\varrho + \frac{2^{\mu-1}\tau_\mu(2H)^{\mu/2}\varpi^{\mu(H+1)}W_\Delta}{\mu+1}\Big(\mathbb{E}_{2,2}\Big(\|\mathbb{D}\|\varpi^2\Big)\Big)^\mu \\
&\quad+(1+\varrho)W_3\} \\
&\leq 5^{\mu-1}\Big\{\theta(\varpi) + \varrho\Big(2^{\mu-1}W_2 + W_3\Big) + W_3\Big\},
\end{aligned}$$

where

$$\theta(\ell) = \left(\mathbb{E}_2\left(\|\mathbb{D}\|(\ell-\zeta)^2\right)\right)^{\mu}\mathbf{E}\|\psi\|_C^{\mu} + \left(\ell\mathbb{E}_{2,2}\left(\|\mathbb{D}\|\ell^2\right)\right)^{\mu}\mathbf{E}\|\psi'\|_C^{\mu}$$
$$+\|\mathbb{D}\|^{\mu}\zeta^{\mu}\left(\ell\mathbb{E}_{2,2}\left(\|\mathbb{D}\|\ell^2\right)\right)^{\mu}\mathbf{E}\|\psi\|_C^{\mu}$$
$$+\frac{2^{\mu-1}\tau_{\mu}(2H)^{\mu/2}\ell^{\mu(H+1)}W_{\Delta}}{\mu+1}\left(\mathbb{E}_{2,2}\left(\|\mathbb{D}\|\ell^2\right)\right)^{\mu}.$$

As a result, from (6), we obtain $\mathcal{L}_1\aleph + \mathcal{L}_2\Im \in \mathcal{T}_{\varrho}$ for some ϱ sufficiency large.

Step 2. We show that $\mathcal{L}_1 : \mathcal{T}_{\varrho} \longrightarrow \mathcal{Q}$ is a contraction. For each $\ell \in \mp$ and $\aleph, \Im \in \mathcal{T}_{\varrho}$, using (7) and (**G2**), we obtain

$$\mathbf{E}\|(\mathcal{L}_1\aleph)(\ell) - (\mathcal{L}_1\Im)(\ell)\|^{\mu}$$
$$= \mathbf{E}\left\|\int_0^{\ell}\mathcal{M}_{\zeta}(\mathbb{D}(\ell-\zeta-\varsigma))[\hbar(\varsigma,\aleph(\varsigma)) - \hbar(\varsigma,\Im(\varsigma))]\mathrm{d}\varsigma\right\|^{\mu}$$
$$\leq \mathbf{E}\|\aleph-\Im\|_{\mathcal{Q}}^{\mu}\int_0^{\ell}\left((\ell-\varsigma)\mathbb{E}_{2,2}\left(\|\mathbb{D}\|(\ell-\varsigma)^2\right)\right)^{\mu}U_{\hbar}(\varsigma)\mathrm{d}\varsigma$$
$$\leq W_3\|\aleph-\Im\|_{\mathcal{Q}}^{\mu}.$$

As we can see from (6), noting $W_3 < 1$, that $\mathcal{L}_1$ is a contraction mapping.

Step 3. We show that $\mathcal{L}_2 : \mathcal{T}_{\varrho} \longrightarrow \mathcal{Q}$ is a continuous compact operator. First, we verify the continuity of $\mathcal{L}_2$. Consider $\aleph_n \in \mathcal{T}_{\varrho}$ with $\aleph_n \longrightarrow \aleph$ as $n \longrightarrow \infty$ in $\mathcal{T}_{\varrho}$. Thus, using Lebesgue's dominated convergence theorem and (8), we obtain, for each $\ell \in \mp$,

$$\mathbf{E}\|(\mathcal{L}_2\aleph_n)(\ell) - (\mathcal{L}_2\aleph)(\ell)\|^{\mu}$$
$$\leq \tau_{\mu}(2H)^{\mu/2}\varpi^{\mu H-1}\int_0^{\ell}\|\mathcal{M}_{\zeta}(\mathbb{D}(\ell-\zeta-\varsigma))\|^{\mu}\mathbf{E}\|\Delta(\varsigma,\aleph_n(\varsigma)) - \Delta(\varsigma,\aleph(\varsigma))\|_{L_2^0}^{\mu}\mathrm{d}\varsigma$$
$$\leq \tau_{\mu}(2H)^{\mu/2}\varpi^{\mu H-1}\int_0^{\ell}\left((\ell-\varsigma)\mathbb{E}_{2,2}\left(\|\mathbb{D}\|(\ell-\varsigma)^2\right)\right)^{\mu}U_{\Delta}(\varsigma)$$
$$\times\|\aleph_n-\aleph\|_{\mathcal{Q}}^{\mu}\mathrm{d}\varsigma \longrightarrow 0, \text{ as } n \longrightarrow \infty.$$

This proves the continuity of $\mathcal{L}_2 : \mathcal{T}_{\varrho} \longrightarrow \mathcal{Q}$. Thereafter, we show that $\mathcal{L}_2$ is uniformly bounded on $\mathcal{T}_{\varrho}$. For each $\ell \in \mp$, $\aleph \in \mathcal{T}_{\varrho}$, we have

$$\|\mathcal{L}_2\aleph\|_{\mathcal{Q}}^{\mu} = \sup_{\ell\in\mp}\mathbf{E}\|(\mathcal{L}_2\aleph)(\ell)\|^{\mu}$$
$$\leq \sup_{\ell\in\mp}\left\{\mathbf{E}\left\|\int_0^{\ell}\mathcal{M}_{\zeta}(\mathbb{D}(\ell-\zeta-\varsigma))\Delta(\varsigma,\aleph(\varsigma))\mathrm{d}Z_H(\varsigma)\right\|^{\mu}\right\}$$
$$\leq 2^{\mu-1}W_2\varrho + \frac{2^{\mu-1}\tau_{\mu}(2H)^{\mu/2}\varpi^{\mu(H+1)}W_{\Delta}}{\mu+1}\left(\mathbb{E}_{2,2}\left(\|\mathbb{D}\|\varpi^2\right)\right)^{\mu},$$

this indicates that, on $\mathcal{T}_{\varrho}$, $\mathcal{L}_2$ is uniformly bounded. Showing that $\mathcal{L}_2$ is equicontinuous is still necessary. For each $\ell_2, \ell_3 \in \mp, 0 < \ell_2 < \ell_3 \leq \varpi$ and $\aleph \in \mathcal{T}_{\varrho}$, using (8), we obtain

$$(\mathcal{L}_2\aleph)(\ell_3) - (\mathcal{L}_2\aleph)(\ell_2)$$
$$= \int_0^{\ell_3}\mathcal{M}_{\zeta}(\mathbb{D}(\ell_3-\zeta-\varsigma))\Delta(\varsigma,\aleph(\varsigma))\mathrm{d}Z_H(\varsigma)$$
$$-\int_0^{\ell_2}\mathcal{M}_{\zeta}(\mathbb{D}(\ell_2-\zeta-\varsigma))\Delta(\varsigma,\aleph(\varsigma))\mathrm{d}Z_H(\varsigma)$$
$$= \Psi_1 + \Psi_2,$$

where
$$\Psi_1 = \int_{\ell_2}^{\ell_3} \mathcal{M}_\zeta(\mathbb{D}(\ell_3 - \zeta - \varsigma))\Delta(\varsigma, \aleph(\varsigma))dZ_H(\varsigma),$$
and
$$\Psi_2 = \int_0^{\ell_2} [\mathcal{M}_\zeta(\mathbb{D}(\ell_3 - \zeta - \varsigma)) - \mathcal{M}_\zeta(\mathbb{D}(\ell_2 - \zeta - \varsigma))]\Delta(\varsigma, \aleph(\varsigma))dZ_H(\varsigma).$$
Thus
$$\begin{aligned}\mathbf{E}\|(\mathcal{L}_2\aleph)(\ell_3) - (\mathcal{L}_2\aleph)(\ell_2)\|^\mu &= \mathbf{E}\|\Psi_1 + \Psi_2\|^\mu \\ &\leq 2^{\mu-1}\{\mathbf{E}\|\Psi_1\|^\mu + \mathbf{E}\|\Psi_2\|^\mu\}. \end{aligned} \tag{12}$$
Now, we can check $\|\Psi_r\| \longrightarrow 0$ as $\ell_2 \longrightarrow \ell_3$, when $r = 1, 2$. For Ψ_1, we obtain
$$\begin{aligned}\mathbf{E}\|\Psi_1\|^\mu &= \mathbf{E}\left\|\int_{\ell_2}^{\ell_3} \mathcal{M}_\zeta(\mathbb{D}(\ell_3 - \zeta - \varsigma))\Delta(\varsigma, \aleph(\varsigma))dZ_H(\varsigma)\right\|^\mu \\ &\leq \tau_\mu(2H)^{\mu/2}(\ell_3 - \ell_2)^{\mu H-1}\int_{\ell_2}^{\ell_3} \mathbf{E}\|\mathcal{M}_\zeta(\mathbb{D}(\ell - \zeta - \varsigma))\Delta(\varsigma, \aleph(\varsigma))\|_{L_2^0}^\mu d\varsigma \\ &\leq 2^{\mu-1}\tau_\mu(2H)^{\mu/2}(\ell_3 - \ell_2)^{\mu H-1} \\ &\quad\times\left\{\varrho\int_{\ell_2}^{\ell_3}\left((\ell - \varsigma)\mathbb{E}_{2,2}\left(\|\mathbb{D}\|(\ell - \varsigma)^2\right)\right)^\mu U_\Delta(\varsigma)d\varsigma \right.\\ &\quad\left.+\frac{(\ell_3 - \ell_2)^{\mu+1}W_\Delta}{\mu+1}\left(\mathbb{E}_{2,2}\left(\|\mathbb{D}\|(\ell_3 - \ell_2)^2\right)\right)^\mu\right\} \longrightarrow 0, \text{ as } \ell_2 \longrightarrow \ell_3.\end{aligned}$$
For Ψ_2, we obtain
$$\begin{aligned}&\mathbf{E}\|\Psi_2\|^\mu \\ &= \mathbf{E}\left\|\int_0^{\ell_2}[\mathcal{M}_\zeta(\mathbb{D}(\ell_3 - \zeta - \varsigma)) - \mathcal{M}_\zeta(\mathbb{D}(\ell_2 - \zeta - \varsigma))]\Delta(\varsigma, \aleph(\varsigma))dZ_H(\varsigma)\right\|^\mu \\ &\leq \tau_\mu(2H)^{\mu/2}\ell_2^{\mu H-1} \\ &\quad\times\int_0^{\ell_2}\mathbf{E}\|[\mathcal{M}_\zeta(\mathbb{D}(\ell_3 - \zeta - \varsigma)) - \mathcal{M}_\zeta(\mathbb{D}(\ell_2 - \zeta - \varsigma))]\Delta(\varsigma, \aleph(\varsigma))\|_{L_2^0}^\mu d\varsigma \\ &\leq 2^{\mu-1}\tau_\mu(2H)^{\mu/2}\ell_2^{\mu H-1} \\ &\quad\times\left\{\varrho\int_0^{\ell_2}\|\mathcal{M}_\zeta(\mathbb{D}(\ell_3 - \zeta - \varsigma)) - \mathcal{M}_\zeta(\mathbb{D}(\ell_2 - \zeta - \varsigma))\|^\mu U_\Delta(\varsigma)d\varsigma\right. \\ &\quad\left.+W_\Delta\int_0^{\ell_2}\|\mathcal{M}_\zeta(\mathbb{D}(\ell_3 - \zeta - \varsigma)) - \mathcal{M}_\zeta(\mathbb{D}(\ell_2 - \zeta - \varsigma))\|^\mu d\varsigma\right\} \\ &\leq 2^{\mu-1}\tau_\mu(2H)^{\mu/2}\ell_2^{\mu H-1} \\ &\quad\times\left\{\varrho\|U_\Delta\|_{L^{r_2}(\mp,\mathbb{R}^+)}\right. \\ &\quad\times\left(\int_0^{\ell_2}\left(\|\mathcal{M}_\zeta(\mathbb{D}(\ell_3 - \zeta - \varsigma)) - \mathcal{M}_\zeta(\mathbb{D}(\ell_2 - \zeta - \varsigma))\|^\mu\right)^{r_1}\right)^{1/r_1}d\varsigma \\ &\quad\left.+W_\Delta\int_0^{\ell_2}\|\mathcal{M}_\zeta(\mathbb{D}(\ell_3 - \zeta - \varsigma)) - \mathcal{M}_\zeta(\mathbb{D}(\ell_2 - \zeta - \varsigma))\|^\mu d\varsigma\right\}.\end{aligned}$$
From (4), knowing that $\mathcal{M}_\zeta(\mathbb{D}\ell)$ is uniformly continuous for $\ell \in \mp$, we obtain
$$\|\mathcal{M}_\zeta(\mathbb{D}(\ell_3 - \zeta - \varsigma)) - \mathcal{M}_\zeta(\mathbb{D}(\ell_2 - \zeta - \varsigma))\| \longrightarrow 0, \text{ as } \ell_2 \longrightarrow \ell_3.$$
Therefore, we have $\|\Psi_r\| \longrightarrow 0$ as $\ell_2 \longrightarrow \ell_3$, when $r = 1, 2$, which leads, via (12), to
$$\mathbf{E}\|(\mathcal{L}_2\aleph)(\ell_3) - (\mathcal{L}_2\aleph)(\ell_2)\|^\mu \longrightarrow 0, \quad \text{as } \ell_2 \longrightarrow \ell_3,$$

for all $\aleph \in \mathcal{T}_\varrho$. Then, $\mathcal{L}_2$ is compact on $\mathcal{T}_\varrho$ via the Arzelà-Ascoli theorem (see [40]). As a result, $F\aleph = \mathcal{L}_1\aleph + \mathcal{L}_2\aleph$ has a fixed point $\aleph$ in $\mathcal{T}_\varrho$, in accordance with Lemma 6. Furthermore, $\aleph$ is also a solution of (2) and $(\mathcal{L}_1\aleph + \mathcal{L}_2\aleph)(\varpi) = \aleph_1$. Therefore, (2) has a mild solution. This completes the proof. □

Next, we verify the Hyers–Ulam stability via Grönwall's inequality lemma approach.

Theorem 2. *If the assumptions of Theorem 1 are satisfied, then the system (2) has Ulam–Hyers stability.*

Proof. Assume that $\aleph$ is the unique solution of (2) and $\Pi \in C(\mp, \mathbb{R}^n)$ is a solution of the inequality (5) with the aid of Theorem 1. Then

$$\begin{aligned}\aleph(\ell) = {} & \mathcal{H}_\zeta(\mathbb{D}(\ell-\zeta))\psi(0) + \mathcal{M}_\zeta(\mathbb{D}(\ell-\zeta))\psi'(0) \\ & -\mathbb{D}\int_{-\zeta}^{0} \mathcal{M}_\zeta(\mathbb{D}(\ell-2\zeta-\varsigma))\psi(\varsigma)\mathrm{d}\varsigma \\ & +\int_0^\ell \mathcal{M}_\zeta(\mathbb{D}(\ell-\zeta-\varsigma))\hbar(\varsigma,\aleph(\varsigma))\mathrm{d}\varsigma \\ & +\int_0^\ell \mathcal{M}_\zeta(\mathbb{D}(\ell-\zeta-\varsigma))\Delta(\varsigma,\aleph(\varsigma))\mathrm{d}Z_H(\varsigma).\end{aligned}$$

Based on Remark 1, then

$$\Pi''(\ell) = -\mathbb{D}\Pi(\ell-\zeta) + \Delta(\ell,\Pi(\ell))\mathrm{d}Z_H(\ell) + \hbar(\ell,\Pi(\ell)) + \mathcal{E}(\ell), \quad \ell \in \mp,$$

can be expressed as

$$\begin{aligned}\Pi(\ell) = {} & \mathcal{H}_\zeta(\mathbb{D}(\ell-\zeta))\psi(0) + \mathcal{M}_\zeta(\mathbb{D}(\ell-\zeta))\psi'(0) \\ & -\mathbb{D}\int_{-\zeta}^{0} \mathcal{M}_\zeta(\mathbb{D}(\ell-2\zeta-\varsigma))\psi(\varsigma)\mathrm{d}\varsigma \\ & +\int_0^\ell \mathcal{M}_\zeta(\mathbb{D}(\ell-\zeta-\varsigma))\Delta(\varsigma,\Pi(\varsigma))\mathrm{d}Z_H(\varsigma) \\ & +\int_0^\ell \mathcal{M}_\zeta(\mathbb{D}(\ell-\zeta-\varsigma))\hbar(\ell,\Pi(\ell))\mathrm{d}\varsigma \\ & +\int_0^\ell \mathcal{M}_\zeta(\mathbb{D}(\ell-\zeta-\varsigma))\mathcal{E}(\varsigma)\mathrm{d}\varsigma.\end{aligned}$$

In the same manner as in the proof of Theorem 1 and, as a consequence of (9), we have

$$\begin{aligned}& \mathbf{E}\|\Pi(\ell)-\aleph(\ell)\|^\mu \\ & \leq 3^{\mu-1}\Bigg\{\mathbf{E}\left\|\int_0^\ell \mathcal{M}_\zeta(\mathbb{D}(\ell-\zeta-\varsigma))[\Delta(\varsigma,\Pi(\varsigma))-\Delta(\varsigma,\aleph(\varsigma))]\mathrm{d}Z_H(\varsigma)\right\|^\mu \\ & +\mathbf{E}\left\|\int_0^\ell \mathcal{M}_\zeta(\mathbb{D}(\ell-\zeta-\varsigma))[\hbar(\ell,\Pi(\ell))-\hbar(\ell,\aleph(\ell))]\mathrm{d}\varsigma\right\|^\mu \\ & +\mathbf{E}\left\|\int_0^\ell \mathcal{M}_\zeta(\mathbb{D}(\ell-\zeta-\varsigma))\mathcal{E}(\varsigma)\mathrm{d}\varsigma\right\|^\mu\Bigg\}\end{aligned}$$

$$
\begin{aligned}
&\leq 3^{\mu-1}\Big\{\tau_\mu(2H)^{\mu/2}\varpi^{\mu H-1}\int_0^\ell\Big((\ell-\varsigma)\mathbb{E}_{2,2}\Big(\|\mathbb{D}\|(\ell-\varsigma)^2\Big)\Big)^\mu U_\Delta(\varsigma) \\
&\qquad\qquad\qquad\qquad\qquad\times\mathbf{E}\|\Pi(\varsigma)-\aleph(\varsigma)\|^\mu d\varsigma \\
&\quad+\int_0^\ell\Big((\ell-\varsigma)\mathbb{E}_{2,2}\Big(\|\mathbb{D}\|^\mu(\ell-\varsigma)^2\Big)\Big)^\mu U_\hbar(\varsigma)\mathbf{E}\|\Pi(\varsigma)-\aleph(\varsigma)\|^\mu d\varsigma \\
&\quad+\int_0^\ell\Big((\ell-\varsigma)\mathbb{E}_{2,2}\Big(\|\mathbb{D}\|(\ell-\varsigma)^2\Big)\Big)^\mu\mathbf{E}\|\mathcal{E}(\varsigma)\|^\mu d\varsigma\Big\} \\
&\leq\int_0^\ell\Big((\ell-\varsigma)\mathbb{E}_{2,2}\Big(\|\mathbb{D}\|(\ell-\varsigma)^2\Big)\Big)^\mu\Big(3^{\mu-1}\tau_\mu(2H)^{\mu/2}\varpi^{\mu H-1}U_\Delta(\varsigma)+3^{\mu-1}U_\hbar(\varsigma)\Big) \\
&\qquad\qquad\qquad\qquad\qquad\times\mathbf{E}\|\Pi(\varsigma)-\aleph(\varsigma)\|^\mu d\varsigma \\
&\quad+\frac{3^{\mu-1}\varpi^{\mu+1}\kappa}{\mu+1}\Big(\mathbb{E}_{2,2}\Big(\|\mathbb{D}\|\varpi^2\Big)\Big)^\mu.
\end{aligned}
$$

Applying Grönwall's inequality (Lemma 5), we obtain

$$
\mathbf{E}\|\Pi(\ell)-\aleph(\ell)\|^\mu\leq\frac{3^{\mu-1}\varpi^{\mu+1}\kappa}{\mu+1}\Big(\mathbb{E}_{2,2}\Big(\|\mathbb{D}\|\varpi^2\Big)\Big)^\mu\exp\Big(3^{\mu-1}(W_2+W_3)\Big),
$$

which implies that

$$
\mathbf{E}\|\Pi(\ell)-\aleph(\ell)\|^\mu\leq W\kappa,
$$

where

$$
W:=\frac{3^{\mu-1}\varpi^{\mu+1}}{\mu+1}\Big(\mathbb{E}_{2,2}\Big(\|\mathbb{D}\|\varpi^2\Big)\Big)^\mu\exp\Big(3^{\mu-1}(W_2+W_3)\Big).
$$

Therefore, there exists W, which satisfies Definition 2. This ends the proof. □

4. An Example

Consider the following nonlinear stochastic delay system driven by the Rosenblatt process:

$$
\begin{aligned}
\aleph''(\ell)+\mathbb{D}\aleph(\ell-0.5)&=\hbar(\ell,\aleph(\ell))+\Delta(\ell,\aleph(\ell))\frac{dZ_H(\ell)}{d\ell},\quad\text{for }\ell\in\mp:=[0,1],\\
\aleph(\ell)\equiv\psi(\ell),\ \aleph'(\ell)&\equiv\psi'(\ell)\quad\text{for }-0.5\leq\ell\leq0,
\end{aligned}
\tag{13}
$$

where

$$
\aleph(\ell)=\begin{pmatrix}\aleph_1(\ell)\\ \aleph_2(\ell)\end{pmatrix},\quad\mathbb{D}=\begin{pmatrix}1&0\\0&0\end{pmatrix},
$$

and

$$
\hbar(\ell,\aleph(\ell))=\begin{pmatrix}(\sin\ell)\aleph_1(\ell)\\(\sin\ell)\aleph_2(\ell)\end{pmatrix},\quad\Delta(\ell,\aleph(\ell))=\begin{pmatrix}\frac{\sqrt{\ell}e^{-\ell}}{4}\aleph_1(\ell)\\ \frac{\sqrt{\ell}e^{-\ell}}{4}\aleph_2(\ell)\end{pmatrix}.
$$

Next, by choosing $\mu=r_1=r_2=2$, we obtain

$$
\begin{aligned}
\mathbf{E}\|\Delta(\ell,\aleph)-\Delta(\ell,\Im)\|^2_{L_2^0}&=\left(\frac{\sqrt{\ell}e^{-\ell}}{4}\right)^2\Big[(\aleph_1(\ell)-\Im_1(\ell))^2+(\aleph_2(\ell)-\Im_2(\ell))^2\Big]\\
&=\frac{\ell e^{-2\ell}}{16}\mathbf{E}\|\aleph-\Im\|^2\\
&\leq\frac{1}{16}\mathbf{E}\|\aleph-\Im\|^2
\end{aligned}
$$

for all $\ell\in\mp$, and $\aleph(\ell),\Im(\ell)\in\mathbb{R}^2$. We set $U_\Delta(\ell)=1/16$, such that $U_\Delta\in L^2(\mp,\mathbb{R}^+)$ in (**G1**), we have

$$
\|U_\Delta\|_{L^2(\mp,\mathbb{R}^+)}=\left(\int_0^1\left[\frac{1}{16}\right]^2d\varsigma\right)^{\frac{1}{2}}=0.0625.
$$

Thus, selecting $H = 0.75$ and $\tau_\mu = 1.15$, we get

$$W_2 = \frac{\tau_\mu (2H)^{\mu/2} \varpi^{\mu(H+1)-\frac{1}{r_2}}}{(\mu r_1 + 1)^{\frac{1}{r_1}}} \left(\mathbb{E}_{2,2}\left(\|\mathbb{D}\|\varpi^2\right)\right)^\mu \|U_\Delta\|_{L^{r_2}(\mp,\mathbb{R}^+)} = 0.065.$$

Furthermore, we have

$$\begin{aligned}\mathbf{E}\|\hbar(\ell,\aleph) - \hbar(\ell,\Im)\|^2 &= \sin^2 \ell\left[(\aleph_1(\ell) - \Im_1(\ell))^2 + (\aleph_2(\ell) - \Im_2(\ell))^2\right] \\ &= U_\hbar(\ell)\mathbf{E}\|\aleph - \Im\|^2.\end{aligned}$$

We set $U_\hbar(\ell) = \sin^2 \ell$, such that $U_\hbar \in L^2(\mp, \mathbb{R}^+)$ in (**G2**), we have

$$\|U_\hbar\|_{L^2(\mp,\mathbb{R}^+)} = \left(\int_0^1 \sin^4 \varsigma d\varsigma\right)^{\frac{1}{2}} = 0.35217.$$

Hence

$$W_3 = \frac{\varpi^{\mu+\frac{1}{r_1}}}{(\mu r_1 + 1)^{\frac{1}{r_1}}} \left(\mathbb{E}_{2,2}\left(\|\mathbb{D}\|\varpi^2\right)\right)^\mu \|U_\hbar\|_{L^{r_2}(\mp,\mathbb{R}^+)} = 0.21752.$$

Finally, we calculate that

$$2^{\mu-1} W_2 + W_3 = 0.3475 < 1,$$

which follows that all the assumptions of Theorems 1 and 2 hold. Therefore, the system (13) has a unique mild solution $\aleph$, and is Hyers–Ulam stable.

5. Conclusions

In this work, based on fixed point theory, we used the solutions of (2) to prove the existence and uniqueness of solutions. After that, we derived the Hyers–Ulam stability results using the delayed matrix functions and Grönwall's inequality. Finally, we verified the theoretical results by providing an example with a numerical simulation, which showed that our results applied to not only all non-singular matrices, but also all singular and arbitrary matrices, not necessarily squares. This is a novel study to prove the well-posedness and Hyers–Ulam stability of (2) using the delayed matrix functions.

In this study, further studies will focus on the obtained results to ascertain the existence and Hyers–Ulam stability of different types of stochastic delay systems, such as fractional or impulsive fractional stochastic delay systems driven by the Rosenblatt process.

Author Contributions: Conceptualization, G.A., M.H., R.U. and A.M.E.; data curation, G.A., M.H. and A.M.E.; formal analysis, G.A., R.U, M.H. and A.M.E.; software, A.M.E.; supervision, M.H.; validation, G.A., M.H. and A.M.E.; visualization, G.A., M.H., R.U. and A.M.E.; writing—original draft, A.M.E.; writing—review & editing, G.A., M.H. and A.M.E.; investigation, M.H. and A.M.E.; methodology, G.A., M.H., and A.M.E.; funding acquisition, G.A. All authors have read and agreed to the published version of the manuscript.

Funding: Princess Nourah bint Abdulrahman University Researchers Supporting Project number (PNURSP2024R45), Princess Nourah bint Abdulrahman University, Riyadh, Saudi Arabia.

Data Availability Statement: Data are contained within the article.

Acknowledgments: Princess Nourah bint Abdulrahman University Researchers Supporting Project number (PNURSP2024R45), Princess Nourah bint Abdulrahman University, Riyadh, Saudi Arabia.

Conflicts of Interest: The authors declare no conflict of interest.

References

1. Rajivganthi, C.; Thiagu, K.; Muthukumar, P.; Balasubramaniam, P. Existence of solutions and approximate controllability of impulsive fractional stochastic differential systems with infinite delay and Poisson jumps. *Appl. Math.* **2015**, *60*, 395–419. [CrossRef]
2. Øksendal, B. *Stochastic Differential Equations: An Introduction with Applications*; Springer: Berlin/Heidelberg, Germany, 2003.
3. Muthukumar, P.; Rajivganthi, C. Approximate controllability of stochastic nonlinear third-order dispersion equation. *Internat. J. Robust Nonlinear Control* **2014**, *24*, 585–594. [CrossRef]
4. Ahmed, H.M. Semilinear neutral fractional stochastic integro-differential equations with nonlocal conditions. *J. Theoret. Probab.* **2015**, *28*, 667–680. [CrossRef]
5. El-Borai, M.M.; EI-Nadi, K.E.S.; Fouad, H.A. On some fractional stochastic delay differential equations. *Comput. Math. Appl.* **2010**, *59*, 1165–1170. [CrossRef]
6. Da Prato, G.; Zabczyk, J. *Stochastic Equations in Infinite Dimensions*; Cambridge University Press: Cambridge, UK, 2014.
7. Diop, M.A.; Ezzinbi, K.; Lo, M. Asymptotic stability of impulsive stochastic partial integrodifferential equations with delays. *Stochastics* **2014**, *86*, 696–706. [CrossRef]
8. Sakthivel, R.; Revathi, P.; Ren, Y. Existence of solutions for nonlinear fractional stochastic differential equations. *Nonlinear Anal.* **2013**, *81*, 70–86. [CrossRef]
9. Dhanalakshmi, K.; Balasubramaniam, P. Stability result of higher-order fractional neutral stochastic differential system with infinite delay driven by Poisson jumps and Rosenblatt process. *Stoch. Anal. Appl.* **2020**, *38*, 352–372. [CrossRef]
10. Khusainov, D.Y.; Shuklin, G.V. Linear autonomous time-delay system with permutation matrices solving. *Stud. Univ. Zilina. Math. Ser.* **2003**, *17*, 101–108.
11. Khusainov, D.Y.; Diblík, J.; Růžičková, M.; Lukáčová, J. Representation of a solution of the Cauchy problem for an oscillating system with pure delay. *Nonlinear Oscil.* **2008**, *11*, 276–285. [CrossRef]
12. Elshenhab, A.M.; Wang, X.T. Representation of solutions for linear fractional systems with pure delay and multiple delays. *Math. Meth. Appl. Sci.* **2021**, *44*, 12835–12850. [CrossRef]
13. Elshenhab, A.M.; Wang, X.T. Representation of solutions of linear differential systems with pure delay and multiple delays with linear parts given by non-permutable matrices. *Appl. Math. Comput.* **2021**, *410*, 1–13. [CrossRef]
14. Sathiyaraj, T.; Wang, J.; O'Regan, D. Controllability of stochastic nonlinear oscillating delay systems driven by the Rosenblatt distribution. *Proc. R. Soc. Edinb. Sect. A* **2021**, *151*, 217–239. [CrossRef]
15. Elshenhab, A.M.; Wang, X.T. Controllability and Hyers–Ulam stability of differential systems with pure delay. *Mathematics* **2022**, *10*, 1248. [CrossRef]
16. Elshenhab, A.M.; Wang, X.T.; Bazighifan, O.; Awrejcewicz, J. Finite-time stability analysis of linear differential systems with pure delay. *Mathematics* **2022**, *10*, 539. [CrossRef]
17. Liang, C.; Wang, J.; O'Regan, D. Controllability of nonlinear delay oscillating systems. *Electron. J. Qual. Theory Differ. Equ.* **2017**, *2017*, 1–18. [CrossRef]
18. Gao, F.; Oosterlee, C.W.; Zhang, J. A deep learning-based Monte Carlo simulation scheme for stochastic differential equations driven by fractional Brownian motion. *Neurocomputing* **2024**, *574*, 127245. [CrossRef]
19. Feng, J.; Wang, X.; Liu, Q.; Li, Y.; Xu, Y. Deep learning-based parameter estimation of stochastic differential equations driven by fractional Brownian motions with measurement noise. *Commun. Nonlinear Sci. Numer. Simul.* **2023**, *127*, 107589. [CrossRef]
20. El-Borai, M.M.; El-Nadi, K.E.S.; Ahmed, H.M.; El-Owaidy, H.M.; Ghanem, A.S.; Sakthivel, R. Existence and stability for fractional parabolic integro-partial differential equations with fractional Brownian motion and nonlocal condition. *Cogent Math. Stat.* **2018**, *5*, 1460030. [CrossRef]
21. Rosenblatt, M. Independence and dependence. *Proc. Berkeley Symp. Math. Statist. Probab.* **1961**, *2*, 431–443.
22. Shen, G.J.; Ren, Y. Neutral stochastic partial differential equations with delay driven by Rosenblatt process in a Hilbert space. *J. Korean Stat. Soc.* **2015**, *4*, 123–133. [CrossRef]
23. Maejima, M.; Tudor, C.A. Selfsimilar processes with stationary increments in the second Wiener chaos. *Probab. Math. Stat.* **2012**, *32*, 167–186.
24. Almarri, B.; Wang, X.; Elshenhab, A.M. Controllability of Stochastic Delay Systems Driven by the Rosenblatt Process. *Mathematics* **2022**, *10*, 4223. [CrossRef]
25. Almarri, B.; Elshenhab, A.M. Controllability of Fractional Stochastic Delay Systems Driven by the Rosenblatt Process. *Fractal. Fract.* **2022**, *6*, 664. [CrossRef]
26. Shen, G.; Sakthivel, R.; Ren, Y.; Li, M. Controllability and stability of fractional stochastic functional systems driven by Rosenblatt process. *Collect. Math.* **2020**, *71*, 63–82. [CrossRef]
27. Maejima, M.; Tudor, C.A. On the distribution of the Rosenblatt process. *Stat. Probab. Lett.* **2013**, *83*, 1490–1495. [CrossRef]
28. Tudor, C.A. Analysis of the Rosenblatt process. *ESAIM Probab. Stat.* **2008**, *12*, 230–257. [CrossRef]
29. Lakhel, E.H.; McKibben, M. Controllability for time-dependent neutral stochastic functional differential equations with Rosenblatt process and impulses. *Int. J. Control Autom. Syst.* **2019**, *17*, 286–297. [CrossRef]
30. Ulam, S. *A Collection of Mathematical Problem*; Interscience: New York, NY, USA, 1960.
31. Hyers, D.H. On the stability of the linear functional equation. *Proc. Natl. Acad. Sci. USA* **1941**, *27*, 222–224. [CrossRef] [PubMed]

32. Li, S.; Shu, L.; Shu, X.B.; Xu, F. Existence and Hyers-Ulam stability of random impulsive stochastic functional differential equations with finite delays. *Stochastics* **2019**, *91*, 857–872. [CrossRef]
33. Selvam, A.; Sabarinathan, S.; Pinelas, S.; Suvitha, V. Existence and Stability of Ulam–Hyers for Neutral Stochastic Functional Differential Equations. *B. Iran. Math. Soc.* **2024**, *50*, 1–18. [CrossRef]
34. Anguraj, A.; Ravikumar, K.; Nieto, J.J. On stability of stochastic differential equations with random impulses driven by Poisson jumps. *Stochastics* **2021**, *93*, 682–696. [CrossRef]
35. Danfeng, L.; Xue, W.; Tomás, C.; Quanxin, Z. Ulam–Hyers stability of Caputo-type fractional fuzzy stochastic differential equations with delay. *Commun. Nonlinear Sci. Numer. Simul.* **2023**, *121*, 107229.
36. Mattuvarkuzhali, C.; Balasubramaniam, P. p^{th} Moment stability of fractional stochastic differential inclusion via resolvent operators driven by the Rosenblatt process and Poisson jumps with impulses. *Stochastics* **2020**, *92*, 1157–1174. [CrossRef]
37. Rus, I.A. Ulam stability of ordinary differential equations. *Stud. Univ. Babeş-Bolyai Math.* **2009**, *54*, 125–133.
38. Kilbas, A.A.; Srivastava, H.M.; Trujillo, J.J. *Theory and Applications of Fractional Differential Equations*; Elsevier Science BV: Amsterdam, The Netherlands, 2006.
39. Hale, J.K. *Ordinary Differential Equations*; Wiley: New York, NY, USA, 1969.
40. Smart, D.R. *Fixed Point Theorems*; University Press: Cambridge, UK, 1980.

mathematics

MDPI

Article

Distributed Interval Observers with Switching Topology Design for Cyber-Physical Systems

Junchao Zhang, Jun Huang * and Changjie Li

The School of Mechanical and Electrical Engineering, Soochow University, Suzhou 215325, China; 2129402025@stu.suda.edu.cn (J.Z.); 20225229049@stu.suda.edu.cn (C.L.)
* Correspondence: jhuang@suda.edu.cn

Abstract: In this paper, the distributed interval estimation problem for networked Cyber-Physical systems suffering from both disturbances and noise is investigated. In the distributed interval observers, there are some connected interval observers built for the corresponding subsystems. Then, due to the communication burden in Cyber-Physical systems, we consider the case where the communication among distributed interval observers is switching topology. A novel approach that combines L_∞ methodology with reachable set analysis is proposed to design distributed interval observers. Finally, the performance of the proposed distributed interval observers with switching topology is verified through a simulation example.

Keywords: cyber-physical systems; distributed interval observers; reachable set analysis; switching topology; L_∞ technique

MSC: 93D05

1. Introduction

Cyber-Physical Systems (CPSs) are the combinations of physical procedures, high-efficiency computation, communication, and effective control defined by [1]. Architecturally, from [2], a typical CPS can be divided into three layers, which are composed of the sensing layer, the network layer, and the control layer. The development of distributed sensing and networking technologies such as [3,4] has enabled omnipresent sensing and computing capabilities. This has led to the implementation of CPSs in large-scale networks. CPSs are widely used in industrial informatics [5] manufacturing [6], healthcare [7], electrical grids [8], and so on. State estimation and observer design are crucial research areas in CPSs. Ref. [9] used a sliding mode observer and integrated the event-triggered mechanism to estimate the state of CPSs from sensor measurements. Ref. [10] introduced a security estimator combined with a Kalman filter to improve the practical performance of state estimation for CPSs. Ref. [11] accomplished state estimation and resilient control of CPSs using finite time observer techniques and switching schemes. It should be noted that the state estimation for CPSs with bounded disturbance and noise has not been investigated sufficiently.

On the other hand, disturbances and noise always exist in real systems, and the interval observer serves as a powerful estimator of upper and lower bounds for uncertain systems with disturbances and noise. In [12], the concept, as well as the framework of interval observer, were presented. Using the monotone system theory, Refs. [13,14] proposed the approach of coordinate transformation that serves as an efficient strategy to reduce the strict conditions for interval observer design. In recent years, the set-membership estimation method was applied effectively in interval observer design. A two-step interval observer design methodology that combines reachability analysis with robust observer was first presented in [15]. In [15], the H_∞ method and reachable set analysis are combined to design interval observers that eliminate the effects of perturbations and noise on the system. At the

Citation: Zhang, J.; Huang, J.; Li, C. Distributed Interval Observers with Switching Topology Design for Cyber-Physical Systems. *Mathematics* **2024**, *12*, 163. https://doi.org/10.3390/math12010163

Academic Editor: Sergey Ershkov

Received: 30 November 2023
Revised: 25 December 2023
Accepted: 29 December 2023
Published: 4 January 2024

same time, the L_∞ (or L_2) method is also a powerful, robust property, and it is widely used in control and observation fields, such as [16–18]. It has been recently shown by [19–21] that the reachability analysis estimation method can not only enhance the accuracy of estimation but also increase the design freedom. Concurrently, in the context of distributed systems, several recent studies have been conducted on distributed interval observers, such as [22–24]. Ref. [22] designed distributed interval observers based on the monotone system approach for multiagent systems. Ref. [23] considered a distributed interval observers design problem for a class of linear time-invariant systems with uncertainty. At the same time, Ref. [24] improved distributed interval observers by using a set-membership estimation approach. However, the topology of interval observer in the aforementioned work [22–24] is fixed, and it is usually switching with respect to time in practice.

In light of the above discussion, for the problem of state estimation for uncertain CPSs, we apply distributed interval observers to such systems. Since the state estimation of CPSs with bounded disturbances and noise has not been investigated sufficiently, it is meaningful to study the state estimation problem for uncertain CPSs. Each interval observer has two types of observer gain: one is obtained by using the traditional observer design method, and the other one is determined by employing neighborhood information. Considering the practical communication problem of the network layer in CPSs, we suppose that the communication among distributed interval observers is described by switching topology. There are three main challenges: one is to design L_∞ observers with optimal performance for networked CPSs, another is to construct a reachable set analysis framework for CPSs, which then gives upper and lower bounds on the state of the system, and the last is to solve the switching topology problem among the observers. The contributions of this paper are summarized in aspects below.

(1) A distributed interval observer methodology for CPSs is proposed. Compared with the monotone system method, the estimation accuracy is greatly improved by using the two-step method. The L_∞ technique is used to deal with the effects of uncertainty in observer design.
(2) The switching topology with average dwell time (ADT) among distributed interval observers is taken into account and is more closely aligned with the actual system. It can also reduce the communication burden of CPSs.

This paper is structured as follows, with the rest of the paper starting with the graph theory, system model, and some basics presented in Section 2. In Section 3, the optimal robust observer is designed using the L_∞ technique. In order to complete the interval estimation, a reachability analysis methodology is used to design the distributed interval observer. In Section 4, the paper simulates a networked CPS with four Unmanned Aerial Vehicles (UAV) models to illustrate the effectiveness of the distributed interval observer. Finally, Section 5 concludes the paper.

Notation: For a matrix of real symmetry $E \in R^{N\times N}, E \succ 0$ demonstrates that E is positive definite, while $E \prec 0$ demonstrates that E is negative and $He(E) = E + E^T$. The maximal (minimum) eigenvalue of the matrix Q is denoted by $\lambda_{max}(Q)(\lambda_{min}(Q))$. The norm L_2 of the vector v is represented by $\|v\|_2$. In other words, $\|v\|_2 = \sqrt{v^Tv}$. Similarly, the norm L_∞ of the vector v is denoted by $\|v\|_\infty$. The symbol $*$ means the term that can arise due to symmetry in the symmetric matrices.

2. Preliminaries

2.1. Graph Theory

For a digraph G which has N vertices, $\mathcal{A} = [a_{ij}] \in R^{N\times N}$ is called the adjacency matrix. The weight associated with the edge $(i,j) \in S$ that connects node i to node j is called a_{ij}, and $\mathcal{A}$ is provided by $a_{ij} = 0$. The length path from vertex i to vertex j is made up of $t+1$ distinct vertices with consecutive vertices adjacent to each other. If there is a path connecting any two vertices of the graph $\mathcal{G}$, the graph $\mathcal{G}$ is considered connected. $\mathcal{L} = \mathcal{D} - \mathcal{A}$ is defined as Laplacian matrix of a graph $\mathcal{G}$. $\mathcal{D}$ is referred to as a degree matrix of $\mathcal{G}$. The Laplacian matrix $\mathcal{L}$ of a connected graph has a single zero eigenvalue, and 1_N

is the associated eigenvector. Furthermore, $0 = \lambda_1(\mathcal{G}) \leq \cdots \leq \lambda_N(\mathcal{G})$ if $\mathcal{G}$ is connected, where $\lambda_i(\mathcal{G})$ $(i = 1, 2, \cdots, N)$ is the eigenvalue of a Laplacian matrix $\mathcal{L}$.

Lemma 1 ([25]). *For a strongly connected graph $\mathcal{G}$, denote $t_i (i = 1, 2, \cdots, N)$ as the left eigenvector with a 0 eigenvalue, and $T = diag\{t_1, t_2, \cdots\cdots, t_N\}$, then, we can obtain $T\mathcal{L} + \mathcal{L}^T T \geq 0$.*

Lemma 2 ([26]). *Suppose that graph $\mathcal{G}$ is a strongly connected and balanced graph, and the algebraic connectivity of $\mathcal{G}$ is defined by $a(\mathcal{L}) = \min_{t^T x = 0, x \neq 0} \frac{x^T (T\mathcal{L} + \mathcal{L}^T T) x}{2x^T T x}$ where $T = diag\{t_1, t_2, \cdots\cdots, t_N\}$. Then, we can obtain $a(\mathcal{L}) = \lambda_{min}(\frac{He(\mathcal{L})}{2})$.*

2.2. System Model

For a networked CPS, consider a network with N subsystems. The following is the ith subsystem with disturbances and noise:

$$\begin{aligned} x_i(k+1) &= Ax_i(k) + Bu_i(k) + p_i(k), \\ y_i(k) &= Cx_i(k) + q_i(k), \end{aligned} \tag{1}$$

where $x_i(k) \in R^n$ is the state, $u_i(k) \in R^n$ is the control input, $y_i(k) \in R^m$ is the output, $p_i(k) \in R^n$ is the disturbance, $q_i(k) \in R^m$ is the noise. $A \in R^{n \times n}$, $B \in R^{n \times n}$ and $C \in R^{m \times n}$ are constant matrices.

In the following, it is supposed that the communication topology of the subsystem is strongly connected. The global dynamic system of (1) can be given:

$$\begin{aligned} x(k+1) &= \overline{A}x(k) + \overline{B}u(k) + p(k), \\ y(k) &= \overline{C}x(k) + q(k), \end{aligned} \tag{2}$$

where $x = [x_1^T, \cdots, x_N^T]^T$, $u = [u_1^T, \cdots, u_N^T]^T$, $y = [y_1^T, \cdots, y_N^T]^T$, $q = [q_1^T, \cdots, q_N^T]^T$, $p = [p_1^T, \cdots, p_N^T]^T$, $\overline{A} = diag\{\underbrace{A, \cdots, A}_{N}\}$, $\overline{B} = diag\{\underbrace{B, \cdots, B}_{N}\}$ and $\overline{C} = diag\{\underbrace{C, \cdots, C}_{N}\}$.

Since the information can be received by a single subsystem from its neighborhood, we consider the case of switching topology of the observer system and then present $\phi(k)$, a stepwise constant function that takes values from the finite collection $S = \{1, \ 2, \ \cdots, \ N\}$. The observer dynamics of ith subsystem are:

$$\hat{x}_i(k+1) = A\hat{x}_i(k) + Bu_i(k) + L_i(y_i(k) - C\hat{x}_i(k)) + \kappa_{\phi(k)} M_i \sum_{j=1}^{N} a_{ij}^{\phi(k)} (\hat{x}_j(k) - \hat{x}_i(k)), \tag{3}$$

where $\kappa_{\phi(k)}$ is the coupled gain that needs to be designed and M_i and L_i are observer gains of the ith subsystem, and $a_{ij}^{\phi(k)}$ represents the connectivity weight from subsystem i to subsystem j at moment k.

By subtracting (1) from (3), we obtain the error dynamics of a single subsystem:

$$\begin{aligned} e_i(k) &= x_i(k) - \hat{x}_i(k), \\ e_i(k+1) &= (A - L_i C)e_i(k) - \kappa_{\phi(k)} M_i \sum_{j=1}^{N} a_{ij}^{\phi(k)} (\hat{x}_j(k) - \hat{x}_i(k)) + D_i d_i(k), \end{aligned} \tag{4}$$

with $D_i = \begin{bmatrix} I & -L_i \end{bmatrix}$ and $d_i(k) = \begin{bmatrix} p_i(k) \\ q_i(k) \end{bmatrix}$.

Then we can obtain the dynamic system of the global observer

$$\begin{aligned} \hat{x}(k+1) &= \overline{A}\hat{x}(k) + \overline{B}u(k) + \overline{L}_{\phi(k)}(y(k) - \overline{C}\hat{x}(k)) \\ &\quad + \kappa_{\phi(k)} \overline{M}_{\phi(k)} (\mathcal{L}_{\phi(k)} \otimes I_N)(x(k) - \hat{x}(k)), \end{aligned} \tag{5}$$

with $\overline{L} = diag\{\underbrace{L_1, \cdots, L_N}_{N}\}$ and $\overline{M} = diag\{\underbrace{M_1, \cdots, M_N}_{N}\}$.

The global error system $e(k) = [e_1^T(k), \cdots, e_N^T(k)]^T$ is as follows

$$e(k+1) = (\overline{A} - \overline{L}_{\phi(k)}\overline{C} - \kappa_{\phi(k)}\overline{M}_{\phi(k)}(\mathcal{L}_{\phi(k)} \otimes I_N))e(k) + \overline{D}d(k), \tag{6}$$

and the compact form of (6) can be writen as

$$e(k+1) = \Gamma_{\phi(k)}e(k) + \overline{D}d(k), \tag{7}$$

where $\overline{D} = diag\{\underbrace{D_1, \cdots, D_N}_{N}\}$ and $\Gamma_{\phi(k)} = \overline{A} - \overline{L}_{\phi(k)}\overline{C} - \kappa_{\phi(k)}\overline{M}_{\phi(k)}(\mathcal{L}_{\phi(k)} \otimes I_N)$.

Definition 1 ([27]). *If the following condition holds,*

$$\|e(k)\|_2 \leq \varkappa\sqrt{\|d\|_\infty^2 + V(0)\omega^k}, \tag{8}$$

where $\varkappa > 0$ and $0 < \omega < 1$, $V(0) = e^T(0)P_{\phi(k)}e(0)$ and $P_{\phi(k)} \succ 0$. Then the observer (5) is a L_∞ robust observer for system (2).

Definition 2 ([28]). *Let $N_\psi(k_1, k_2)$ be the switching times of $\psi(k)$ across the range $[k_1, k_2)$. If*

$$N_\psi(k_1, k_2) \leq N_0 + \frac{k_2 - k_1}{\tilde{\tau}}, \tag{9}$$

for given $\tilde{\tau} > 0$ and $N_0 \geq 0$, $\tilde{\tau}$ is the average dwell time (ADT) of the switching signal $\psi(k)$. In this paper, we let $N_0 = 0$.

Definition 3 ([29]). *The definition of an α-dimensional zonotope is as follow*

$$\Omega = \nu \oplus HY^\alpha = \nu + Hz, z \in Y^\alpha, \tag{10}$$

where $\nu \in R^\iota$ represents a given vector, $H \in R^{\iota \times \alpha}$ represents a given matrix, Y^α is a unitary box made up of α unitary intervals and Y^α is a unitary interval. In the sequel, the zonotope Ω is represented as $\langle \nu, H \rangle$ for the sake of simplicity.

Lemma 3 ([30]). *The following equation is satisfied for a zonotope defined in (10):*

$$\langle \nu_1, H_1 \rangle \oplus \langle \nu_2, H_2 \rangle = \langle \nu_1 + \nu_2, [H_1, H_2] \rangle,$$
$$W \odot \langle \nu, H \rangle = \langle W\nu, WH \rangle,$$
$$\langle \nu, H \rangle \subseteq \langle \nu, \overline{H} \rangle,$$

where H_1 and H_2 represent the shape matrices of each of the zonotopes, and ν_1 and ν_2 are their centers. $\overline{H} \in R^{N \times N}$ means a diagonal matrix with $\overline{H}_{M,M} = \sum_{i=1}^{\alpha} |H_{M,i}|, M = 1, ..., \rho$.

Remark 1. *$\overline{H}$ can be expressed in the form shown below:*

$$\overline{H} = \begin{bmatrix} \sum_{i=1}^{\alpha} |H_{1,i}| & \cdots & 0 \\ \vdots & \ddots & \vdots \\ 0 & \cdots & \sum_{i=1}^{\alpha} |H_{\rho,i}| \end{bmatrix}. \tag{11}$$

If $\alpha > \rho$, then $\langle \nu, H \rangle \subseteq \langle \nu, \overline{H} \rangle$ is applied to reduce the order of high order zonotopes.

Remark 2. *For zonotopes $\Omega_M \subset R^N, M = 1, ..., \rho$, the Minkowski sum of them is*

$$\bigoplus_{M=1}^{\rho} \Omega_M = \Omega_1 \oplus \Omega_2 \oplus \cdots \oplus \Omega_\rho. \tag{12}$$

Definition 4 ([19]). *For an α-order zonotope, there is an interval hull Ω that could contain Ω in its entirety:*

$$\Omega \subset Box(\Omega) = [a, b], \tag{13}$$

with $a = [a_1, \cdots, a_\alpha]^T$, $b = [b_1, \cdots, b_\alpha]^T$ and $Box(\cdot)$ stands for the interval hull. For any zonotope, the interval hull is the smallest interval vector.

Lemma 4 ([29]). *If $\Omega = \langle v, H \rangle$, the components of its interval hull are*

$$\begin{cases} a_i = v_i - \sum_{j=0}^{\alpha} |H_{ij}|, i = 1, \ldots, \iota, \\ b_i = v_i + \sum_{j=0}^{\alpha} |H_{ij}|, i = 1, \ldots, \iota. \end{cases} \tag{14}$$

Lemma 5 ([29]). *Given zonotopes Ω_i, $i = 1, 2, \ldots, m$*

$$Box(\bigoplus_{m=1}^{\rho} \Omega_m) = \bigoplus_{m=1}^{\rho} (Box(\Omega_m)). \tag{15}$$

Lemma 6 ([31]). *For the given symmetric matrix $\begin{bmatrix} \mathbb{A} & \mathbb{B} \\ \mathbb{B}^T & \mathbb{C} \end{bmatrix}$, the following inequalities are equivalent:*

(1) $\begin{bmatrix} \mathbb{A} & \mathbb{B} \\ \mathbb{B}^T & \mathbb{C} \end{bmatrix} \prec 0,$
(2) $\mathbb{C} \prec 0; \mathbb{A} - \mathbb{B}\mathbb{A}^{-1}\mathbb{B}^T \prec 0,$
(3) $\mathbb{A} \prec 0; \mathbb{C} - \mathbb{B}^T\mathbb{C}^{-1}\mathbb{B} \prec 0.$

Assumption 1. *The initial state of the ith subsystem satisfies the following condition*

$$\underline{x}_i(0) \le x_i(0) \le \overline{x}_i(0). \tag{16}$$

Assumption 2. *The disturbances and output noise in system (6) are bounded, which are:*

$$\|d(k)\|_2 \le \|d\|_\infty, \tag{17}$$

where $\|d\|_\infty$ is a constant.

Assumption 3. *The initial state, disturbance and noise, and initial error are represented by $x(0)$, $d(0)$ and $e(0)$, which can be wrapped as in the equations below:*

$$\begin{aligned} x(0) &\in \langle v_0, H_0 \rangle = \mathcal{X}(0), \\ d(0) &\in \langle 0, E_d \rangle = \mathcal{D}(0), \\ e(0) &\in \langle 0, H_0 \rangle = \mathcal{E}(0), \end{aligned} \tag{18}$$

where H_0 is given matrix and $E_d = diag\{d_i\}$.

3. Main Results

In this part, we provide sufficient conditions for the observer with L_∞ property for CPS. Then, a reachability analysis methodology is used to design the distributed interval observer.

Theorem 1. *Let η and θ be two given constants with $0 < \eta < 1$ and $\theta > 1$. If there exists a constant χ and matrices $P_n \succ 0 \in R^{N\times N}$, $P_m \succ 0 \in R^{N\times N}$ such that*

$$\begin{cases} min \quad \chi^2, \\ subject \quad to: \\ \begin{bmatrix} -\eta P_n & * & * \\ 0 & -\chi^2 I_N & * \\ P_n\overline{A} - W_n\overline{C} - \kappa_n Q_n(\mathcal{L}_n \otimes I_N) & P_n\overline{D} & -P_n \end{bmatrix} \prec 0, \\ P_n \prec \theta P_m, \\ \dfrac{1}{\tilde{\tau}} + \dfrac{ln\eta}{ln\theta} < 0, \\ \kappa_n > \dfrac{1}{a(\mathcal{L}_n)}, \end{cases} \tag{19}$$

where $W_n = P_n\overline{L}_n, Q_n = P_n\overline{M}_n$, $n \neq m, \forall n, m \in S$ and $\tilde{\tau}$ satisfying ADT. Then (5) *is a robust L_∞ observer for system* (3).

Proof. Please see the Appendix A. □

Remark 3. *In Theorem 1, the parameter κ_n depends on $a(\mathcal{L}_n)$. The algebraic connectivity $a(\mathcal{L}_n)$ of a graph tends to increase with graph stability. As the stability of the graph increases, there will be a greater range of options for κ_n.*

Remark 4. *In practice, the ADT $\tilde{\tau}$ and the disturbance attenuation level χ^2 stand for the performance of the observer. Owing to the fact that the ADT $\dfrac{1}{\tilde{\tau}} + \dfrac{ln\eta}{ln\theta} < 0$ depends on η and θ, it is necessary to select suitable values for η and θ to minimize the ADT $\tilde{\tau}$.*

After completing the design of the optimal L_∞ observer, we need to construct the interval observer by designing the interval hull that can completely wrap the system disturbances, noise, and errors. The real-time error from (5) can be wrapped by the zonotope as

$$e(k) \in \langle v_0, H(k)\rangle = \mathcal{E}(k). \tag{20}$$

Then, we add the reachability analysis methodology to the distributed L_∞ observer designed in Theorem 1 and then give the following interval observer:

$$\begin{cases} \overline{x} = \hat{x} + \overline{e} \\ \underline{x} = \hat{x} + \underline{e} \end{cases} \tag{21}$$

where $\overline{e}$ and $\underline{e}$ are the upper and lower bounds of $e(k)$.

Theorem 2. *An interval estimate of the system state is provided by* (22)*, if Assumption 3 holds, then the error $e(k)$ has the following upper and lower bounds:*

$$\begin{bmatrix}\underline{e}(k), & \overline{e}(k)\end{bmatrix} = Box((\prod_{n=0}^{k-1} \Gamma_n)\mathcal{E}(0)) \oplus \bigoplus_{n=0}^{k-1} Box((\prod_{m=1}^{k-1} \Gamma_m)\Gamma_n^{-1} D\mathcal{D}), \tag{22}$$

where $Box(\mathcal{E}(0)) = \begin{bmatrix}\underline{e}(0), & \overline{e}(0)\end{bmatrix}$.

Proof. Based on Assumption 3, we can obtain $\mathcal{E}(1)$,

$$\mathcal{E}(1) = \Gamma_{\phi(0)}\mathcal{E}(0) \oplus D\mathcal{D}. \tag{23}$$

When $k = 1$, then we can obtain

$$\begin{aligned}\mathcal{E}(2) &= \Gamma_{\phi(1)}\mathcal{E}(1) \oplus \mathcal{DD} \\ &= \Gamma_{\phi(1)}\Gamma_{\phi(0)}\mathcal{E}(0) \oplus \Gamma_{\phi(1)}\mathcal{DD} \oplus \mathcal{DD}.\end{aligned} \tag{24}$$

Iterating the above process yields

$$\mathcal{E}(k+1) = (\prod_{n=0}^{k-1} \Gamma_n)\mathcal{E}(0) \oplus \bigoplus_{n=0}^{k-1}(\prod_{m=1}^{k-1} \Gamma_m)\Gamma_n^{-1}\mathcal{DD}. \tag{25}$$

Then the interval hull below describes the set $\mathcal{E}(k)$

$$\mathcal{E}(k) = \left[\underline{e}(k), \quad \overline{e}(k)\right] = Box((\prod_{n=0}^{k-1} \Gamma_n)\mathcal{E}(0)) \oplus \bigoplus_{n=0}^{k-1} Box((\prod_{m=1}^{k-1} \Gamma_m)\Gamma_n^{-1}\mathcal{DD}). \tag{26}$$

□

The proof of Theorem 2 is now completed.

Based on Theorems 1 and 2, the design of a distributed interval observer with switching topology can be implemented by Algorithm 1.

Remark 5. *According to Theorem 1, the robust L_∞ observer is designed to reduce the effect of outside disturbance and output noise. It is obvious that a bounded interval hull exists based on the result of Theorem 1, which can include errors and disturbances. Then, using the given interval hull $\mathcal{E}(0)$ as a starting point, Theorem 2 gives the interval hull $\mathcal{E}(k)$. We propose a reachable set analysis technique by combining Theorem 1 with Theorem 2.*

Algorithm 1: Algorithm for designing the distributed interval observer with switching topology.

(1) Model CPSs with given disturbances and noise.
(2) Design distributed observers for subsystems.
(3) Select the appropriate κ_n according to the switching topology.
(4) Solve the LMI problem in Theorem 1 using the information of the bounds of disturbances
and noise.
(5) Calculate the gains of the observers by $\overline{M}_n = P_n^{-1}Q_n, \overline{L}_n = P_n^{-1}W_n$.
(6) Determine the ADT by given η and θ.
(7) Obtain the zonotopes of the initial value according to (18)
(8) Transform the zonotope at the moment $k = n$ into an interval hull starting from $n = 0$.
(9) Iterate the interval hull in step (6).
(10) Obtain the interval hull at the moment $k = n + 1$.
(11) The interval observer is obtained through (21).

Remark 6. *If the modeling uncertainty is taken into consideration, the new model of this paper is as follows:*

$$\begin{aligned}x_i(k+1) &= (A + \Delta A(k))x_i(k) + (B + \Delta B(k))u_i(k) + p_i(k), \\ y_i(k) &= (C + \Delta C(k))x_i(k) + q_i(k),\end{aligned}$$

where $\Delta A(k), \Delta B(k)$ and $\Delta C(k)$ represent the uncertainty of the system, respectively. Then, we may use the norm-bounded condition on $\Delta A(k), \Delta B(k)$ and $\Delta C(k)$ to design the L_∞ interval observer. However, it is not an easy task to construct the interval hull of the corresponding error system since *the time-varying terms are contained. In the near future, we will deal with systems with model uncertainty with the method proposed in this paper.*

Remark 7. *In Theorems 1 and 2, limited by the current knowledge of authors, this paper only gives sufficient conditions for observer design. In the future, we will study the necessary conditions for the design of interval observers, and in conjunction with this paper, we will give the necessary and sufficient conditions for the design of interval observers.*

4. Simulation

Among CPSs, UAVs have recently achieved widespread application. In this section, we simulate through a networked CPSs with four UAVs.

Referring to [11], the dynamical system of each UAV is as follows

$$\begin{aligned} \begin{bmatrix} \dot{\alpha}_i(t) \\ \dot{\beta}_i(t) \end{bmatrix} &= A \begin{bmatrix} \alpha_i(t) \\ \beta_i(t) \end{bmatrix} + Bu_i(k) + p_i(k), \\ y_i(k) &= C \begin{bmatrix} \alpha_i(t) \\ \beta_i(t) \end{bmatrix} + q_i(k), \end{aligned} \tag{27}$$

where the pitch rate and angle of attack of each UAV are indicated by $\beta_i(t)$ and $\alpha_i(t)$. The schematic of each UAV is shown in Figure 1, and the detailed derivation of the dynamics model is omitted here.

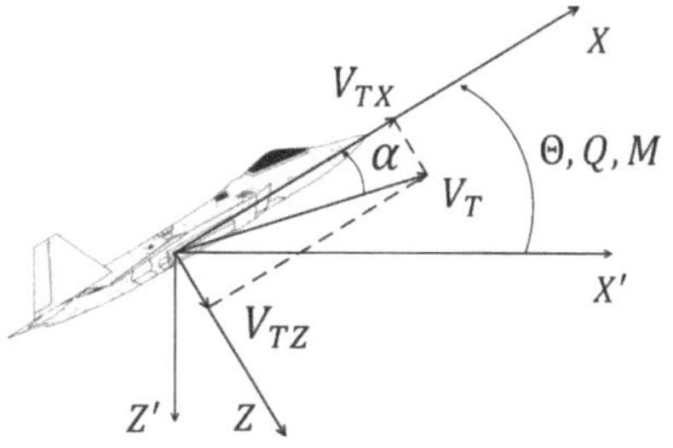

Figure 1. Longitudinal axis system of UAV.

Table 1 displays the output noise and the external disturbance borrowed from [11], and the partial matrix values are as follows:

$$A = \begin{bmatrix} 0.8825 & 0.0987 \\ -0.8458 & 0.9122 \end{bmatrix}, B = \begin{bmatrix} -0.0194 & -0.0036 \\ -1.9290 & -0.3803 \end{bmatrix}, C = \begin{bmatrix} 1 & 0.2 \end{bmatrix}.$$

Table 1. The output noise and the external disturbance.

Subsystem	Output Noise	External Disturbance
1	$0.1\cos(k)$	$0.2 + 0.1\cos(0.5k)$
2	$0.1\sin(k)$	$0.1\cos(0.1k)$
3	$0.01\cos(k)$	$\cos(0.2\pi k) + 0.3\sin(0.2\pi k)$
4	$0.05\sin(k)$	$\cos(0.3\pi k) + 0.1\sin(0.3\pi k)$

For illustrative purposes, Figure 2 shows a switching communication topology $\mathcal{G}_1$,$\mathcal{G}_2$ and $\mathcal{G}_3$ with four UAVs. Figure 3 displays the change in the switching signal $\phi(k)$. Then, from Figure 2, we can give the corresponding matrix $\mathcal{L}_n$:

$$\mathcal{L}_1 = \begin{bmatrix} 1 & -1 & 0 & 0 \\ -1 & 3 & -1 & -1 \\ 0 & -1 & 2 & -1 \\ 0 & -1 & -1 & 2 \end{bmatrix}, \mathcal{L}_2 = \begin{bmatrix} 1 & -1 & 0 & 0 \\ 0 & 2 & -1 & -1 \\ 0 & 0 & 1 & -1 \\ -1 & -1 & 0 & 2 \end{bmatrix}, \mathcal{L}_3 = \begin{bmatrix} 1 & 0 & 0 & -1 \\ 0 & 1 & 0 & -1 \\ 0 & 0 & 1 & -1 \\ -1 & -1 & -1 & 3 \end{bmatrix}.$$

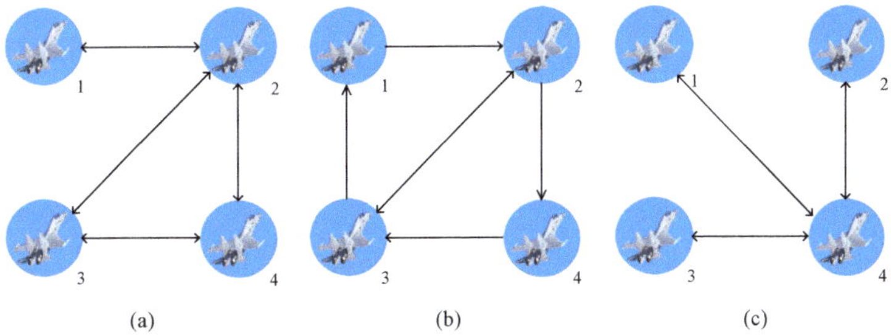

Figure 2. Three switching communication topology of UAVs.

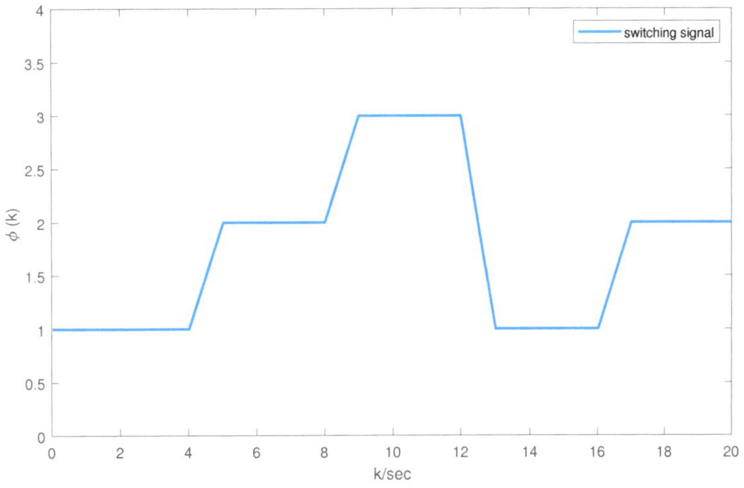

Figure 3. The change in the switching signal of distributed interval observers.

From Lemmas 1 and 2, we can obtain $a(\mathcal{L}_1) = a(\mathcal{L}_2) = a(\mathcal{L}_3) = 1$, then we have $\kappa > 1$. In this simulation, $\kappa_1 = 1.8, \kappa_2 = 2.3, \kappa_3 = 3.2$ are chosen. We can determine L_n and M_n by solving (19), the observer gains for 1th subsystem are listed below

$$L_1 = \begin{bmatrix} 47.8095 & -46.2386 \end{bmatrix}, L_2 = \begin{bmatrix} -9.5528 & 47.7273 \end{bmatrix}, L_3 = \begin{bmatrix} 17.8244 & -11.9289 \end{bmatrix}.$$

$$M_1 = \begin{bmatrix} -41.7480 & -36.5506 \\ -1.8874 & -6.9585 \end{bmatrix}, M_2 = \begin{bmatrix} 10.3302 & 8.4984 \\ -79.5140 & -71.1991 \end{bmatrix}, M_3 = \begin{bmatrix} -11.7589 & -7.5478 \\ -6.1219 & -9.0834 \end{bmatrix}.$$

The peak estimation error of Angle of Attack and Pitch Rate of 1th subsystem are 0.4184 and 0.7668, respectively. The disturbance attenuation level χ^2 and ADT $\tilde{\tau}$ are as follows

$$\chi^2 = 0.4984, \tilde{\tau} \leq 0.1763.$$

Below are the results of the numerical simulation. The states of the original systems and the upper and lower observers are depicted in Figures 4 and 5. x_{ij} are the original states of the subsystems. u_{ij} and v_{ij} reflect the bounds produced via the interval hull technique used in this paper and the monotone system method in [32], where i denotes the ith subsystem and j denotes the jth state. Figure 6 shows the error system for a single subsystem, eu_{ij}, and ev_{ij} reflect the observation error through the method used in this paper and the monotone system method in [32].

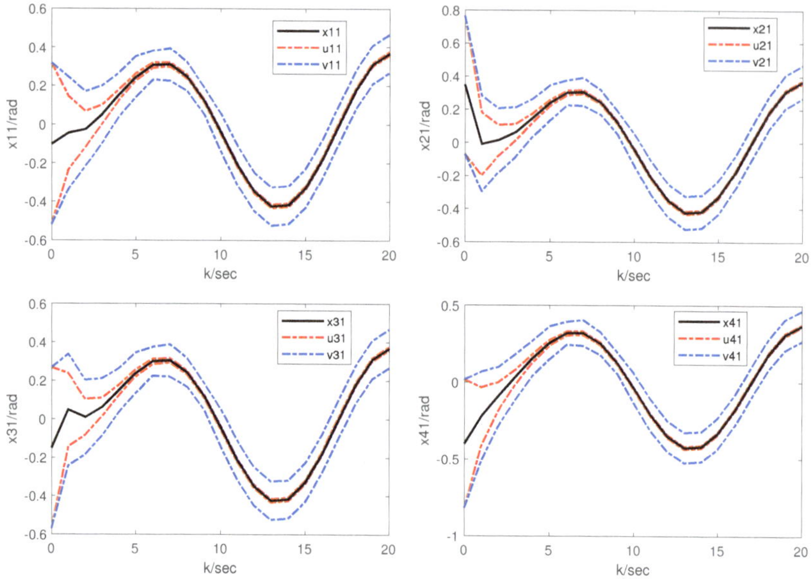

Figure 4. Angle of attack and interval estimates of UAVs.

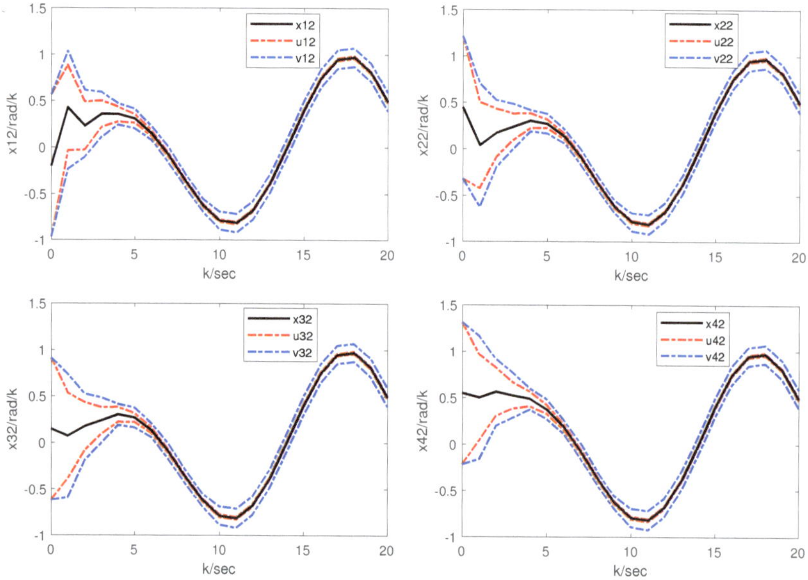

Figure 5. Pitch rate and interval estimates of UAVs.

It is evident that the states of CPS are completely surrounded by those of the upper and lower observers. From Figures 4–6, it can be seen that the distributed interval observer designed in this paper has higher estimation accuracy compared to the traditional monotone system approach in [32]. We design the optimal robust observer using the L_∞ technique, which reduces the design requirements of the observer, unlike the H_∞ technique applied in [24]. Different from [23] and most of the work, we consider for the first time the case of switching topology among distributed interval observers. Thus, the proposed distributed interval observers design method for CPSs is effective and feasible.

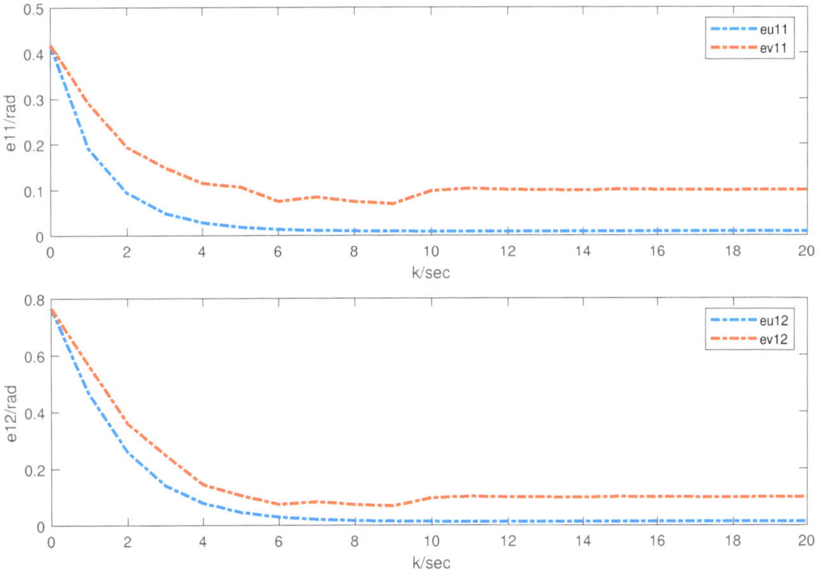

Figure 6. Observation error of angle of attack and pitch rate of UAVs.

Remark 8. *For networked CPSs with perturbations and noise, we propose a class of distributed interval observer design methods with switching topology that combine the design of L_∞ observers with interval hulls. This class of methods eliminates the need to consider the error system to be Schur and, therefore, eliminates the need to use coordinate transformation methods, significantly reducing the conservatism of the estimation. It can be seen from the observer form* (3)*, sufficient conditions* (19)*, and the proof of Theorem 1. In addition, the estimation accuracy is better than that in [32], as it can also be seen from* (22).

5. Conclusions

In this paper, a distributed interval estimation method for uncertain CPSs is investigated. Due to the communication burden of networked CPSs, we consider the case of switching topology among distributed interval observers. To improve the accuracy of the estimation, a reachability analysis is introduced in conjunction with the L_∞ technique. Finally, the validity of the main results of this paper is verified by one example. In the future, we may focus our research on the distributed interval estimation of the attacked CPSs.

Author Contributions: Methodology, J.Z.; validation, C.L.; writing—original draft preparation, J.Z.; writing—review and editing, J.H.; supervision, J.H. All authors have read and agreed to the published version of the manuscript.

Funding: This research received no external funding.

Data Availability Statement: Data are contained within the article.

Conflicts of Interest: The authors declare no conflict of interest.

Appendix A

Proof. Define $\mathcal{A} = \begin{bmatrix} -\eta P_n & 0 \\ 0 & -\chi^2 I_N \end{bmatrix}$, $\mathcal{B} = \begin{bmatrix} P_n\overline{A} - W_n\overline{C} - \kappa_n Q_n(\mathcal{L}_n \otimes I_N) \\ P_n\overline{D} \end{bmatrix}$, $\mathcal{C} = -P_n$. Using Lemma 6 and the fact that $\mathcal{C} \prec 0$, we can determine that

$$\mathcal{A} - \mathcal{B}\mathcal{A}^{-1}\mathcal{B}^T \prec 0. \tag{A1}$$

Substituting $Q_n = P_n\overline{M}_n, W_n = P_n\overline{L}_n$ into (5), we obtain

$$G = \begin{bmatrix} \Gamma_n^T P_n \Gamma_n - \eta P_n & * \\ \overline{D}^T P_n \Gamma_n & \overline{D}^T P_n \overline{D} - \chi^2 I_N \end{bmatrix} \prec 0. \tag{A2}$$

where $\Gamma_n = \overline{A} - \overline{L}_n\overline{C} - \kappa_n\overline{M}_n(\mathcal{L}_n \otimes I_N)$. It follows from (A2) that

$$\begin{bmatrix} e^T(k) & d^T(k) \end{bmatrix} \begin{bmatrix} \Gamma_n^T P_n \Gamma_n - \eta P_n & * \\ \overline{D}^T P_n \Gamma_n & \overline{D}^T P_n \overline{D} - \chi^2 I_N \end{bmatrix} \begin{bmatrix} e(k) \\ d(k) \end{bmatrix} \prec 0. \tag{A3}$$

Thus, (A3) implies that

$$e^T(k)(\Gamma_n^T P_n \Gamma_n - \eta P_n)e(k) + d^T(k)(\overline{D}^T P_n \overline{D} - \chi^2 I_N)d(k) + 2d^T(k)\overline{D}^T P_n \Gamma_n e(k) < 0. \tag{A4}$$

We then choose the following Lyapunov function

$$V_n(k) = e^T(k) P_n e(k). \tag{A5}$$

Thus,

$$\begin{aligned} \triangle V_n(k) =& V_n(k+1) - V_n(k) \\ =& e^T(k)(\Gamma_n^T P_n \Gamma_n - \eta P_n)e(k) + d^T(k)(\overline{D}^T P_n \overline{D} - \chi^2 I_N)d(k) \\ &+ 2d^T(k)\overline{D}^T P_n \Gamma_n e(k). \end{aligned} \tag{A6}$$

Further simplification of (A6) yields

$$\triangle V_n(k) < (\eta - 1)e^T(k) P_n e(k) + \chi^2 d^T(k) d(k), \tag{A7}$$

which means

$$V_n(k+1) < \eta V_n(k) + \chi^2 d^T(k) d(k). \tag{A8}$$

Then consider the interval $[k_\sigma, k)$. Iterating (A8) yields

$$V_n(k) < \eta^{k-k_\sigma} V_n(k_\sigma) + \chi^2 \sum_{\varrho=0}^{k-k_\sigma-1} \eta^\varrho d^T(k-1-\varrho) d(k-1-\varrho). \tag{A9}$$

Suppose that $\phi(k_{\sigma-1}) = m$, using $P_n \prec \theta P_m$, then

$$\begin{aligned} V_n(k) <& \eta^{k-k_\sigma} V_n(k_\sigma) + \chi^2 \sum_{\varrho=0}^{k-k_\sigma-1} \eta^\varrho d^T(k-1-\varrho) d(k-1-\varrho) \\ <& \eta^{k-k_\sigma+1}\theta V_m(k_\sigma - 1) + \eta^{k-k_\sigma}\theta\chi^2 d^T(k_\sigma - 1) d(k_\sigma - 1) \\ &+ \chi^2 \sum_{\varrho=0}^{k-k_\sigma-1} \eta^l d^T(k-1-\varrho) d(k-1-\varrho). \end{aligned} \tag{A10}$$

Iterating (A10) yields the inequality below

$$V_n(k) < \eta^k \theta^{N_\phi(0,k)} V_{\phi(0)}(0) + \chi^2 \sum_{\varrho=0}^{k-k_\sigma-1} \eta^\varrho \theta^{N_\phi(k-\varrho,k)} d^T(k-1-\varrho)d(k-1-\varrho). \tag{A11}$$

In view of $\theta > 1$ and $N_\phi(0,k) \geq N_\phi(k-\varrho,k)$, we have

$$\begin{aligned} &\chi^2 \sum_{\varrho=0}^{k-k_\sigma-1} \eta^\varrho \theta^{N_\phi(k-\varrho,k)} d^T(k-1-\varrho)d(k-1-\varrho) \\ &< \chi^2 \sum_{\varrho=0}^{k-k_\sigma-1} \eta^\varrho \theta^{N_\phi(0,k)} d^T(k-1-\varrho)d(k-1-\varrho). \end{aligned} \tag{A12}$$

It follows from (A11) and (A12) that

$$V_n(k) < \theta^{N_\phi(0,k)}[\eta^k V_{\phi(0)}(0) + \chi^2 \sum_{\varrho=0}^{k-k_\sigma-1} \eta^l d^T(k-1-\varrho)d(k-1-\varrho)]. \tag{A13}$$

Due to $V_n(k) \geq \underline{\lambda}\|e(k)\|_2^2$, the error $e(k)$ in (6) satisfies

$$\begin{aligned} \|e(k)\|_2^2 &\leq \frac{1}{\underline{\lambda}(P_n)} V_n(k), \\ \|e(k)\|_2^2 &\leq \frac{\theta^{N_\phi(0,k)}}{\underline{\lambda}(P_n)} (\eta^k V_{\phi(0)}(0) + \chi^2 \sum_{\varrho=0}^{k-1} \eta^\varrho \|d(k)\|_\infty^2). \end{aligned} \tag{A14}$$

By Definition 1, the proof of Theorem 1 is completed. □

References

1. Cao, X.; Cheng, P.; Chen, J.; Sun, Y. An online optimization approach for control and communication codesign in networked cyber-physical systems. *IEEE Trans. Ind. Inform.* **2012**, *9*, 439–450. [CrossRef]
2. Gunes, V.; Peter, S.; Givargis, T.; Vahid, F. A survey on concepts, applications, and challenges in cyber-physical systems. *KSII Trans. Internet Inf. Syst.* **2014**, *8*, 4242–4268.
3. Kim, K.D.; Kumar, P.R. Cyber–physical systems: A perspective at the centennial. *Proc. IEEE* **2012**, *100*, 1287–1308.
4. Cai, S.; Lau, V.K. Zero MAC latency sensor networking for cyber-physical systems. *IEEE Trans. Signal Process.* **2018**, *66*, 3814–3823. [CrossRef]
5. Sangaiah, A.K.; Siarry, P. Cognitive Brain-Inspired Cyber-Physical Systems in Industrial Informatics. *Front. Neurorobot.* **2022**, *16*, 926538. [CrossRef]
6. Mitchell, R.; Chen, R. Modeling and analysis of attacks and counter defense mechanisms for cyber physical systems. *IEEE Trans. Reliab.* **2015**, *65*, 350–358. [CrossRef]
7. Verma, R. Smart city healthcare cyber physical system: Characteristics, technologies and challenges. *Wirel. Pers. Commun.* **2022**, *122*, 1413–1433. [CrossRef]
8. Hozdić, E.; Butala, P. Concept of socio-cyber-physical work systems for Industry 4.0. *Tehnički Vjesn.* **2020**, *27*, 399–410.
9. Wu, C.; Hu, Z.; Liu, J.; Wu, L. Secure estimation for cyber-physical systems via sliding mode. *IEEE Trans. Cybern.* **2018**, *48*, 3420–3431. [CrossRef]
10. Chang, Y.H.; Hu, Q.; Tomlin, C.J. Secure estimation based Kalman filter for cyber–physical systems against sensor attacks. *Automatica* **2018**, *95*, 399–412. [CrossRef]
11. Zhang, C.L.; Yang, G.H.; Lu, A.Y. Resilient observer-based control for cyber-physical systems under denial-of-service attacks. *Inf. Sci.* **2021**, *545*, 102–117. [CrossRef]
12. Gouzé, J.L.; Rapaport, A.; Hadj-Sadok, M.Z. Interval observers for uncertain biological systems. *Ecol. Model.* **2000**, *133*, 45–56. [CrossRef]
13. Raïssi, T.; Efimov, D.; Zolghadri, A. Interval state estimation for a class of nonlinear systems. *IEEE Trans. Autom. Control* **2011**, *57*, 260–265. [CrossRef]
14. Mazenc, F.; Dinh, T.N.; Niculescu, S.I. Interval observers for discrete-time systems. *Int. J. Robust Nonlinear Control* **2014**, *24*, 2867–2890. [CrossRef]

15. Tang, W.; Wang, Z.; Wang, Y.; Raïssi, T.; Shen, Y. Interval estimation methods for discrete-time linear time-invariant systems. *IEEE Trans. Autom. Control* **2019**, *64*, 4717–4724. [CrossRef]
16. Zhou, X.; Wang, Z.; Wang, J. Automated Ground Vehicle Path-Following: A Robust Energy-to-Peak Control Approach. *IEEE Trans. Intell. Transp. Syst.* **2022**, *23*, 14294–14305. [CrossRef]
17. Zhang, W.; Chen, Y.; Gao, H. Energy-to-peak control for seismic-excited buildings with actuator faults and parameter uncertainties. *J. Sound Vib.* **2011**, *330*, 581–602. [CrossRef]
18. Feng, J.; Han, K. Robust full- and reduced-order energy-to-peak filtering for discrete-time uncertain linear systems. *Signal Process.* **2015**, *108*, 183–194. [CrossRef]
19. Huang, J.; Che, H.; Raïssi, T.; Wang, Z. Functional interval observer for discrete-time switched descriptor systems. *IEEE Trans. Autom. Control* **2021**, *67*, 2497–2504. [CrossRef]
20. Guo, S.; Ren, W.; Ahn, C.K.; Wen, C.; Lam, H.K. Reachability analysis-based interval estimation for discrete-time Takagi–Sugeno fuzzy systems. *IEEE Trans. Fuzzy Syst.* **2021**, *30*, 1981–1992. [CrossRef]
21. Ren, W.; Guo, S. State and Faults Interval Estimations for Discrete-time Linear Systems. *Int. J. Control Autom. Syst.* **2023**, *21*, 2303–2312. [CrossRef]
22. Zhang, Z.H.; Yang, G.H. Distributed fault detection and isolation for multiagent systems: An interval observer approach. *IEEE Trans. Syst. Man Cybern. Syst.* **2018**, *50*, 2220–2230. [CrossRef]
23. Li, D.; Chang, J.; Chen, W.; Raïssi, T. IPR-based distributed interval observers design for uncertain LTI systems. *ISA Trans.* **2022**, *121*, 147–155. [CrossRef] [PubMed]
24. Huang, J.; Zhang, H.; Raïssi, T. Distributed interval estimation methods for multiagent systems. *IEEE Syst. J.* **2022**, *17*, 1843–1852. [CrossRef]
25. Yu, W.; Chen, G.; Cao, M.; Kurths, J. Second-order consensus for multiagent systems with directed topologies and nonlinear dynamics. *IEEE Trans. Syst. Man Cybern. Part B (Cybern.)* **2009**, *40*, 881–891.
26. Li, Z.; Liu, X.; Fu, M.; Xie, L. Global H_∞ consensus of multi-agent systems with Lipschitz non-linear dynamics. *IET Control Theory Appl.* **2012**, *6*, 2041–2048. [CrossRef]
27. Briat, C. Robust stability and stabilization of uncertain linear positive systems via integral linear constraints: L1-gain and L_∞-gain characterization. *Int. J. Robust Nonlinear Control* **2013**, *23*, 1932–1954. [CrossRef]
28. Zhao, X.; Zhang, L.; Shi, P.; Liu, M. Stability of switched positive linear systems with average dwell time switching. *Automatica* **2012**, *48*, 1132–1137. [CrossRef]
29. Alamo, T.; Bravo, J.M.; Camacho, E.F. Guaranteed state estimation by zonotopes. *Automatica* **2005**, *41*, 1035–1043. [CrossRef]
30. Combastel, C. Zonotopes and Kalman observers: Gain optimality under distinct uncertainty paradigms and robust convergence. *Automatica* **2015**, *55*, 265–273. [CrossRef]
31. Boyd, S.; El Ghaoui, L.; Feron, E.; Balakrishnan, V. *Linear Matrix Inequalities in System and Control Theory*; SIAM: Philadelphia, PA, USA, 1994.
32. Zhang, H.; Huang, J.; He, S. Fractional-order interval observer for multiagent nonlinear systems. *Fractal Fract.* **2022**, *6*, 355. [CrossRef]

mathematics

Article

Combination of Functional and Disturbance Observer for Positive Systems with Disturbances

Lanai Huang [1], Xudong Zhao [2,*], Fengyu Lin [3] and Junfeng Zhang [3]

[1] School of Control Science and Engineering, Dalian University of Technology, Dalian 116024, China
[2] School of Artificial Intelligence, Dalian University of Technology, Dalian 116024, China
[3] School of Automation, Hangzhou Dianzi University, Hangzhou 310018, China
* Correspondence: xudongzhao@dlut.edu.cn

Abstract: This technique note proposes two classes of functional and disturbance observers for positive systems with structural and non-structural disturbances, respectively. A positive functional observer is first proposed for positive systems by introducing the estimation of disturbance to the observer. By developing the disturbance observer technique, a positive disturbance observer is designed to supply the estimation of disturbance in the functional observer. Then, a new unknown input observer is constructed for positive systems. A matrix decomposition method is employed to design the observer gains. All conditions are described in terms of linear programming. The corresponding algorithms are addressed for computing the presented conditions. Finally, two examples are provided to verify the effectiveness of the theoretical findings.

Keywords: functional observer; disturbance observer; positive systems; linear programming

MSC: 93C28

check for updates

Citation: Huang, L.; Zhao, X.; Lin, F.; Zhang, J. Combination of Functional and Disturbance Observer for Positive Systems with Disturbances. *Mathematics* **2023**, *11*, 200. https://doi.org/10.3390/math11010200

Academic Editors: Huaizhong Zhao and António Lopes

Received: 18 November 2022
Revised: 13 December 2022
Accepted: 23 December 2022
Published: 30 December 2022

1. Introduction

Observer is a popular technology for estimating the system state when the state is unmeasured [1,2]. For linear systems, linear matrix inequalities can be directly used for dealing with the observer design [3]. The observer technique has also been widely applied for nonlinear systems [4], time-varying systems [5,6], stochastic systems [7], hybrid systems [8,9], etc. Disturbance is a key factor when describing a control system. It is also inevitable for a system to receive some affection from disturbances. Generally speaking, structural and non-structural disturbances are two wide classes of disturbances in practice. For the observation problem of a system with disturbances, the first idea is to propose an observer such that the corresponding error is bounded [10] or the corresponding error system is robustly stable with respect to the disturbances [11]. It is clear that such observers cannot estimate the system state accurately. The error between the state of the observer and the state of the original system depends on the disturbance. The other idea is to design an unknown input observer to eliminate the influence of the disturbance on the observer [12,13]. For the observation of a system with structural disturbance, the strategy is to design a disturbance observer [14] to supply the state observer. Specifically, it is a state observer constructed by replacing the disturbance with the state of disturbance observer [15].

Nonnegativity is a common property of many quantities in real systems. For example, the density of material in physical systems, economic indicators in social systems, the population of people and insect biologic systems, and the water storage capacity in water systems are always nonnegative. Positive systems are naturally utilized to describe such dynamic process with nonnegativity [16,17]. Some significant achievements have also been presented in stability [18,19], observation [20], control synthesis [21,22], etc.

Positive systems have many distinct features that are different from general systems. Copositive Lyapunov functions are more suitable for positive systems than the Lyapunov functions with quadratic form [23,24]. Linear programming is more powerful for dealing with the computation issues of positive systems than linear matrix inequalities [19,25,26]. Luenberger-type observer of positive systems and the corresponding interval observer were proposed in [20] by virtue of linear programming. It is required that the observer of positive systems is also positive since the negative value part of an observer cannot estimate the nonnegative state of positive systems. State-bounded functional observers of positive systems were also designed in [27–29]. In existing results on positive systems, the gain performances-based observer is commonly used for dealing with the observation of positive systems with disturbances [30,31]. However, few efforts are devoted to the asymptotic observation of positive systems with disturbances. The disturbance observer and unknown input observer are two new issues to positive systems [32,33]. Developing the disturbance and unknown input observers of general systems to positive systems is not an easy work. First, how to establish new frameworks on disturbance and unknown input observers? As stated above, positive systems have distinct research approaches from general systems. Therefore, existing observer frameworks cannot be easily developed for positive systems. New linear observer frameworks are expected for positive systems. Second, the positivity of the observer is a difficult issue. Due to the essential positivity of positive systems, the observer of positive systems should also be positive [20,27–31]. This issue is complex for investigating positive systems. For the simultaneous state and disturbance observer, how to reach the positivity requirement is key to the corresponding design. The introduction of disturbance observer increases the difficulty of the design. Third, the disturbance and unknown input observers are full new topics for positive systems. The disturbance observer design of positive systems is distinct from the one of general systems. How to connect the state observer and disturbance observer and how to transform the corresponding conditions into linear form are two key issues.

This paper will design two kinds of observers: One is disturbance observer for positive systems with structural disturbance and the other is unknown input observer for positive systems with non-structural disturbance. First, a functional observer is designed for positive systems, which uses the estimated disturbance to replace the original disturbance. Meanwhile, a positive disturbance observer is proposed to estimate the disturbance. The observer gain matrices are designed based on matrix decomposition technique. All the presented conditions are computed via linear programming. Then, an unknown input observer is proposed for positive systems with non-structural disturbance. A nonlinear programming algorithm is proposed for computing the presented conditions. The rest of the paper is organized as follows. Section 2 introduces the preliminaries, Section 3 presents main design approaches, Section 4 gives two examples, and Section 5 concludes the paper.

Notations. Let $\Re$ (or $\Re_+$), $\Re^n$ (or $\Re^n_+$), and $\Re^{n\times m}$ be the sets of (nonnegative) real numbers, n-dimensional (nonnegative) vectors and $n \times m$ matrices, respectively. For a matrix $A = [a_{ij}] \in \Re^{n\times n}$, $A \succeq 0$ ($\succ 0$) and $A \preceq 0$ ($\prec 0$) mean that $a_{ij} \geq 0$ ($a_{ij} > 0$) and $a_{ij} \leq 0$ ($a_{ij} < 0$) $\forall i, j = 1, \ldots, n$. Similarly, $A \succeq B$ ($A \preceq B$) means that $a_{ij} \geq b_{ij}$ ($a_{ij} \leq b_{ij}$) $\forall i, j = 1, \ldots, n$. A matrix is called Metzler if all its off-diagonal elements are nonnegative. A matrix I_n denotes the n-dimensional identity matrix. Denote by $\mathbf{1}_n = (\underbrace{1, 1, \ldots, 1}_{n})^\top$, $\mathbf{1}_n^{(i)} = (\underbrace{0, \ldots, 0}_{i-1}, 1, 0, \ldots, 0)^\top$, and $\mathbf{1}_{n\times n}$ is a matrix with all elements being 1.

2. Preliminaries

Consider the following continuous-time system:

$$\begin{aligned} \dot{x}(t) &= Ax(t) + Bu(t) + Ew(t), \\ y(t) &= Cx(t) + Dw(t), \end{aligned} \tag{1}$$

where $x(t) \in \Re^n, u(t) \in \Re^m, w(t) \in \Re^r_+, y(t) \subset \Re^s$ represent the system state, the input, the disturbance, and the output, respectively. Suppose that A is Metzler and $B \succeq 0$, $C \succeq 0, D \succeq 0, E \succeq 0$ in system (1).

Definition 1 ([16,17]). *A system is said to be positive if all states and outputs are nonnegative for any nonnegative initial conditions and nonnegative inputs and external disturbances.*

Lemma 1 ([16,17]). *System (1) is positive if and only if A is Metzler and $B \succeq 0, C \succeq 0$, $D \succeq 0, E \succeq 0$.*

Noting the assumptions on system (1), it is easy to derive that the system (1) is positive.

Lemma 2 ([16,17]). *For a continuous-time positive system $\dot{x}(t) = Ax(t)$, the following statements are equivalent:*

(i) The system is stable.
(ii) The system matrix A is Hurwitz.
(iii) There exists a vector $v \succ 0$ such that $A^\top v \prec 0$.

Lemma 3 ([16,17]). *For a positive system, the state is non-positive for any non-positive initial conditions.*

Lemma 4 ([16,17]). *Matrix A is Metzler if and only if there is a positive constant γ such that $A + \gamma I \succeq 0$.*

3. Main Results

We mainly consider the observer design of two classes of systems: One contains structural disturbance and the other one refers to non-structural disturbance. For the structural disturbance, simultaneous state and disturbance observers are designed. For the non-structural disturbance, a new unknown input observer will be proposed.

3.1. Structural Disturbance

Assume that the disturbance is structural, that is, it is dependent on an exogenous system:

$$\begin{aligned} \dot{\xi}(t) &= \Upsilon \xi(t), \\ w(t) &= \Gamma \xi(t), \end{aligned} \tag{2}$$

where $\xi(t) \in \Re^r_+$ is the state of the exogenous system, $\Gamma \succeq 0, \Gamma \in \Re^{r\times r}$, and $\Upsilon \in \Re^{r\times r}$ is a Metzler matrix. By Lemma 1, the exogenous system is positive. Thus, the disturbance observer design can be achieved by estimating the state $\xi(t)$.

Define the linear functional:

$$\eta(t) = Tx(t), \tag{3}$$

where $\eta(t) \in \Re^o$ is the state to be estimated and $T \succeq 0, T \in \Re^{o\times n}$. This implies that a functional observer with respect to the state will be designed later.

The state functional observer of system (1) is designed as:

$$\dot{\hat{\eta}}(t) = G\hat{\eta}(t) + TBu(t) + M\hat{w}(t) + L_c y(t), \tag{4}$$

where $\hat{\eta}(t) \in \Re^o$ is the observer state, $\hat{w}(t) \in \Re^r$ is the estimate of the disturbance signal and $\hat{w}(t) = \Gamma\hat{\xi}(t)$, and $G \in \Re^{o\times o}, M \in \Re^{o\times r}, L_c \in \Re^{o\times s}$ are the observer gains to be designed. The disturbance observer is constructed as:

$$\dot{\hat{\xi}}(t) = H\hat{\xi}(t) + F\hat{\eta}(t) + L_d y(t), \tag{5}$$

where $\hat{\xi}(t) \in \Re^r$ is the state of the disturbance observer and $H \in \Re^{r\times r}, F \in \Re^{r\times o}, L_d \in \Re^{r\times s}$ are the observer gains to be designed.

Denote by the errors $e(t) = \eta(t) - \hat{\eta}(t)$ and $\sigma(t) = \xi(t) - \hat{\xi}(t)$. Then,

$$\begin{aligned} \dot{e}(t) &= (TA - L_cC)x(t) - G\hat{\eta}(t) + (TE\Gamma - L_cD\Gamma)\xi(t) - M\Gamma\hat{\xi}(t), \\ \dot{\sigma}(t) &= (\Upsilon - L_dD\Gamma)\xi(t) - H\hat{\xi}(t) - F\hat{\eta}(t) - L_dCx(t). \end{aligned} \tag{6}$$

It is well known that it is impossible to estimate a nonnegative variable using a non-positive variable. Therefore, the observer of positive systems is also positive. By Lemma 1, the positivity of the disturbance observer (5) is reached by virtue of the conditions: (i) H is Metzler, (ii) $F \succeq 0$, and (iii) $L_d \succeq 0$. In the literature [6,8], some equations were introduce to transform the system (6) into an error system with variables $e(t)$ and $\sigma(t)$. For example, the equations $TA - L_cC = GT, TB = Q, TE\Gamma - L_cD\Gamma = M\Gamma$, and $\Upsilon - L_dD\Gamma = H$ are imposed on the state error dynamic in (6). Noting the facts $F \succeq 0, L_d \succeq 0, T \succeq 0$, and $C \succeq 0$, the relation $FT + L_dC = 0$ does not hold. Thus, the term $-F\hat{\eta}(t) - L_dCx(t)$ in (6) can not be transformed into the error term $e(t)$. This implies that the positivity of (5) contradicts with the stability of (6). The following theorem will solve the mentioned problems.

Theorem 1. *If there exist constants $\delta_1 > 0, \delta_2 > 0, \delta_3 > 0, \alpha > 0$, positive $\Re^o$ vectors $v_1, z_g^{(i)}, z_g, z_f^{(i)}, z_f$, and positive $\Re^r$ vector v_2 such that*

$$TA\mathbf{1}_o^\top v_1 - \sum_{i=1}^{o} \mathbf{1}_o^{(i)} z_c^{(i)\top} C - \sum_{i=1}^{o} \mathbf{1}_o^{(i)} z_g^{(i)\top} T + \delta_1 T = 0, \tag{7a}$$

$$TE\mathbf{1}_o^\top v_1 - \sum_{i=1}^{o} \mathbf{1}_o^{(i)} z_c^{(i)\top} D \succeq 0, \tag{7b}$$

$$\Upsilon\mathbf{1}_r^\top v_2 - \sum_{i=1}^{r} \mathbf{1}_r^{(i)} z_d^{(i)\top} D\Gamma + \delta_3 I_r \succeq 0, \tag{7c}$$

$$\sum_{i=1}^{r} \mathbf{1}_r^{(i)} z_f^{(i)\top} T + \sum_{i=1}^{r} \mathbf{1}_r^{(i)} z_d^{(i)\top} C = 0, \tag{7d}$$

and

$$\sum_{i=1}^{o} \mathbf{1}_o^{(i)} z_g^{(i)\top} - \delta_1 I_o + \delta_2 I_o \succeq 0, \tag{8a}$$

$$z_g - \alpha v_1 + z_f \prec 0, \tag{8b}$$

$$\Gamma^\top E^\top T^\top v_1 + \Gamma^\top D^\top z_c + \Upsilon^\top v_2 - \Gamma^\top D^\top z_d, \tag{8c}$$

$$\alpha \mathbf{1}_o^\top v_1 \leq \delta_1, \tag{8d}$$

$$z_g^{(i)} \preceq z_g,\ i = 1, 2, \ldots, o, \tag{8e}$$

$$z_f^{(i)} \preceq z_f,\ i = 1, 2, \ldots, r, \tag{8f}$$

$$z_c^{(i)} \preceq z_c, z_d^{(i)} \succeq z_d,\ i = 1, 2, \ldots, s, \tag{8g}$$

hold, then under the observer gain matrices

$$\begin{aligned} G = \frac{\sum_{i=1}^{o} \mathbf{1}_o^{(i)} z_g^{(i)\top} - \delta_1 I_o}{\mathbf{1}_o^\top v_1}, F = \frac{\sum_{i=1}^{r} \mathbf{1}_r^{(i)} z_f^{(i)\top}}{\mathbf{1}_r^\top v_2}, L_c = \frac{\sum_{i=1}^{o} \mathbf{1}_o^{(i)} z_c^{(i)\top}}{\mathbf{1}_o^\top v_1}, L_d = \frac{\sum_{i=1}^{r} \mathbf{1}_r^{(i)} z_d^{(i)\top}}{\mathbf{1}_r^\top v_2}, \\ M = TE - \frac{\sum_{i=1}^{o} \mathbf{1}_o^{(i)} z_c^{(i)\top} D}{\mathbf{1}_o^\top v_1}, H = \Upsilon - \frac{\sum_{i=1}^{r} \mathbf{1}_r^{(i)} z_d^{(i)\top} D\Gamma}{\mathbf{1}_r^\top v_2}, \end{aligned} \tag{9}$$

and the initial conditions satisfy $e(0) \preceq 0$ and $\sigma(0) \preceq 0$, the observers (4) and (5) are positive, and the error system (6) is stable.

Proof. First, we prove the positivity of the observers (4) and (5). From (7a,d) and (9), we have

$$\begin{aligned} TA - L_cC - GT &= 0, \\ TE - L_cD - M &= 0, \\ \Upsilon - L_dD\Gamma - H &= 0. \end{aligned} \tag{10}$$

Then, (6) can be transformed into

$$\begin{pmatrix} \dot{e}(t) \\ \dot{\sigma}(t) \end{pmatrix} = \begin{pmatrix} G & M\Gamma \\ F & H \end{pmatrix} \begin{pmatrix} e(t) \\ \sigma(t) \end{pmatrix}. \tag{11}$$

From (8a) and (7c), it follows that

$$\begin{aligned} &\frac{\sum_{i=1}^{o} \mathbf{1}_o^{(i)} z_g^{(i)\top} - \delta_1 I_o}{\mathbf{1}_o^\top v_1} + \frac{\delta_2}{\mathbf{1}_o^\top v_1} I_o \succeq 0, \\ &\Upsilon - \frac{\sum_{i=1}^{p} \mathbf{1}_p^{(i)} z_d^{(i)\top}}{\mathbf{1}_p^\top v_2} + \frac{\delta_3}{\mathbf{1}_p^\top v_2} I_r \succeq 0. \end{aligned} \tag{12}$$

Together with (9) gives that $G + \frac{\delta_2}{\mathbf{1}_o^\top v_1} I_o \succeq 0$ and $H + \frac{\delta_3}{\mathbf{1}_p^\top v_2} I_p \succeq 0$, which imply that G and H are Metzler by Lemma 4. By (7b), $M \succeq 0$. It is also easy to know $F \succeq 0$. By Lemma 1, the system (12) is positive. Since $e(0) \preceq 0$ and $\sigma(0) \preceq 0$, then $e(t) \preceq 0$ and $\sigma(t) \preceq 0$. Thus, the observers (4) and (5) are positive.

First, we have

$$G^\top v_1 + F^\top v_2 = \frac{\sum_{i=1}^{o} z_g^{(i)} \mathbf{1}_o^{(i)\top} v_1 - \delta_1 v_1}{\mathbf{1}_o^\top v_1} + \frac{\sum_{i=1}^{p} z_f^{(i)} \mathbf{1}_p^{(i)\top} v_2}{\mathbf{1}_p^\top v_2}. \tag{13}$$

Together with (8e), (8f), and (8d) gives

$$G^\top v_1 + F^\top v_2 \preceq z_g - \frac{\delta_1}{\mathbf{1}_o^\top v_1} v_1 + z_f \preceq z_g - \alpha v_1 + z_f. \tag{14}$$

By (8b), $G^\top v_1 + F^\top v_2 \prec 0$. Then, it follows from (9) that

$$\begin{aligned} \Gamma^\top M^\top v_1 + H^\top v_2 &= \Gamma^\top E^\top T^\top v_1 + \Gamma^\top D^\top \frac{\sum_{i=1}^{o} z_c^{(i)} \mathbf{1}_o^{(i)\top} v_1}{\mathbf{1}_o^\top v_1} + \Upsilon^\top v_2 \\ &\quad - \Gamma^\top D^\top \frac{\sum_{i=1}^{p} z_d^{(i)} \mathbf{1}_p^{(i)\top} v_2}{\mathbf{1}_p^\top v_2}. \end{aligned} \tag{15}$$

By (8g) and

$$\Gamma^\top M^\top v_1 + H^\top v_2 \preceq \Gamma^\top E^\top T^\top v_1 + \Gamma^\top D^\top z_c + \Upsilon^\top v_2 - \Gamma^\top D^\top z_d \prec 0. \tag{16}$$

Then,

$$\begin{pmatrix} G & M\Gamma \\ F & H \end{pmatrix}^\top \begin{pmatrix} v_1 \\ v_2 \end{pmatrix} \prec 0. \tag{17}$$

By Lemma 2, the matrix $\begin{pmatrix} G & M\Gamma \\ F & H \end{pmatrix}$ is Hurwitz. Then, $e(t)$ and $\sigma(t)$ converge to zero with $t \to \infty$, that is, $\hat{\eta}(t) \to Tx(t)$ and $\hat{\xi}(t) \to \xi(t)$. □

Remark 1. *In [20], the Luenberger observer of positive systems was proposed in terms of linear programming. Following the linear programming technique in [20], Theorem 1 is the first attempt to introduce simultaneous functional and disturbance observers for positive systems. Under the designed observers (4) and (5), the error system (6) is positive and asymptotically stable. In existing literature [30,31],* L_1/ℓ_1 *gain stability was used for positive systems to assess the performance of observer, which can only reduce the influence of disturbance to a bounded range and cannot achieve accurate observation. Under the observer (5), the asymptotic stability of system (6) can be reached rather than gain stability. Such kind of observer can be used in the systems with high precision or high system performance, and has potential applications in practical systems.*

In Theorem 1, two key conditions $e(0) \preceq 0$ *and* $\sigma(0) \preceq 0$ *are imposed on the system (6). Together with the fact that system (11) is positive,* $e(t) \preceq 0$ *and* $\sigma(t) \preceq 0$ *hold* $\forall t \geq 0$ *by Lemma 3. This implies that* $\eta(t) \preceq \hat{\eta}(t)$ *and* $\xi(t) \preceq \hat{\xi}(t)$*. Thus, the state observer (4) and the disturbance* *observer (5) are positive since $\eta(t) \succeq 0$ and $\xi(t) \succeq 0$. In most literature [20,27–31], the error $e(t)$ was required to be nonnegative. Under such a case, it is hard to guarantee the positivity of the disturbance observer (5). To solve this problem, Theorem 1 changes the nonnegativity condition as non-positivity condition. Such a strategy smooths the development of the positive disturbance observer.*

Remark 2. *In [27–29], the functional observer of positive systems had been investigated. However, few efforts are devoted to the simultaneous functional and disturbance observers of positive systems. For positive systems with disturbances, the current observer design can only obtain the gain performance-based state estimation [30,31]. Up to now, the disturbance observer issue is full open in the field of positive systems. There are three difficulties for the issue. How to construct a new framework on the error system of simultaneous functional and disturbance observers? How to guarantee the positivity of the functional and disturbance observers? How to design the functional and disturbance observers gains of positive systems via linear programming? Theorem 1 establishes a new linear framework on the disturbance observer of positive systems.*

3.2. Non-Structural Disturbance

In the last subsection, the disturbance is assumed to be structural. A dynamic system is introduced to describe the disturbance. In this subsection, the dynamic disturbance system is removed, that is, the disturbance is non-structural. This object of this subsection is to propose an unknown input observer for system (1) with non-structural disturbance.

For the convenience of the design, we introduce an additional transformation:

$$\hat{\eta}(t) = \zeta(t) + Wy(t), \tag{18}$$

where $\hat{\eta}(t) \in \Re^o$ is the estimate of $\eta(t)$, $\zeta(t) \in \Re^o$ is an additional state, and $W \in \Re^{o\times s}$. It is clear that the estimate state $\hat{\eta}(t)$ is dependent on the state $\zeta(t)$. Thus, one only needs to design the dynamics of $\zeta(t)$. The corresponding dynamics is designed as:

$$\dot{\zeta}(t) = G\zeta(t) + Qu(t) + Ly(t), \tag{19}$$

where $G \in \Re^{o\times o}, Q \in \Re^{o\times m}, L \in \Re^{o\times s}$ are the observer gains to be designed.

Firstly, consider the case: $y(t) = Cx(t)$. Denote $e(t) = \eta(t) - \hat{\eta}(t)$. Then

$$\begin{aligned} \dot{e}(t) &= T\dot{x}(t) - \dot{\zeta}(t) - W\dot{y}(t) \\ &= ((T-WC)A - LC)x(t) + ((T-WC)B - Q)u(t) \\ &\quad + (T-WC)Ew(t) - G\zeta(t) \\ &= ((T-WC)A - LC + GWC)x(t) + ((T-WC)B - Q)u(t) \\ &\quad + (T-WC)Ew(t) - G\hat{\eta}(t). \end{aligned} \tag{20}$$

Theorem 2. *If there exist constants $\delta_1 > 0, \delta_2 > 0, \alpha > 0$, $\Re^o$ vectors $v \succ 0, z_g^{(i)} \succ 0, z_g \succ 0$, $\Re^r$ vectors $z_w^{(i)}, z_w$, $\Re^m$ vector $z_q^{(i)}$, and $\Re^s$ vector $z_c^{(i)}$ such that*

$$\begin{aligned} &TA - \textstyle\sum_{i=1}^{o} \mathbf{1}_o^{(i)} z_w^{(i)\top} CA - \sum_{i=1}^{o} \mathbf{1}_o^{(i)} z_c^{(i)\top} C + (\sum_{i=1}^{o} \mathbf{1}_o^{(i)} z_g^{(i)\top} \\ &-\delta_1 I_o) \textstyle\sum_{i=1}^{o} \mathbf{1}_o^{(i)} z_w^{(i)\top} C - \sum_{i=1}^{o} \mathbf{1}_o^{(i)} z_g^{(i)\top} T + \delta_1 T = 0, \end{aligned} \tag{21a}$$

$$TB - \textstyle\sum_{i=1}^{o} \mathbf{1}_o^{(i)} z_w^{(i)\top} CB - \sum_{i=1}^{o} \mathbf{1}_o^{(i)} z_q^{(i)\top} = 0, \tag{21b}$$

$$TE - \textstyle\sum_{i=1}^{o} \mathbf{1}_o^{(i)} z_w^{(i)\top} CE = 0, \tag{21c}$$

$$\textstyle\sum_{i=1}^{o} \mathbf{1}_o^{(i)} z_g^{(i)\top} - \delta_1 I_o + \delta_2 I_o \succeq 0, \tag{21d}$$

$$z_g \mathbf{1}_o^\top v - \delta_1 v \prec 0, \tag{21e}$$

$$z_g^{(i)} \preceq z_g,\ i = 1,2,\ldots,o, \tag{21f}$$

hold, then under the observer gain matrices

$$\begin{aligned} W = \sum_{i=1}^{o} \mathbf{1}_o^{(i)} z_w^{(i)\top}, G = \sum_{i=1}^{o} \mathbf{1}_o^{(i)} z_g^{(i)\top} - \delta_1 I_o, \\ Q = \sum_{i=1}^{o} \mathbf{1}_o^{(i)} z_q^{(i)\top}, L = \sum_{i=1}^{o} \mathbf{1}_o^{(i)} z_c^{(i)\top}, \end{aligned} \tag{22}$$

and the initial condition satisfying $e(0) \preceq 0$, the observer state $\hat{\eta}(t)$ is nonnegative and the error system (20) is stable.

Proof. By (21a) and (22), it follows that $(T - WC)A - LC - GT = 0$. Using (21b) and (22) yields $(T - WC)B - Q = 0$. Using (21c) and (22) gives $(T - WC)E = 0$. Then, (20) becomes

$$\dot{e}(t) = Ge(t). \tag{23}$$

By (21d), it holds that $G + \delta_2 I_o \succeq 0$, which follows that G is Metzler by Lemma 4. By Lemma 2, the system (23) is positive. Since $e(0) \preceq 0$, then $e(t) \preceq 0$. That is to say, $\eta(t) \preceq \hat{\eta}(t)$. Owing to the nonnegative property of $\eta(t)$, $\hat{\eta}(t) \succeq 0$. It is not hard to obtain

$$G^\top v = \sum_{i=1}^{o} z_g^{(i)} \mathbf{1}_o^{(i)\top} v - \delta_1 v. \tag{24}$$

By (21f), (24) is transformed into

$$G^\top v \preceq \sum_{i=1}^{o} z_g \mathbf{1}_o^{(i)\top} v - \delta_1 v = z_g \mathbf{1}_o^\top v - \delta_1 v. \tag{25}$$

Using (21e), $G^\top v \prec 0$. Then, $e(t) \to 0$ with $t \to \infty$. □

Remark 3. *The literature [19,20,25,27,28] had investigated the observer issues of positive systems. In these literature, a commonly used approach is that the positivity of the observer is achieved by imposing some conditions on the observer matrices. Take (18) and (19) for example. In order to guarantee the positivity of the observer state $\hat{\eta}(t)$, two classes of conditions are required: The first one is that G is Metzler, $Q \succeq 0$, and $L \succeq 0$, and the second one is $W \succeq 0$. The first one is to guarantee the positivity of (19) and the second one is to achieve the positivity of $\hat{\eta}(t)$. These conditions are rigorous and hard to be guaranteed. In Theorem 2, a new design approach is presented. The restrictions on W, Q, and L are removed. Moreover, a design framework on the observer gains is constructed in (22). The conditions in (21) are solvable in terms of linear programming. These increase the reliability of the design in Theorem 2.*

Next, consider the case $y(t) = Cx(t) + Dw(t)$. Then, the equation (20) can be rewritten as

$$\begin{aligned} \dot{e}(t) &= ((T - WC)A - LC + GWC)x(t) + ((T - WC)B - Q)u(t) \\ &\quad + ((T - WC)E - LD + GWD)w(t) - G\hat{\eta}(t) - WD\dot{w}(t). \end{aligned} \tag{26}$$

Theorem 3. *If there exist constants $\delta_1 > 0, \delta_2 > 0, \alpha > 0$, $\Re^o$ vectors $v \succ 0, z_g^{(i)} \succ 0, z_g \succ 0$, $\Re^r$ vectors $z_w^{(i)}, z_w$, $\Re^m$ vector $z_q^{(i)}$, and $\Re^s$ vector $z_c^{(i)}$ such that*

$$\sum_{i=1}^{o} \mathbf{1}_o^{(i)} z_w^{(i)\top} D = 0, \tag{27a}$$

$$\begin{aligned} TA - \sum_{i=1}^{o} \mathbf{1}_o^{(i)} z_w^{(i)\top} CA - \sum_{i=1}^{o} \mathbf{1}_o^{(i)} z_c^{(i)\top} C + (\sum_{i=1}^{o} \mathbf{1}_o^{(i)} z_g^{(i)\top} - \\ \delta_1 I_o) \sum_{i=1}^{o} \mathbf{1}_o^{(i)} z_w^{(i)\top} C - \sum_{i=1}^{o} \mathbf{1}_o^{(i)} z_g^{(i)\top} T + \delta_1 T = 0, \end{aligned} \tag{27b}$$

$$TB - \sum_{i=1}^{o} \mathbf{1}_o^{(i)} z_w^{(i)\top} CB - \sum_{i=1}^{o} \mathbf{1}_o^{(i)} z_q^{(i)\top} = 0, \tag{27c}$$

$$TE - \sum_{i=1}^{o} \mathbf{1}_o^{(i)} z_w^{(i)\top} CE - \sum_{i=1}^{o} \mathbf{1}_o^{(i)} z_c^{(i)\top} D = 0, \tag{27d}$$

$$\sum_{i=1}^{o} \mathbf{1}_o^{(i)} z_g^{(i)\top} - \delta_1 I_o + \delta_2 I_o \succeq 0, \tag{27e}$$

$$z_g \mathbf{1}_o^\top v - \delta_1 v \prec 0, \tag{27f}$$

$$z_g^{(i)} \preceq z_g,\ i = 1, 2, \ldots, o, \tag{27g}$$

hold, then under the observer gain matrices (22) and the initial condition satisfying $e(0) \preceq 0$, the observer state $\hat{\eta}(t)$ is nonnegative and the error system (21) is stable.

Proof. From (27a) and (22), it is clear that $WD = 0$. By (27b) and (22), it follows that $(T - WC)A - LC - GT = 0$. Using (27c) and (22) yields $(T - WC)B - Q = 0$. Using (27d) and (22) gives $(T - WC)E - LD + GWD = 0$. Then, (26) becomes (23). By (21e), it holds that $G + \delta_2 I_o \succeq 0$, which follows that G is Metzler by Lemma 4. By Lemma 2, the system (23) is positive. Since $e(0) \preceq 0$, then $e(t) \preceq 0$. This implies, $\eta(t) \preceq \hat{\eta}(t)$. Due to the nonnegative property of $\eta(t)$, $\hat{\eta}(t) \succeq 0$.

By (27f) and (27g), one can obtain $G^\top v \prec 0$, which implies $e(t) \to 0$ with $t \to \infty$. □

The conditions (27b) and (27f) are nonlinear. Then, the nonlinear programming toolbox in Matlab can be directly used for dealing with the conditions.

Remark 4. *As the early attempt on the observer design of positive systems with unknown input, the literature [12,13] proposed the functional observer and the disturbance observer for positive systems, respectively. However, there still exist some open issues to the observer design of positive systems. First, existing results are concerned with the disturbance-free output, i.e., $y(t) = Cx(t)$. Indeed, the output will contain the disturbance when the dynamics of the system contains disturbance. Therefore, it is unreasonable to ignore the disturbance in the output. Second, linear (nonlinear) programming is more effective for dealing with the issues of positive systems than linear matrix inequalities. In [13], linear matrix inequalities were employed for computing the corresponding conditions. This will increase the complexity of the design. Linear programming has been verified to be more suitable for positive systems [16,17,19,20,23–25,27]. Third, a unified is needed to the observer gain design. In [12,13], the observer gains were computed based on some algorithms. However, there are no unified framework on these gains. Thus, it limits the further extension of the proposed design. To further present a unified observer design approach and overcome existing open issues, Theorems 2 and 3 are presented. The presented framework has potential applications in the related issues of positive systems.*

4. Illustrative Examples

In recent years, urban water supply and water resources management have become a hot topic with the rapid development of cities. In some large cities such as Paris, Barcelona, Hangzhou, etc., the large water pipes are constructed to meet the city's water demand and facilitate water resource management. In literature [34,35], a state-space model with disturbance was established for water systems, and corresponding control methods were designed to achieve effective control of water systems and improve the management ability of water resources. Considering the positivity of water flow in the water systems, the literature [36] studied the robust model predictive controllers of the water system by using positive system theory. The main physical quantities considered in the water system studied in the literature [34–36] include the water capacity in the tank, the water flow operated by the actuator (pump station or valve), and the flow generated by the disturbances (water demands or rainwater flow). Based on the models described in the literature [34–36], a virtual water tank of the water systems is as shown in Figure 1 and the state-space model can be established under the form (1), where $x(t)$ is the volume of all the tanks at the tth time instant, $u(t)$ is the manipulated flows through the actuators (pumps and valves), $y(t)$ is the outputs of sensors network, and $w(t)$ represents the vector of the value of water demands or rainwater flow. Here, we assume that the disturbance $w(t)$ is generated by a structural system (2) in Example 1, and the disturbance $w(t)$ is non-structural in Example 2.

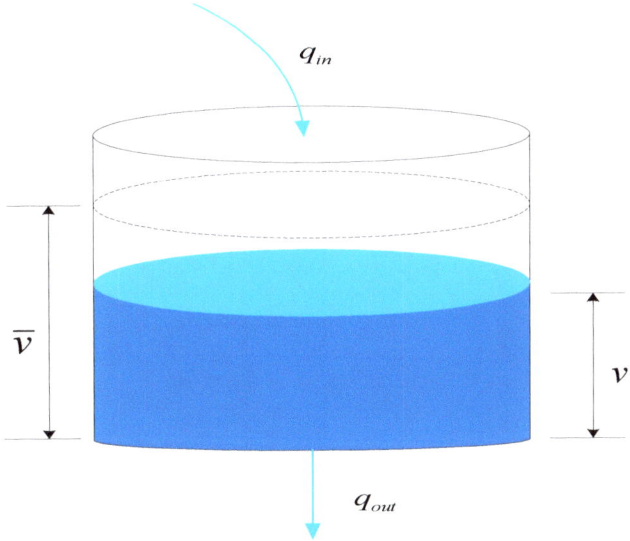

Figure 1. The virtual tank.

Example 1. *Consider system (1) with*

$$A = \begin{pmatrix} -2.50 & 0.35 & 0.30 \\ 0.52 & -1.98 & 0.58 \\ 0.38 & 0.40 & -2.28 \end{pmatrix}, B = \begin{pmatrix} 0.88 & 0.56 \\ 0.59 & 0.90 \\ 0.66 & 0.55 \end{pmatrix},$$
$$C = \begin{pmatrix} 1.23 & 0.95 \\ 0.98 & 1.15 \\ 1.10 & 0.86 \end{pmatrix}^{\top}, D = \begin{pmatrix} 0.85 & 0.78 \\ 0.78 & 0.88 \end{pmatrix}, E = \begin{pmatrix} 0.78 & 0.68 \\ 0.69 & 0.70 \\ 0.56 & 0.65 \end{pmatrix}.$$

Give the structural disturbance system (2) with

$$\Upsilon = \begin{pmatrix} -0.51 & 0.41 \\ 0.45 & -0.52 \end{pmatrix}, \Gamma = \begin{pmatrix} 1.10 & 0.10 \\ 0.10 & 0.13 \end{pmatrix}.$$

By Theorem 1, one can obtain the corresponding gain matrices:

$$G = \begin{pmatrix} -2.5497 & 0.2793 & 0.2142 \\ 0.3502 & -2.1804 & 0.3714 \\ 0.3305 & 0.3018 & -2.3818 \end{pmatrix}, F = \begin{pmatrix} 0.0049 & 0.0043 & 0.0036 \\ 0.0052 & 0.0045 & 0.0038 \end{pmatrix},$$
$$L_c = \begin{pmatrix} 0.0400 & 0.0364 \\ 0.1416 & 0.1070 \\ 0.0929 & 0.0705 \end{pmatrix}, L_d = \begin{pmatrix} -0.0031 & -0.0026 \\ -0.0035 & -0.0025 \end{pmatrix}.$$

Give $u(t) = 200e^{-0.05t}(|\sin(0.2\pi t)| \; |\cos(0.15\pi t)|)^{\top}$. *Under different initial conditions, the state trajectories of* $\eta(t)$ *and the observed signal* $\hat{\eta}(t)$ *are shown in Figure 2. The corresponding error signal* $e(t) = \eta(t) - \hat{\eta}(t)$ *is given in Figure 3. It can be observed from Figure 2 that all observer states* $(\hat{\eta}_1(t), \hat{\eta}_2(t), \hat{\eta}_3(t))$ *remain in positive orthant when the conditions are satisfied in Theorem 1. Moreover, it can be obtained that the observer errors* $e(t)$ *asymptotically converge to zero from Figure 3. Figures 2 and 3 show that the state and disturbance observers design for system (1) with structural disturbance is effective.*

In order to prove that the state observer obtained by Theorem 1 has a good performance, another input $u'(t) = 10{,}000e^{-0.05t}(|\sin(0.2\pi t)| \; |\cos(0.15\pi t)|)^{\top}$ *is given and the simulation results obtained are shown in Figure 4. By comparing Figure 2 with Figure 4, it can be found that the simultaneous state and disturbance observers designed in Theorem 1 are all effective for different inputs.*

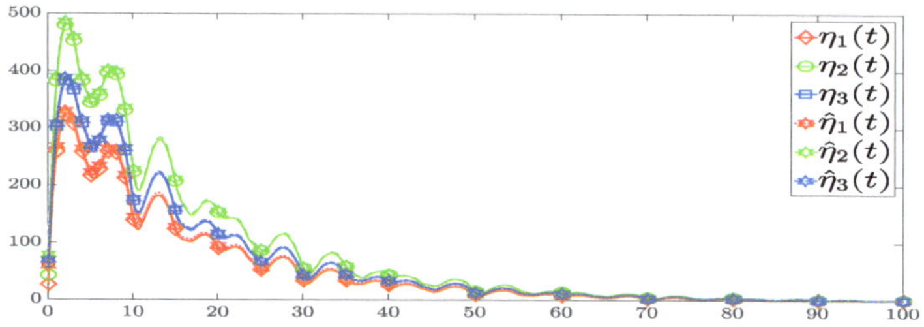

Figure 2. The state trajectories of system (1).

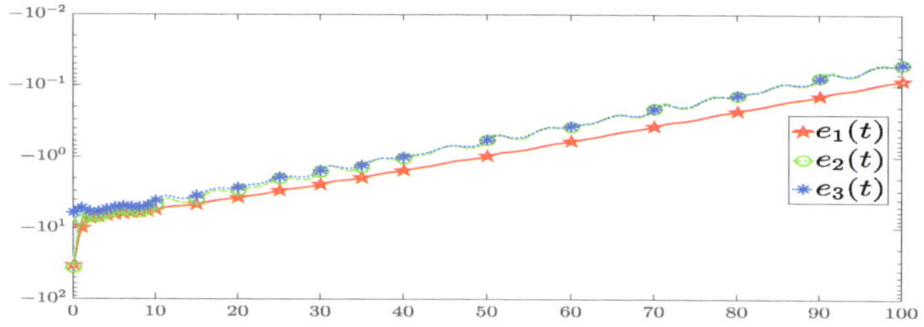

Figure 3. The corresponding error trajectories of system (1).

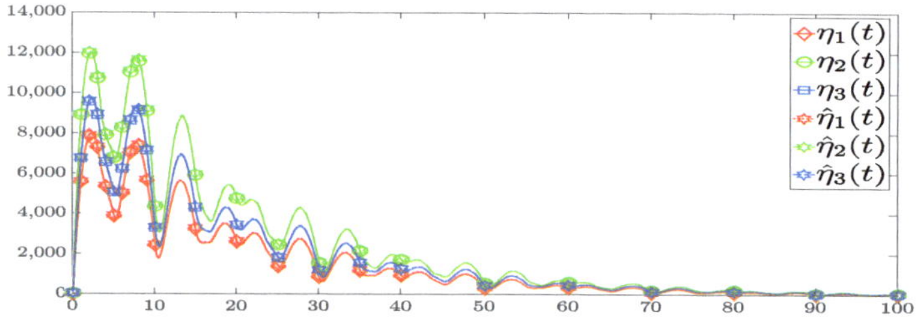

Figure 4. The state trajectories of system (1) with input $u'(t)$.

Example 2. *Consider the system (1) with*

$$A = \begin{pmatrix} -13.08 & 5.08 & 5.30 \\ 5.62 & -12.38 & 4.50 \\ 4.98 & 4.68 & -11.98 \end{pmatrix}, \ B = \begin{pmatrix} 0.38 & 0.56 \\ 0.39 & 0.60 \\ 0.46 & 0.35 \end{pmatrix},$$
$$C = \begin{pmatrix} 0.68 & 0.39 \\ 0.35 & 0.41 \\ 0.32 & 0.33 \end{pmatrix}^{\top}, \ E = \begin{pmatrix} 0.1950 & 0.1755 \\ 0.1800 & 0.1620 \\ 0.1400 & 0.1260 \end{pmatrix}.$$

By Theorem 2, the gain matrices are:

$$G = \begin{pmatrix} -10.7958 & 0.0070 \\ 0.0115 & -10.2686 \end{pmatrix}, \ Q = \begin{pmatrix} 0.0099 & 0.1400 \\ 0.0050 & 0.0213 \end{pmatrix},$$
$$W = \begin{pmatrix} 6.2873 & -5.2919 \\ -1.4874 & 3.1204 \end{pmatrix}, \ L = \begin{pmatrix} 1.3829 & 0.0125 \\ 0.9604 & 0.0256 \end{pmatrix}.$$

Give $u(t) = 100e^{-0.05t}|\sin(0.2\pi t)|\ |\cos(0.15\pi t)|^\top$ *and* $w(t) = 75e^{-0.05t}|\cos(0.1\pi t)|\ |\sin(0.15\pi t)|^\top$. *Under differential initial conditions, the state trajectories of* $\eta(t)$ *and the observed signal* $\hat{\eta}(t)$ *are depicted in Figure 5. The corresponding error signal* $e(t) = \eta(t) - \hat{\eta}(t)$ *is shown in Figure 6. It can be seen from Figure 5 that all observer states* ($\hat{\eta}_1(t)$,$\hat{\eta}_2(t)$,$\hat{\eta}_3(t)$) *remain in positive orthant when the conditions are satisfied in Theorem 2. Besides, it can be obtained that the observer errors* $e(t)$ *asymptotically converge to zero from Figure 6. Figures 5 and 6 show that the unknown input observer design for system (1) with non-structural disturbance is effective.*

Different input and disturbance with $u'(t) = 10{,}000e^{-0.05t}(|\sin(0.2\pi t)|\ |\cos(0.15\pi t)|)^\top$ *and* $w'(t) = 7500e^{-0.05t}(|\cos(0.1\pi t)|\ |\sin(0.15\pi t)|)^\top$ *are re-selected for simulation. The simulation results are shown in Figure 7. By comparing Figure 5 with Figure 7, it can be found that the unknown input observer designed in Theorem 2 is effective for different input and disturbance.*

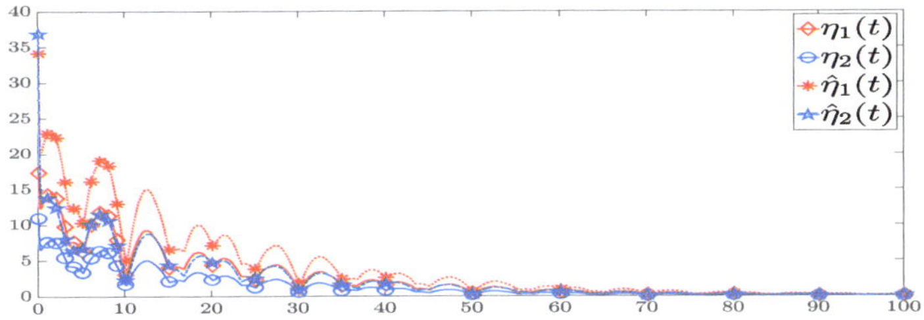

Figure 5. The state trajectories of system (1).

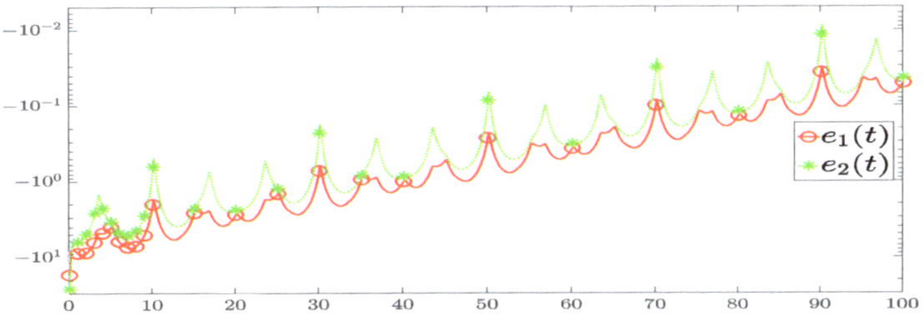

Figure 6. The corresponding error trajectories of system (1).

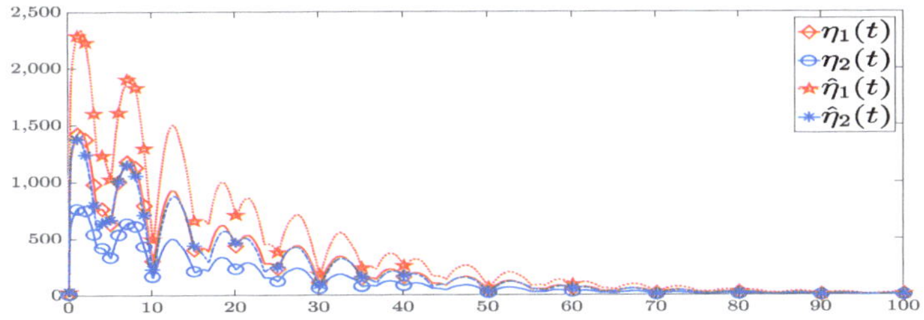

Figure 7. The state trajectories of system (1) with input $u'(t)$ and disturbance $w'(t)$.

5. Conclusions

This paper proposes two classes of observers for positive systems with disturbance. One is for the structural disturbance and the other one is the non-structural disturbance. A novel designed approach without additional conditions on the observer gain matrices is introduced, which removes the limitation of gain performance and thus improves the accuracy of the observer. The observer frameworks proposed in this paper are universal to positive systems with structural/non-structural disturbance and the proposed design method can provide valuable reference for the control synthesis of positive systems. In addition, linear programming is used to solve the presented conditions, which greatly reduces the computational complexity. In future work, it will be interesting to develop symmetry observer [37] and sliding mode observer [38] to positive systems.

Author Contributions: Conceptualization, X.Z. and J.Z.; Methodology, L.H., X.Z., F.L. and J.Z.; Software, L.H. and F.L.; Validation, X.Z.; Formal analysis, X.Z.; Investigation, L.H. and J.Z.; Resources, X.Z.; Writing—original draft, L.H.; Writing—review & editing, F.L. and J.Z.; Visualization, F.L.; Funding acquisition, J.Z. All authors have read and agreed to the published version of the manuscript.

Funding: This work was supported by the National Nature Science Foundation of China (62073111), the Fundamental Research Funds for the Provincial Universities of Zhejiang (GK229909299001-010 and GK219909299001-002), and Graduate Scientific Research Foundation of Hangzhou Dianzi University (CXJJ2022153 and CXJJ2022163).

Institutional Review Board Statement: Not applicable.

Informed Consent Statement: Not applicable.

Data Availability Statement: Not applicable.

Conflicts of Interest: The authors declare no conflict of interest. The funders had no role in the design of the study; in the collection, analyses, or interpretation of data; in the writing of the manuscript, or in the decision to publish the results.

References

1. Luenberger, D.G. Observing the state of a linear system. *IEEE Trans. Mil. Electron.* **1964**, *8*, 74–80. [CrossRef]
2. Fortmann, T.; Williamson, D. Design of low-order observers for linear feedback control laws. *IEEE Trans. Autom. Control* **1972**, *17*, 301–308. [CrossRef]
3. Lien, C.H. Robust observer-based control of systems with state perturbations via LMI approach. *IEEE Trans. Autom. Control* **2004**, *49*, 1365–1370. [CrossRef]
4. Rajamani, R. Observers for Lipschitz nonlinear systems. *IEEE Trans. Autom. Control* **1998**, *43*, 397–401. [CrossRef]
5. Efimov, D.; Raïssi, T.; Chebotarev, S.; Zolghadri, A. Interval state observer for nonlinear time varying systems. *Automatica* **2013**, *49*, 200–205. [CrossRef]
6. Tranninger, M.; Seeber, R.; Zhuk, S.; Steinberger, M.; Horn, M. Detectability analysis and observer design for linear time varying systems. *IEEE Control Syst. Lett.* **2019**, *4*, 331–336. [CrossRef]
7. Liu, Y.; Zhang, J. Reduced-order observer-based control design for nonlinear stochastic systems. *Syst. Control Lett.* **2004**, *52*, 123–135. [CrossRef]
8. Eltag, E.; Aslam, M.S.; Chen, Z. Functional observer-based T-S fuzzy systems for quadratic stability of power system synchronous generator. *Int. J. Fuzzy Syst.* **2020**, *22*, 172–180. [CrossRef]
9. Zhao, X.; Liu, H.; Zhang, J.; Li, H. Multiple-mode observer design for a class of switched linear systems. *IEEE Trans. Autom. Sci. Eng.* **2013**, *12*, 272–280. [CrossRef]
10. Chen, M.S.; Chen, C.C. Robust nonlinear observer for Lipschitz nonlinear systems subject to disturbances. *IEEE Trans. Autom. Control* **2007**, *52*, 2365–2369. [CrossRef]
11. Penarrocha, I.; Sanchis, R.; Albertos, P. H_∞ observer design for a class of nonlinear discrete systems. *Eur. J. Control* **2009**, *15*, 157–165. [CrossRef]
12. Guan, Y.; Saif, M. A novel approach to the design of unknown input observers. *IEEE Trans. Autom. Control* **1991**, *36*, 632–635. [CrossRef]
13. Zheng, G.; Bejaranoo, F.J.; Perruquetti, W.; Richard, J.P. Unknown input observer for linear time-delay systems. *Automatica* **2015**, *61*, 35–43. [CrossRef]
14. Chen, W.H. Disturbance observer based control for nonlinear systems. *IEEE/ASME Trans. Mechatron.* **2004**, *9*, 706–710. [CrossRef]
15. Yong, S.Z.; Zhu, M.; Frazzoli, E. Simultaneous input and state estimation for linear time-varying continuous-time stochastic systems. *IEEE Trans. Autom. Control* **2016**, *62*, 2531–2538. [CrossRef]
16. Farina, L.; Rinaldi, S. *Positive Linear Systems: Theory and Applications*; John Wiley & Sons: Hoboken, NJ, USA, 2000.
17. Kaczorek, T. *Positive 1D and 2D Systems*; Springer: London, UK, 2001.
18. Fornasini, E.; Valcher, M.E. Stability and stabilizability criteria for discrete-time positive switched systems. *IEEE Trans. Autom. Control* **2011**, *57*, 1208–1221. [CrossRef]
19. Briat, C. Robust stability and stabilization of uncertain linear positive systems via integral linear constraints: L_1-gain and L-gain characterization. *Int. J. Robust Nonlinear Control* **2013**, *23*, 1932–1954. [CrossRef]
20. Rami, M.A.; Tadeo, F.; Helmke, U. Positive observers for linear positive systems, and their implications. *Int. J. Control* **2011**, *84*, 716–725. [CrossRef]

21. Ebihara, Y.; Peaucelle, D.; Arzelier, D. Analysis and synthesis of interconnected positive systems. *IEEE Trans. Autom. Control* **2016**, *62*, 652–667. [CrossRef]
22. Zhang, J.; Zheng, G.; Feng, Y.; Chen, Y. Event-triggered state-feedback and dynamic output-feedback control of positive Markovian jump systems with intermittent faults. *IEEE Trans. Autom. Control* **2022**. [CrossRef]
23. Blanchini, F.; Colaneri, P.; Valcher, M.E. Co-positive Lyapunov functions for the stabilization of positive switched systems. *IEEE Trans. Autom. Control* **2012**, *57*, 3038–3050. [CrossRef]
24. Knorn, F.; Mason, O.; Shorten, R. On linear co-positive Lyapunov functions for sets of linear positive systems. *Automatica* **2009**, *45*, 1943–1947. [CrossRef]
25. Rami, M.A.; Tadeo, F. Controller synthesis for positive linear systems with bounded controls. *IEEE Trans. Circuits Syst. II Express Briefs* **2007**, *54*, 151–155. [CrossRef]
26. Shao, S.Y.; Chen, M.; Wu, Q.X. Stabilization control of continuous-time fractional positive systems based on disturbance observer. *IEEE Access* **2016**, *4*, 3054–3064. [CrossRef]
27. Li, P.; Lam, J. Positive state-bounding observer for positive interval continuous-time systems with time delay. *Int. J. Robust Nonlinear Control* **2012**, *22*, 1244–1257. [CrossRef]
28. Zaidi, I.; Chaabane, M.; Tadeo, F.; Benzaouia, A. Static state-feedback controller and observer design for interval positive systems with time delay. *IEEE Trans. Circuits Syst. II Express Briefs* **2014**, *62*, 506–510. [CrossRef]
29. Zhang, J.; Zhang, R.; Chen, Y.; Fu, S. Linear programming based dynamic output-feedback controller for positive systems. In Proceedings of the 2017 American Control Conference (ACC), Seattle, WA, USA, 24–26 May 2017; pp. 1281–1290.
30. Qi, W.; Park, J.H.; Zong, G.; Cao, J.; Cheng, J. A fuzzy Lyapunov function approach to positive L_1 observer design for positive fuzzy semi-Markovian switching systems with its application. *IEEE Trans. Syst. Man Cybern. Syst.* **2018**, *51*, 775–785. [CrossRef]
31. Zhang, D.; Zhang, Q.; Du, B. L_1 fuzzy observer design for nonlinear positive Markovian jump system. *Nonlinear Anal. Hybrid Syst.* **2018**, *27*, 271–288. [CrossRef]
32. Arogbonlo, A.; Huynh, V.T.; Oo, A.M.T.; Trinh, H. Functional observers design for positive systems with delays and unknown inputs. *IET Control Theory Appl.* **2020**, *14*, 1656–1661. [CrossRef]
33. Oghbaee, A.; Shafai, B.; Nazari, S. Complete characterisation of disturbance estimation and fault detection for positive. *IET Control Theory Appl.* **2018**, *12*, 883–891. [CrossRef]
34. Ocampo-Martínez, C.; Puig, V.; Cembrano, G.; Creus, R.; Minoves, M. Improving water management efficiency by using optimization-based control strategies: The Barcelona case study. *Water Sci. Technol. Water Supply* **2009**, *9*, 565–575. [CrossRef]
35. Ocampo-Martínez, C.; Puig, V.; Cembrano, G.; Quevedo, J. Application of predictive control strategies to the management of complex networks in the urban water cycle. *IEEE Control Syst. Mag.* **2013**, *3*, 15–41.
36. Zhang, J.; Yang, H.; Li, M.; Wang, Q. Robust model predictive control for uncertain positive time-delay systems. *Int. J. Control. Autom. Syst.* **2019**, *17*, 307–318. [CrossRef]
37. Bonnabel, S.; Martin, P.; Rouchon, P. Symmetry-preserving observers. *IEEE Trans. Auto-Matic Control* **2008**, *53*, 2514–2526. [CrossRef]
38. Edwards, C.; Spurgeon, S.K.; Patton, R.J. Sliding mode observers for fault detection and isolation. *Autom. Autom.* **2000**, *36*, 541–553. [CrossRef]

mathematics

Article

Robust Synchronization of Fractional-Order Chaotic System Subject to Disturbances

Dongya Li [1,*], Xiaoping Zhang [1], Shuang Wang [1] and Fengxiang You [2]

[1] Applied Technology College of Soochow University, Suzhou 215325, China
[2] School of Mechanical and Electrical Engineering, Soochow University, Suzhou 215021, China
* Correspondence: lidongya@suda.edu.cn

Abstract: This paper studies the synchronization problem for a class of chaotic systems subject to disturbances. The nonlinear functions contained in the master and slave systems are assumed to be incremental quadratic constraints. Under some assumptions, a feedback law is designed so that the error system behaves like the H^∞ performance. Meanwhile, the detailed algorithm for computing the incremental multiplier matrix is also given. Finally, one numerical example and one practical example are simulated to show the effectiveness of the designed method.

Keywords: fractional-order; chaotic systems; robust synchronization

MSC: 93-10

Citation: Li, D.; Zhang, X.; Wang, S.; You, F. Robust Synchronization of Fractional-Order Chaotic System Subject to Disturbances. *Mathematics* **2022**, *10*, 4639. https://doi.org/10.3390/math10244639

Academic Editor: Ichiro Tsuda

Received: 16 November 2022
Accepted: 5 December 2022
Published: 7 December 2022

1. Introduction

The chaotic system is a kind of nonlinear system, and the characteristics of the chaotic system behave like the chaotic attractors . The definition of chaos was introduced by [1]. The investigation of the chaotic system has been paid much attention since it plays an important role in areas such as image encryption [2], fault detection [3], neural networks [4], communication security [5], and so on. In practice, the synchronization of master and slave chaotic systems is very essential to secure communication. Thus, synchronization has been studied extensively [6–8]. In [6], the authors used an active nonlinear controller to realize the synchronization of two hybrid chaotic systems, while [7] studied the design of adaptive controller for the purpose of the synchronization. Ref. [8] focused on the time-delay chaotic system, and a feedback law was designed to realize the robust synchronization. There were also other important works on synchronization for chaotic systems [9,10] herein.

On the other hand, the research on fractional-order systems is also a hot topic. The fractional-order system first appeared in a pure mathematical problem [11]. In the context of mathematics, Ref. [12] used the fractal-fractional mathematical model to describe the situation of corona virus, and [13] studied a class of nonlinear delayed corona virus pandemic model, while in [14], an optimal control problem of a nonholonomic macroeconomic system was investigated. Some researchers used the fractional-order system to describe more general practical systems. In fact, a fractional-order differential equation can be more accurate in describing complicated systems than integral-order differential equation. In the aspect of the fractional-order system, there were many interesting works, such as [15–18]. Since it is very powerful in describing more general systems, the study of fractional-order chaotic systems has always been a hot spot. Ref. [19] employed the active control method to investigate the synchronization problem for fractional-order chaotic systems, while, in [20], an adaptive impulsive controller was designed to achieve synchronization . The robust observer design problem for fractional-order chaotic systems was addressed in [21]. Moreover, the stability conditions of a class of impulsive incommensurate chaotic systems were analyzed in [22].

It should be noted that the nonlinear terms considered in the above-mentioned works [19–22] are all Lipschitz. Recently, a more general nonlinearity, called incremental quadratic constraints (IQC), has attracted much attention. It is pointed out in the work [23] that IQC is characterized by the incremental multiplier matrix (IMM) and can include Lipschitz constraints and one-sided Lipschitz constraints. The work [24] presented an observer design method for IQC systems, the results of [24] were extended to chaotic systems, and the secure communication problem was studied in [25]. Ref. [26] designed the controller for IQC systems with external disturbances. However, the robust synchronization of fractional-order chaotic systems under the framework of IQC has been reported rarely.

In the light of the above discussion, this paper considers the synchronization problem of fractional-order chaotic systems whose nonlinearity is described by IQC. The controller is designed by using the output, and the fractional-order stability theory is employed to derive sufficient conditions on robust synchronization. The remainder of the paper is as follows: Section 2 formulates the problem and presents some necessary basics. Section 3 designs the feedback law so that the robust synchronization is achieved. Section 4 suggests an algorithm to compute IMM. Section 5 simulates two examples to illustrate the validity of the designed method.

2. Preliminaries and Problem Statements

Consider the following fractional-order chaotic system:

$$\begin{cases} D_t^{\alpha}x(t) = Ax(t) + Gf(Hx(t)) + D\omega_x(t), \\ z_x(t) = Cx(t), \end{cases} \tag{1}$$

where D_t^{α} is the α-order Caputo derivative with $0 < \alpha < 1$. $A \in R^{n\times n}$, $G \in R^{n\times m}$, $H \in R^{l\times n}$, $D \in R^{n\times s}$, and $C \in R^{q\times n}$ are constant matrices. $x(t) \in R^n$ is the system state, $\omega_x(t) \in R^s$ is the disturbance, and $z_x(t) \in R^q$ is the output. $f(q) : R^l \to R^m$ with $q = Hx(t)$ is the nonlinear function. For the purpose of simplification, the variable t is omitted when necessary. Then, the master system Equation (1) is written as:

$$\begin{cases} D_t^{\alpha}x = Ax + Gf(Hx) + D\omega_x, \\ z_x = Cx. \end{cases} \tag{2}$$

In the above system Equation (2), when the system matrices are given as:

$$A = \begin{bmatrix} -10 & 10 & 0 \\ 28 & -1 & 0 \\ 0 & 0 & -\frac{8}{3} \end{bmatrix}, \ G = \begin{bmatrix} 0 & 0 \\ -1 & 0 \\ 0 & 1 \end{bmatrix}, \ H = \begin{bmatrix} 1 & 0 & 0 \\ 0 & 1 & 0 \\ 0 & 0 & 1 \end{bmatrix},$$

$$f(Hx) = \begin{bmatrix} x_1x_3 \\ x_1x_2 \end{bmatrix}, \ D = \begin{bmatrix} 1 \\ 0 \\ 0 \end{bmatrix}, \ C = \begin{bmatrix} 1 & 0 & 0 \end{bmatrix}.$$

The above fractional-order system behaves the chaos phenomenon. Figure 1 shows the phase plane of the system when $\alpha = 0.98$.

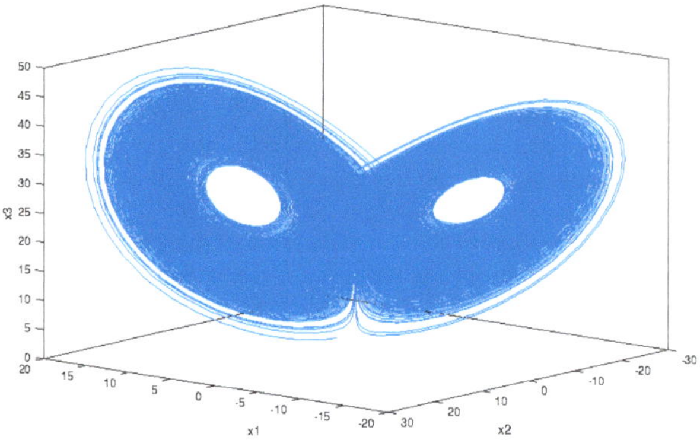

Figure 1. The chaotic behavior of fractional-order chaotic system.

Then, the following slave system is presented:

$$\begin{cases} D_t^a y = Ay + Gf(Hy) + D\omega_y + Bu, \\ z_y = Cy, \end{cases} \tag{3}$$

where ω_y is the disturbance, and u is the controller, which is designed as $u = K(z_y - z_x)$. Let $e = y - x$, the error system is derived as follows:

$$D_t^a e = (A + BKC)e + G[f(Hy) - f(Hx)] + D\Delta\omega. \tag{4}$$

where $\Delta\omega = \omega_y - \omega_x$.

Denote that $\Gamma(\alpha) = \int_0^\infty e^{-t}t^{\alpha-1}dt$. We firstly give the definitions of the fractional-order integral and derivative. More details can be found in [27].

Remark 1. *If the parameter errors exist in the system* (1)*, i.e., A and B will be substituted by* $A + \Delta A(t)$ *and* $B + \Delta B(t)$*. Then, we can use the norm-bounded conditions of* $\Delta A(t)$ *and* $\Delta B(t)$ *to design the controller in* (3)*.*

Definition 1 ([27])**.** *A fractional integral of the function z with order* $\alpha > 0$ *is defined as follows:*

$$I_{t_0}^\alpha z(t) = \frac{1}{\Gamma(\alpha)} \int_{t_0}^t (t-s)^{\alpha-1} z(s)ds.$$

Definition 2 ([27])**.** *A Caputo fractional derivative of the function z with order* $0 < \alpha < 1$ *is defined as follows:*

$$D_t^\alpha z(t) = \frac{1}{\Gamma(\alpha)} \int_0^t (t-s)^{-\alpha} \frac{d}{ds} z(s)ds.$$

Definition 3 ([24])**.** *For a nonlinear function* $\varphi(\theta)$*, if there exists a symmetric matrix W such that*

$$\begin{bmatrix} \theta_1 - \theta_2 \\ \varphi(\theta_1) - \varphi(\theta_2) \end{bmatrix}^T W \begin{bmatrix} \theta_1 - \theta_2 \\ \varphi(\theta_1) - \varphi(\theta_2) \end{bmatrix} \geq 0, \tag{5}$$

then $\varphi(\theta)$ *is IQC, and W is called the IMM for* $\varphi(\theta)$*.*

From [24], it is known that Lispchitz constraints or one-sided Lipschitz constraints are a special case of IQC.

Definition 4 ([28])**.** *The error system* (4) *behaves the* H^∞ *performance, if*

(1) $\Delta\omega = 0, \lim\limits_{t\to\infty} e(t) = 0;$

(2) $\Delta\omega \neq 0$; *under the zero-initial condition, the following inequality holds:*

$$\int_0^\infty e^T(s)e(s)ds < \varsigma \int_0^\infty \Delta\omega^T(s)\Delta\omega(s)ds,$$

where $\varsigma > 0$ *is the disturbance attenuation level.*

Assumption 1. *The nonlinear function* $f(q)$ *in systems (2) and (3) satisfies IQC with IMM M, i.e.,*

$$\begin{bmatrix} q_2 - q_1 \\ f(q_2) - f(q_1) \end{bmatrix}^T M \begin{bmatrix} q_2 - q_1 \\ \Phi_1(q_2) - \Phi_1(q_1) \end{bmatrix} \geq 0, \quad (6)$$

and M has the blocked form as follows:

$$M = \begin{bmatrix} M_{11} & M_{12} \\ M_{21} & M_{22} \end{bmatrix}, \quad (7)$$

where $M_{11} \in \mathbb{R}^{n\times n}$ *and* $M_{22} \in \mathbb{R}^{s\times s}$.

Lemma 1 ([29]). *Let* $V(t) = x^T(t)Px(t)$ *be a continuously differentiable function, then* $D_t^\alpha V(t)$ *satisfies:*

$$D_t^\alpha V(t) \leq (D_t^\alpha e(t))^T Pe(t) + e^T P(D_t^\alpha e(t)). \quad (8)$$

3. Main Results

We first state the following theorem, where sufficient conditions are given so that the controller design is effective.

Theorem 1. *Let Assumption 1 hold. If matrices* $P = P^T > 0$ *and K exist such that*

$$\begin{bmatrix} (A+BKC)^T P + P(A+BKC) + I + H^T M_{11} H & PG + H^T M_{12} & PD \\ G^T P + M_{21} H & M_{22} & 0 \\ D^T P & 0 & -\varsigma I \end{bmatrix} < 0, \quad (9)$$

where $\varsigma > 0$ *is the disturbance attenuation level, then the error system Equation (4) behaves like the* H^∞ *performance , i.e., the robust synchronization of systems Equations (2) and (3) is achieved.*

Proof. Consider the following Lyapunov function candidate:

$$V = e^T Pe. \quad (10)$$

From Lemma 1, the fractional derivative of V is

$$D_t^\alpha V \leq (D_t^\alpha e^T)Pe + e^T PD_t^\alpha e. \quad (11)$$

Thus, along the error dynamics Equation (4), we have

$$\begin{aligned} D_t^\alpha V \leq & e^T[(A+BKC)^T P + P(A+BKC)]e + e^T PD\Delta\omega + \Delta\omega^T D^T Pe \\ & e^T PG[f(Hy) - f(Hx)] + [f(Hy) - f(Hx)]^T G^T Pe. \end{aligned} \quad (12)$$

The proof is divided into two steps according to Definition 4.
(i) $\Delta\omega(t) = 0$. It follows from Equation (12) that

$$\begin{aligned} D_t^\alpha V \leq & e^T[(A+BKC)^T P + P(A+BKC)]e \\ & e^T PG[f(Hy) - f(Hx)] + [f(Hy) - f(Hx)]^T G^T Pe. \end{aligned} \quad (13)$$

Define that $\delta = f(Hy) - f(Hx)$, by using Assumption 1, we have

$$\begin{bmatrix} e \\ \delta \end{bmatrix}^T \begin{bmatrix} H^T & 0 \\ 0 & I \end{bmatrix} M \begin{bmatrix} H & 0 \\ 0 & I \end{bmatrix} \begin{bmatrix} e \\ \delta \end{bmatrix} \geq 0,$$

i.e.,

$$\begin{bmatrix} e \\ \delta \end{bmatrix}^T \begin{bmatrix} H^T M_{11} H & H^T M_{12} \\ M_{21} H & M_{22} \end{bmatrix} \begin{bmatrix} e \\ \delta \end{bmatrix} \geq 0. \tag{14}$$

Denote that $\xi = [e^T \ \delta^T]^T$. Substituting Equation (14) into Equation (13) yields

$$D_t^\alpha V \leq \xi^T \Lambda \xi, \tag{15}$$

where

$$\Lambda = \begin{bmatrix} (A+BKC)^T P + P(A+BKC) + H^T M_{11} H & PG + H^T M_{12} \\ G^T P + M_{21} H & M_{22} \end{bmatrix}.$$

In view of Equation (9), by using the matrix theory, we have

$$\begin{bmatrix} (A+BKC)^T P + P(A+BKC) + I + H^T M_{11} H & PG + H^T M_{12} \\ G^T P + M_{21} H & M_{22} \end{bmatrix} < 0.$$

Thus, $\Gamma < 0$. From Equation (15), we can deduce that $D_t^\alpha V < 0$, which means that $\lim_{t\to\infty} e(t) = 0$.

(ii) $\Delta\omega(t) \neq 0$. Let

$$\begin{aligned} J &= \int_0^\infty e^T(t)e(t)\, dt - \int_0^\infty \varsigma\omega^T(t)\omega(t)\, dt \\ &= \int_0^\infty [e^T(t)e(t) - \varsigma\omega^T(t)\omega(t)]\, dt. \end{aligned} \tag{16}$$

Recall that

$$I_0^1 D_t^\alpha V(e) = I_0^{1-\alpha} I_0^\alpha D_t^\alpha V(e), \tag{17}$$

then

$$I_0^1 D_t^\alpha V(e) = I_0^{1-\alpha}(V(e(t)) - V(e(0))). \tag{18}$$

By using the zero-initial condition $e(0) = 0$, one gets

$$I_0^1 D_t^\alpha V(e) = I_0^{1-\alpha}(V(e(t))). \tag{19}$$

Since $V(e(t)) \geq 0$, we have $I_0^{1-\alpha}(V(e(t))) \geq 0$, i.e., $I_0^1 D_t^\alpha V(e) \geq 0$, which implies that

$$\int_0^\infty D_\alpha^t V(t) dt \geq 0. \tag{20}$$

Thus, it follows from Equations (16) and (20) that

$$J \leq \int_0^\infty [e^T(t)e(t) - \varsigma\omega^T(t)\omega(t) + D_\alpha^t V(t)]\, dt. \tag{21}$$

Denote that $S = e^T e - \varsigma\omega^T\omega + D_\alpha^t V$, together with Equation (12), we have

$$\begin{aligned} S =& e^T e - \varsigma\omega^T\omega + e^T[(A+BKC)^T P + P(A+BKC)]e \\ &+ \delta^T G^T P e + e^T P G \delta + \Delta\omega^T D^T P e + e^T P D \Delta\omega. \end{aligned} \tag{22}$$

Define that $\zeta = [e^T \ \ \delta^T \ \ \Delta\omega^T]^T$. Substituting Equation (14) into Equation (22) yields

$$S \leq \zeta^T \Theta \zeta, \tag{23}$$

where

$$\Theta = \begin{bmatrix} (A+BKC)^T P + P(A+BKC) + I + H^T M_{11} H & PG + H^T M_{12} & PD \\ G^T P + M_{21} H & M_{22} & 0 \\ D^T P & 0 & -\varsigma I \end{bmatrix}.$$

It follows from Equations (9), (19) and (23) that

$$J \leq \int_0^\infty S(t)\, dt < 0, \tag{24}$$

which implies

$$\int_0^\infty e^T(t)e(t)dt - \int_0^\infty \varsigma\omega^T(t)\omega(t)dt < 0, \tag{25}$$

i.e.,

$$\int_0^\infty e^T(t)e(t)dt < \varsigma \int_0^\infty \omega^T(t)\omega(t)dt. \tag{26}$$

Combining (i) with (ii), the proof is completed. □

Remark 2. *The condition Equation* (9) *cannot be solved by the Matlab LMI toolbox since it is not a standard LMI. Here, one solution is suggested to deal with the matrix inequality Equation* (9)*. Let $PBK = Z$; then Equation* (9) *is equivalent to*

$$\begin{bmatrix} A^T P + PA + ZC + C^T Z^T + I + H^T M_{11} H & PG + H^T M_{12} & PD \\ G^T P + M_{21} H & M_{22} & 0 \\ D^T P & 0 & -\varsigma I \end{bmatrix} < 0. \tag{27}$$

Thus, we can use the Matlab LMI toolbox to solve the matrix P from Equation (27)*. Then, if $rank(PB) = rank(PB\ \ Z)$, then $K = (PB)^{+}Z$, where $(PB)^{+}$ is the Moore inverse matrix of PB.*

4. The Determination of IMM M

In Equation (27), the matrix M is essential to the solutions P and K. Thus, we will give a detailed algorithm to compute M in this section. Generally, the nonlinear function $f(q)$ in system Equations (2) and (3) is supposed to be continuously differentiable, and it is characterized by a known set Ω of matrices. For any $q_1, q_2 \in R^l$, there is a matrix Y in Ω such that $f(q_1) - f(q_2) = Y(q_1 - q_2)$. Ω is a known polytope of matrices with vertices Y_1, Y_2, ..., Y_n denoted by $\Omega = co\{Y_1, Y_2, ..., Y_n\}$. By using IQC, one gets

$$\begin{bmatrix} I \\ Y_i \end{bmatrix}^T M \begin{bmatrix} I \\ Y_i \end{bmatrix} \geq 0, \quad i \in I[1, N]. \tag{28}$$

In view of the decomposition form of M, Equation (28) becomes

$$M_{11} + M_{12}Y_i + Y_i^T M_{12}^T + Y_i^T M_{22} Y_i \geq 0, \quad i \in I[1, N].$$

By using the Shur complements, we have

$$\begin{bmatrix} M_{11} + M_{12}Y_i + Y_i^T M_{12}^T & Y_i^T M_{22} \\ M_{22}Y_i & -M_{22} \end{bmatrix} \geq 0, \; i \in I[1, N]. \tag{29}$$

By solving Equation (29), the solution M may be found. However, it is not easy to determine Y_i. In the sequel, we provide a method with to compute these vertex matrices Y_i of $f(q)$ in systems Equations (2) and (3). Since $f(q)$ is continuously differentiable, then

$$\begin{aligned}\frac{\partial f(q)}{\partial q} &= \begin{bmatrix} q_3 & 0 & q_1 \\ q_2 & q_1 & 0 \end{bmatrix} = \begin{bmatrix} x_3 & 0 & x_1 \\ x_2 & x_1 & 0 \end{bmatrix} \\ &= x_1\begin{bmatrix} 0 & 0 & 1 \\ 0 & 1 & 0 \end{bmatrix} + x_2\begin{bmatrix} 0 & 0 & 0 \\ 1 & 0 & 0 \end{bmatrix} + x_3\begin{bmatrix} 1 & 0 & 0 \\ 0 & 0 & 0 \end{bmatrix}.\end{aligned}$$

Thus, the vertices of the polytope Ω can be described by

$$\begin{aligned}Y_1 &= \tau\begin{bmatrix} 0 & 0 & 1 \\ 0 & 1 & 0 \end{bmatrix}, Y_2 = \tau\begin{bmatrix} 0 & 0 & 0 \\ 1 & 0 & 0 \end{bmatrix}, Y_3 = \tau\begin{bmatrix} 1 & 0 & 0 \\ 0 & 0 & 0 \end{bmatrix}, \\ Y_4 &= -\tau\begin{bmatrix} 0 & 0 & 1 \\ 0 & 1 & 0 \end{bmatrix}, Y_5 = -\tau\begin{bmatrix} 0 & 0 & 0 \\ 1 & 0 & 0 \end{bmatrix}, Y_6 = -\tau\begin{bmatrix} 1 & 0 & 0 \\ 0 & 0 & 0 \end{bmatrix},\end{aligned}$$

By using the bounded condition in [30], we solve Equation (29) and have

$$M_{11} = \begin{bmatrix} 10.1989 & 0 & 0 \\ 0 & 7.9306 & 0 \\ 0 & 0 & 7.4134 \end{bmatrix}, M_{12} = \begin{bmatrix} -0.952 & 0 \\ 0 & -1.4586 \\ 0 & 0 \end{bmatrix},$$

$$M_{22} = \begin{bmatrix} -4.0588 & 0 \\ 0 & -4.0836 \end{bmatrix}. \tag{30}$$

Remark 3. *By the definition of IQC, we know that the matrix M may have infinite solutions. To some extent, the computation of M is not an easy task. The detailed procedure depends on the parameter τ. Owing to the technique proposed in [30], the detailed computation process of τ is omitted here.*

5. Numerical Simulation

Example 1. *Consider the system Equation* (2) *with the matrix parameters:*

$$A = \begin{bmatrix} -10 & 10 & 0 \\ 28 & -1 & 0 \\ 0 & 0 & -\frac{8}{3} \end{bmatrix}, G = \begin{bmatrix} 0 & 0 \\ -1 & 0 \\ 0 & 1 \end{bmatrix}, H = \begin{bmatrix} 1 & 0 & 0 \\ 0 & 1 & 0 \\ 0 & 0 & 1 \end{bmatrix},$$

$$f(Hx) = \begin{bmatrix} x_1x_3 \\ x_1x_2 \end{bmatrix}, D = \begin{bmatrix} 1 \\ 0 \\ 0 \end{bmatrix}, C = \begin{bmatrix} 1 & 0 & 0 \end{bmatrix}.$$

By using the results of the fractional order system, if $\alpha \geq 0.98$, the system Equation (2) *behaves like chaotic attractors . Therefore, α is setting as 0.98 in the simulation. We solve IMM for $f(Hx)$, and M is given as shown in Equation* (30)*. The disturbances ω_x and ω_y are as follows:*

$$\omega_x = \omega_y = \begin{cases} 300\sin(10t), 0 \leq t \leq 3s, \\ 0, t > 3s. \end{cases}$$

The initial values of x and y are chosen as

$$x^T(0) = \begin{bmatrix} 5 & -3 & 6 \end{bmatrix}, y^T(0) = \begin{bmatrix} -5 & 6 & -3 \end{bmatrix}.$$

First, we deal with nonlinearity through the IMM algorithm in Section 4, and the matrix M is as follows:

$$M = \begin{bmatrix} 10.1989 & 0 & 0 & -0.9520 & 0 \\ 0 & 7.9306 & 0 & 0 & -1.4586 \\ 0 & 0 & 7.4134 & 0 & 0 \\ -0.9520 & 0 & 0 & -4.0588 & 0 \\ 0 & -1.4586 & 0 & 0 & -4.0836 \end{bmatrix}.$$

We let

$$\int_0^\infty e^T(t)e(t)dt < 5\int_0^\infty \omega^T(t)\omega(t)dt,$$

when $\Delta\omega(t) \neq 0$. Substituting M into Equation (27) yields

$$P = \begin{bmatrix} 6.0584 & -0.5478 & 0.1606 \\ -0.5478 & 2.0698 & 0.4468 \\ 0.1606 & 0.4468 & 2.6717 \end{bmatrix},$$

which can be seen that P is positive definite. Furthermore, we have

$$Z = \begin{bmatrix} 67.2326 \\ -124.5635 \\ -10.4750 \end{bmatrix}.$$

It is obvious that $rank(PB) = rank(PB\ \ Z)$, i.e., $K = (PB)^{\dagger}Z$, where $(PB)^{\dagger}$ is the Moore inverse matrix of PB. Hence, the controller gain can be obtained

$$K = \begin{bmatrix} 5.5223 \\ -59.9667 \\ 5.7749 \end{bmatrix}.$$

During the simulation, we denote that

$$x = \begin{bmatrix} x_1 \\ x_2 \\ x_3 \end{bmatrix}, y = \begin{bmatrix} y_1 \\ y_2 \\ y_3 \end{bmatrix}, e = \begin{bmatrix} e_1 \\ e_2 \\ e_3 \end{bmatrix}.$$

In Figure 2, the trajectories of each state of system Equation (2) and system Equation (3) are shown. In Figure 3, the trajectories of error dynamics $e(t)$ are shown. It can be seen that the trajectories of errors exhibit bounded convergence under the influence of disturbances ω_x and ω_y in the first 3 s. However, after 3 s, when the disturbances disappears, the error will gradually converge to zero. It is consistent with the theoretical results. Thus, the proposed method is valid in this paper.

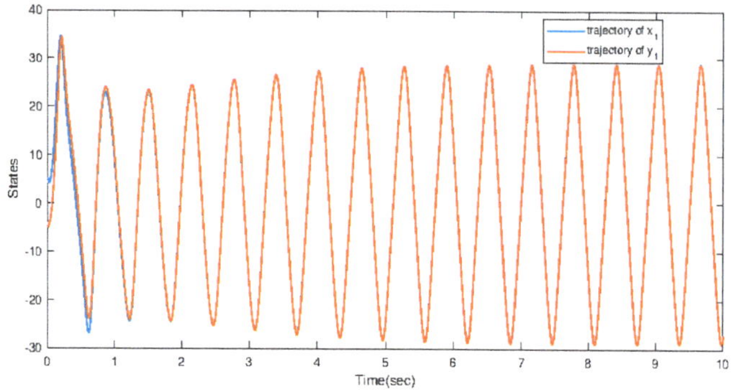

(**a**)

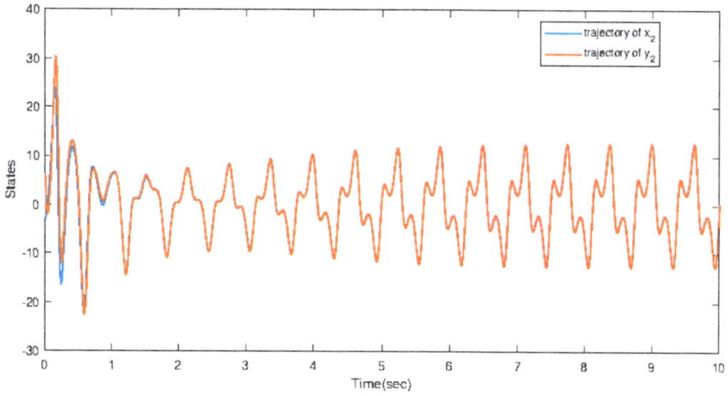

(**b**)

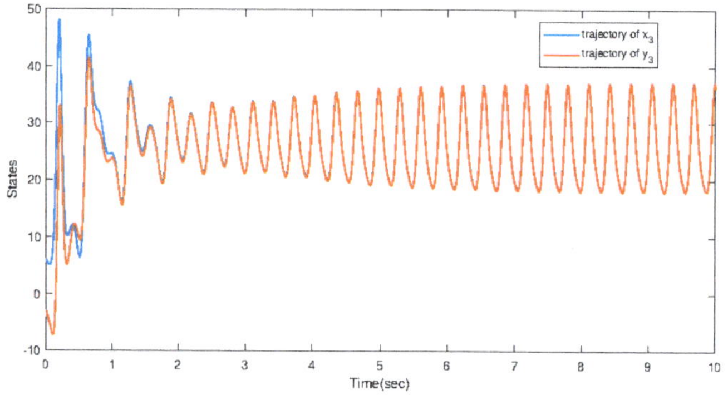

(**c**)

Figure 2. The trajectories of $x(t)$ and $y(t)$, (**a**) the trajectories of x_1 and y_1, (**b**) the trajectories of x_2 and y_2, and (**c**) the trajectories of x_3 and y_3.

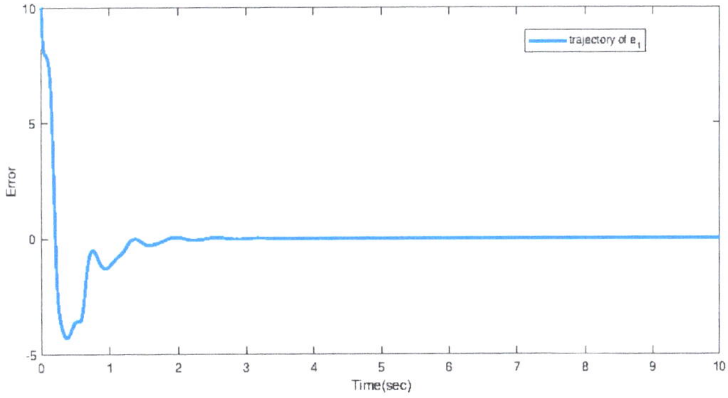

(**a**)

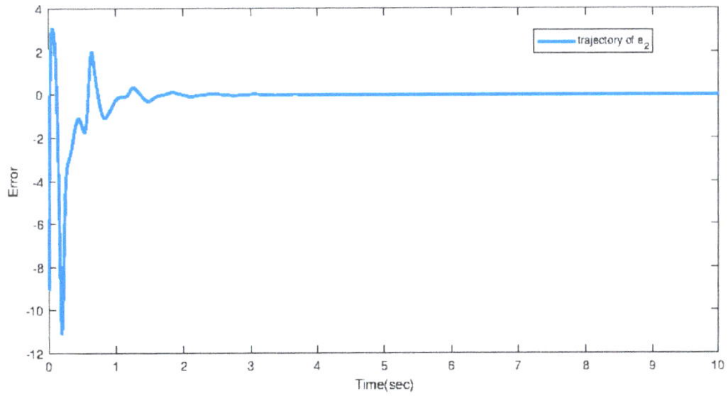

(**b**)

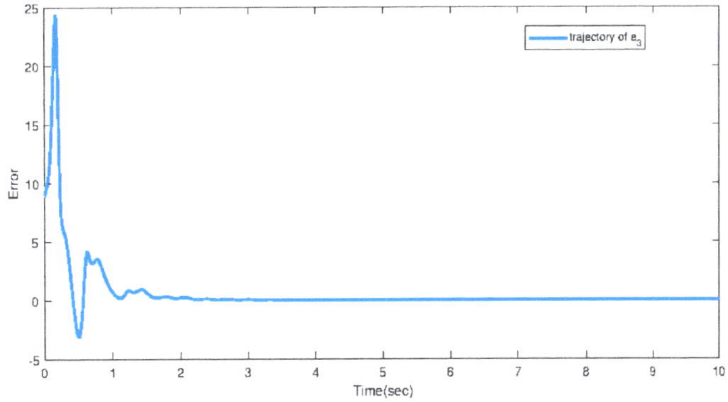

(**c**)

Figure 3. The trajectory of tracking error $e(t)$, (**a**) The trajectory of tracking error e_1, (**b**) the trajectory of tracking error e_2, and (**c**) the trajectory of tracking error e_3.

Example 2. *Fractional-order Lorentz systems can be used to describe a class of circuit systems. In Figure 4, the three state variables x, y, and z are implemented by three channels, respectively, and some of these calculus operations are replaced by operational amplifiers and analog multipliers. The resistors in Figure 4 are* $R_i = 10\ k\Omega; i = 1,2,8,9,12,13,15,16,18,19,21,22;$ $R_4 = 1\ k\Omega;$ $R_5 = 1.55\ M\Omega;$ $R_6 = 62\ M\Omega;$ $R_7 = 2.5\ k\Omega;$ $R_{10} = 3.57\ k\Omega;$ $R_j = 100\ k\Omega; j = 11,14,20;$ *and* $R_{17} = 37.5\ k\Omega;$ *R_3 is adjustable, and the capacitors are* $C_1 = 0.73\ \mu F$, $C_2 = 0.52\ \mu F$, $C_3 = 1.1\ \mu F$, $C_4 = C_5 = 1\ nF$. *By adjusting R_3, different chaotic phenomena can be obtained.*

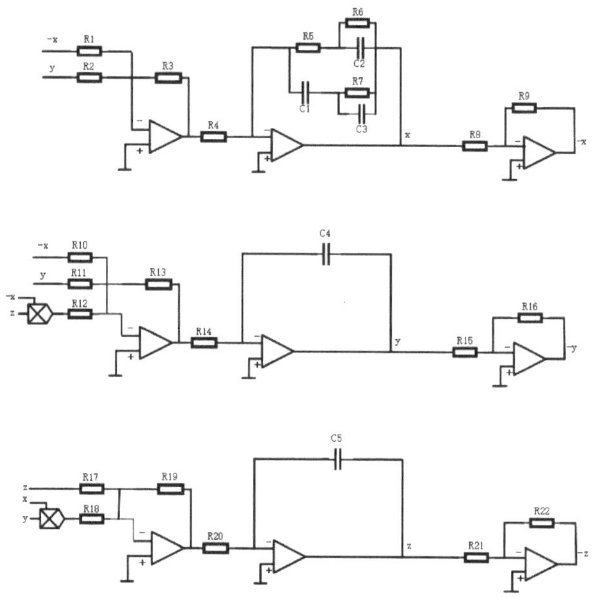

Figure 4. Analog circuit of the fractional-order Lorenz system.

From [31], by analyzing the circuit system in Figure 4, we obtain the following state-space dynamics:

$$\begin{cases} D_t^{\alpha} x(t) = Ax(t) + Gf(Hx(t)) + D\omega_x(t), \\ z_x(t) = Cx(t), \end{cases}$$

where

$$A = \begin{bmatrix} -10 & 10 & 0 \\ 28 & -1 & 0 \\ 0 & 0 & -\frac{8}{3} \end{bmatrix}, G = \begin{bmatrix} 0 & 0 \\ -1 & 0 \\ 0 & 1 \end{bmatrix}, H = \begin{bmatrix} 1 & 0 & 0 \\ 0 & 1 & 0 \\ 0 & 0 & 1 \end{bmatrix},$$

$$f(Hx) = \begin{bmatrix} x_1 x_3 \\ x_1 x_2 \end{bmatrix}, D = \begin{bmatrix} 1 & 0 & 0 \\ 0 & 1 & 0 \\ 0 & 0 & 0 \end{bmatrix}, C = \begin{bmatrix} 1 & 0 & 0 \end{bmatrix}.$$

The disturbance is chosen as:

$$w_x = w_y = \begin{cases} \begin{bmatrix} 10sin(10t) \\ 20sin(10t) \\ 0 \end{bmatrix}, 0 \le t \le 10s, \\ 0, t > 10s. \end{cases}$$

Then, the control gain K is obtained by solving Equation (27)

$$K = \begin{bmatrix} -3.6982 \\ -53.0802 \\ 4.1746 \end{bmatrix},$$

and

$$P = \begin{bmatrix} 1.7077 & -0.5876 & 0.1182 \\ -0.5876 & 1.0806 & 0.3321 \\ 0.1182 & 0.3321 & 2.4589 \end{bmatrix}, Z = \begin{bmatrix} 25.3700 \\ -53.7972 \\ -7.8021 \end{bmatrix}.$$

The state trajectories of three variables in the circuit system are depicted in Figure 5, and the trajectories of the tracking errors between the master–slave circuit systems are shown in Figure 6. The simulation results are consistent with the theoretical results.

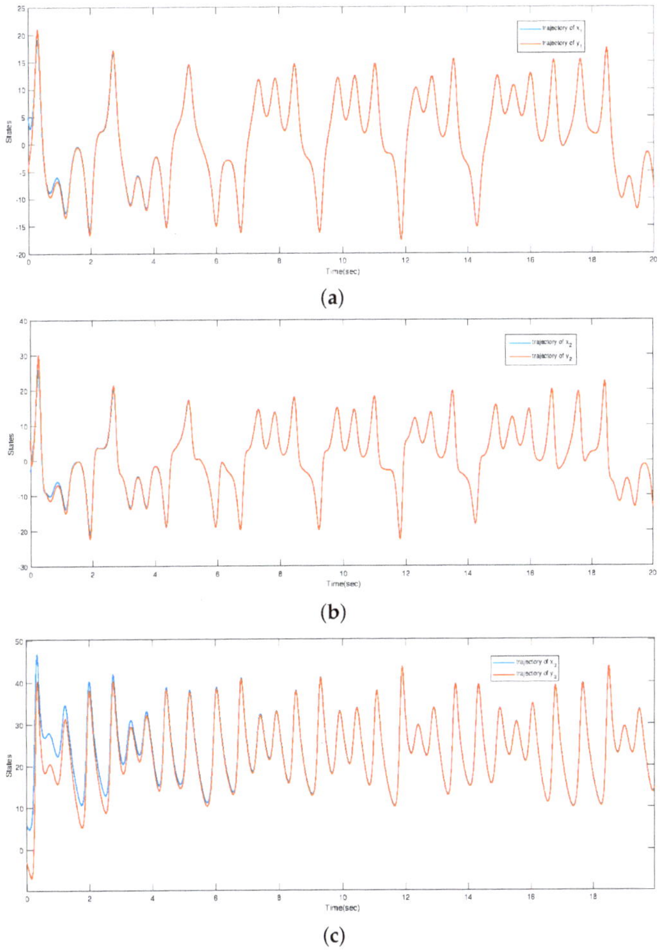

Figure 5. The trajectories of $x(t)$ and $y(t)$, (**a**) the trajectories of x_1 and y_1, (**b**), the trajectories of x_2 and y_2, and (**c**) the trajectories of x_3 and y_3.

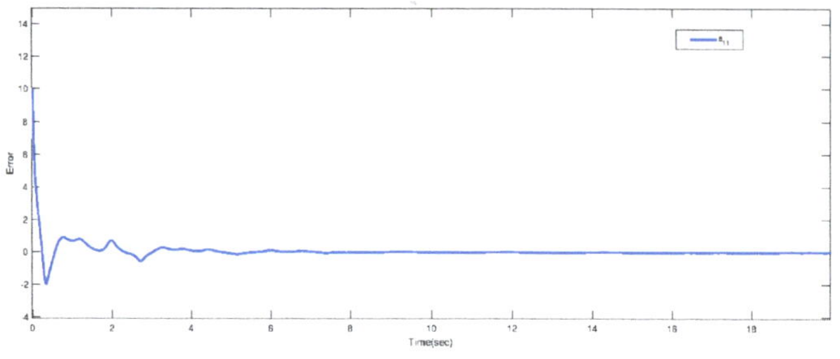

(a)

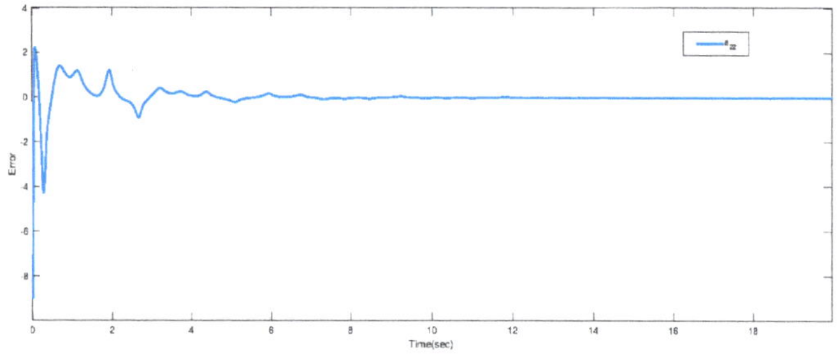

(b)

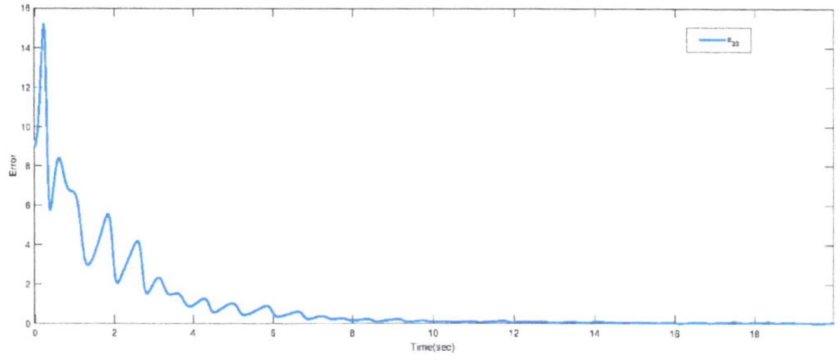

(c)

Figure 6. The trajectory of tracking error $e(t)$, (**a**) the trajectory of tracking error e_1, (**b**) the trajectory of tracking error e_2, and (**c**) the trajectory of tracking error e_3.

Remark 4. *In [18], the sliding mode control problem for fractional-order systems was considered. Unlike [18], the synchronization problem for fractional-order chaotic systems is concerned in this paper. Moreover, compared with [19–22], the nonlinearity in this paper satisfies IQC, and it is a more gerenal description.*

6. Conclusions

The robust synchronization problem of nonlinear fractional-order chaotic systems was investigated in this paper. Under the framework of IQC, the nonlinear function was described by using the IMM and state variables. The detailed computation method for IMM was also presented when the nonlinearity in chaotic system was concerned. Under the sufficient conditions, the controller was designed so that the error system behaved like the H^{∞} performance. At last, two examples were given to verify the validity of the proposed method. In future work, we will focus on the adaptive control or another advanced control for fractional-order chaotic systems.

Author Contributions: Methodology, X.Z.; Software, S.W.; Writing—original draft, D.L.; Supervision, F.Y. All authors have read and agreed to the published version of the manuscript.

Funding: This research received no external funding.

Institutional Review Board Statement: Not applicable.

Informed Consent Statement: Not applicable.

Data Availability Statement: Not applicable.

Conflicts of Interest: The authors declare no conflict of interest.

References

1. Ott, E.; Grebogi, C.; Yorke, J.A. Controlling chaos. *Phys. Rev. Lett.* **1990**, *64*, 1196. [CrossRef] [PubMed]
2. Wang, X.; Teng, L.; Qin, X. A novel colour image encryption algorithm based on chaos. *Signal Process.* **2012**, *92*, 1101–1108. [CrossRef]
3. Du, Y.; Duever, T.A.; Budman, H. Fault detection and diagnosis with parametric uncertainty using generalized polynomial chaos. *Comput. Chem. Eng.* **2015**, *76*, 63–75. [CrossRef]
4. Potapov, A.; Ali, M. Robust chaos in neural networks. *Phys. Lett. A* **2000**, *277*, 310–322. [CrossRef]
5. Cheng, C.J. Robust synchronization of uncertain unified chaotic systems subject to noise and its application to secure communication. *Appl. Math. Comput.* **2012**, *219*, 2698–2712. [CrossRef]
6. Vaidyanathan, S. Hybrid chaos synchronization of Liu and Lü systems by active nonlinear control. In Proceedings of the International Conference on Computational Science, Engineering and Information Technology, Tirunelveli, India, 23–25 September 2011; pp. 1–10.
7. Ahn, C.K.; Jung, S.T.; Kang, S.K.; Joo, S.C. Adaptive H_∞ synchronization for uncertain chaotic systems with external disturbance. *Commun. Nonlinear Sci. Numer. Simul.* **2010**, *15*, 2168–2177. [CrossRef]
8. Zhao, Z.; Lv, F.; Zhang, J.; Du, Y. H_∞ synchronization for uncertain time-delay chaotic systems with one-sided lipschitz nonlinearity. *IEEE Access* **2018**, *6*, 19798–19806. [CrossRef]
9. Pecora, L.M.; Carroll, T.L. Synchronization in chaotic systems. *Phys. Rev. Lett.* **1990**, *64*, 821. . 64.821. [CrossRef]
10. Olusola, O.I.; Vincent, E.; Njah, A.N.; Ali, E. Control and synchronization of chaos in biological systems via backsteping design. *Int. J. Nonlinear Sci.* **2011**, *11*, 121–128.
11. Oldham, K.; Spanier, J. *The Fractional Calculus Theory and Applications of Differentiation and Integration to Arbitrary Order*; Elsevier: Amsterdam, The Netherlands, 1974.
12. Shah, K.; Arfan, M.; Mahariq, I.; Ahmadian, A.; Salahshour, S.; Ferrara, M. Fractal-fractional mathematical model addressing the situation of corona virus in Pakistan. *Results Phys.* **2020**, *19*, 103560. [CrossRef]
13. Raza, A.; Ahmadian, A.; Rafiq, M.; Salahshour, S.; Ferrara, M. An analysis of a nonlinear susceptible-exposed-infected-quarantine-recovered pandemic model of a novel coronavirus with delay effect. *Results Phys.* **2021**, *21*, 103771. . 2020.103771. [CrossRef] [PubMed]
14. Udrişte, C.; Ferrara, M.; Zugrăvescu, D.; Munteanu, F. Controllability of a nonholonomic macroeconomic system. *J. Optim. Theory Appl.* **2012**, *154*, 1036–1054. [CrossRef]
15. Tavazoei, M.S.; Haeri, M. A note on the stability of fractional order systems. *Math. Comput. Simul.* **2009**, *79*, 1566–1576. [CrossRef]
16. Li, M.; Li, D.; Wang, J.; Zhao, C. Active disturbance rejection control for fractional-order system. *ISA Trans.* **2013**, *52*, 365–374. [CrossRef]
17. Zheng, S. Robust stability of fractional order system with general interval uncertainties. *Syst. Control Lett.* **2017**, *99*, 1–8. [CrossRef]
18. Kamal, S.; Sharma, R.K.; Dinh, T.N.; Ms, H.; Bandyopadhyay, B. Sliding mode control of uncertain fractional-order systems: A reaching phase free approach. *Asian J. Control* **2021**, *23*, 199–208. [CrossRef]
19. Agrawal, S.; Srivastava, M.; Das, S. Synchronization of fractional order chaotic systems using active control method. *Chaos Solitons Fractals* **2012**, *45*, 737–752. [CrossRef]
20. Andrew, L.Y.T.; Xian-Feng, L.; Yan-Dong, C.; Hui, Z. A novel adaptive-impulsive synchronization of fractional-order chaotic systems. *Chin. Phys. B* **2015**, *24*, 100502. [CrossRef]
21. N Doye, I.; Salama, K.N.; Laleg-Kirati, T.M. Robust fractional-order proportional-integral observer for synchronization of chaotic fractional-order systems. *IEEE/CAA J. Autom. Sin.* **2019**, *6*, 268–277. [CrossRef]
22. Luo, R.; Su, H. The stability of impulsive incommensurate fractional order chaotic systems with Caputo derivative. *Chin. J. Phys.* **2018**, *56*, 1599–1608. [CrossRef]
23. D'Alto, L.; Corless, M. Incremental quadratic stability. *Numer. Algebra Control Optim.* **2013**, *3*, 175–201. [CrossRef]
24. Açıkmeşe, B.; Corless, M. Observers for systems with nonlinearities satisfying incremental quadratic constraints. *Automatica* **2011**, *47*, 1339–1348. [CrossRef]
25. Zhao, Y.; Zhang, W.; Su, H.; Yang, J. Observer-based synchronization of chaotic systems satisfying incremental quadratic constraints and its application in secure communication. *IEEE Trans. Syst. Man Cybern. Syst.* **2018**, 1–12. [CrossRef]
26. Xu, X.; Açıkmeşe, B.; Corless, M.J. Observer-Based Controllers for Incrementally Quadratic Nonlinear Systems With Disturbances. *IEEE Trans. Autom. Control.* **2021**, *66*, 1129–1143. [CrossRef]
27. Podlubny, I. Fractional Derivatives and Integrals. In *Fractional Differential Equations: An Introduction to Fractional Derivatives*; Elsevier: Amsterdam, The Netherlands, 1998; pp. 41–80.

28. Xiang, W.; Xiao, J.; Iqbal, M.N. Robust observer design for nonlinear uncertain switched systems under asynchronous switching. *Nonlinear Anal. Hybrid Syst.* **2012**, *6*, 754–773. [CrossRef]
29. Zhang, H.; Huang, J.; He, S. Fractional-Order Interval Observer for Multiagent Nonlinear Systems. *Fractal Fract.* **2022**, *6*, 355. [CrossRef]
30. Li, D.; Lu, J.; Wu, X.; Chen, G. Estimating the bounds for the Lorenz family of chaotic systems. *Chaos Solitons Fractals* **2005**, *23*, 529–534. [CrossRef]
31. Jia, H.; Tao, Q.; Chen, Z. Analysis and circuit design of a fractional-order Lorenz system with different fractional orders. *Syst. Sci. Control Eng. Open Access J.* **2014**, *2*, 745–750. [CrossRef]

mathematics

Review

An Overview of Recent Advances in the Event-Triggered Consensus of Multi-Agent Systems with Actuator Saturations

Jing Xu and Jun Huang *

School of Mechanical and Electrical Engineering, Soochow University, Suzhou 215031, China
* Correspondence: cauchyhot@163.com

Abstract: The event-triggered consensus of multi-agent systems received extensive attention in academia and industry perspectives since it ensures all agents eventually converge to a stable state while reducing the utilization of network communication resources effectively. However, the practical limitation of the actuator could lead to a saturation phenomenon, which may degrade the systems or even induce instability. This paper plans to offer a detailed review of some recent results in the event-triggered consensus of multi-agent systems subject to actuator saturation. First, the multi-agent system model with actuator saturation constraints is given, and the basic framework of the event-triggering mechanism is introduced. Second, representative results reported in recent valuable papers are reviewed based on methods for dealing with saturated terms, including low-gain feedback, sector-bounded conditions, and convex hull representations. Finally, some challenging topics worthy of research efforts are dicussed for future research.

Keywords: event-triggered consensus; multi-agent systems; actuator saturation; low-gain feedback; sector-bounded condition; convex hull representation

MSC: 93D50

check for updates

Citation: Xu, J.; Huang, J. An Overview of Recent Advances in the Event-Triggered Consensus of Multi-Agent Systems with Actuator Saturations. *Mathematics* **2022**, *10*, 3879. https://doi.org/10.3390/math10203879

Academic Editor: António Lopes

Received: 28 September 2022
Accepted: 15 October 2022
Published: 19 October 2022

1. Introduction

The investigation on multi-agent systems (MASs) received great attention due to its wide range of application scenarios, including robot team control [1–3], unmanned aerial vehicle formation control [4,5], sensor network control [6–8], power grid control [9], etc. The consensus problem is one of the most popular issues among researchers, which intended to make all agents of MASs converge to the expected state. A key issue about consensus control of MASs is how to design an appropriate control protocol such that the consensus of MASs can be achieved. However, due to large-scale agent actions and the complexity of information exchange in MASs, it is really difficult or even unrealistic to adopt a conventional simple centralized control strategy. In order to investigate the consensus problem of MASs, academia prefers to employ a distributed control method that uses the information exchange between neighboring agents in a shared information network. Over the past decade, some distributed control methods have been presented [10–15].

According to the traditional consensus control protocol, it is widely assumed that control signals can be transmitted to agents in MASs continuously. However, the aforementioned assumption is harsh because it requires the network of the MASs to provide sufficient communication resources, which is difficult in practical environments, especially considering that agents are powered by limited energy devices, such as batteries. Moreover, the communication resources and bandwidth of the MASs are limited at a certain time. Therefore, a suitable distributed control protocol not only needs to ensure the control performance of the system but also needs to consider the limitation of communication resources. One formerly adopted method is time-triggered control (TTC), where the action of information sampling is triggered by preset sampling periodic intervals [16–19].

However, subsequent research studies have found that this method not only consumes excessive communication resources but also behaves poorly in the presence of external interference in systems [20]. On the other hand, if control protocols require frequent information updates, it may lead to detrimental results in the system, such as communication congestion and increased packet loss. It is well-known that communication congestion will deteriorate related performance indicators, such as severe time-delay phenomenon and reduced throughput, thus inevitably destroying the stability, reliability, and rapidity of the system. Therefore, it is of great theoretical and practical significance to design a distributed control protocol that can not only satisfy the control performance to the greatest extent but also simultaneously reduce the communication frequency in the system as much as possible.

The adoption of the event-triggered control (ETC) provides an effective solution to the above problems [21]; since then, it has been a focus of attention from the researchers [22,23]. Distinct from the TTC, the controller update is implemented if the predefined event-triggered mechanism (ETM) is violated, which helps save communication resources [24]. Essentially, the triggered instant of ETC depends on the state change inside the system, while the triggered instant of TTC depends on the time period pre-defined by the designer, which cannot reflect the internal laws of the system. For example, if the system has a stable trend, i.e., the system state changes are quite small, the ETC can significantly reduce the update frequency of information and economize communication resources compared with TTC [24]. Benefiting from this advantage, research studies related to ETC attracted tremendous attention in the last decade. The research on ETC are quite mature, particularly on fault detection approaches, the influence of disturbances, modeling errors, and various uncertainties in the real systems. The event-triggered consensus problem of a fuzzy-basis-dependent event generator and an asynchronous filter of fuzzy Markovian jump systems was investgated in [25]. Djordjevic [26] considered the data-driven optimal controller of hydraulic servo actuators with completely unknown dynamics. An event-triggered observation scheme was considered for a perturbed nonlinear dynamical system in [27]. Moreover, [28] investigated the adaptive neural network fixed-time tracking control issue. In addition to the above results, a number of meaningful results emerged [20,29–37].

It can be seen from the above discussion that ETC has obvious practical significance, i.e., to reduce the utilization of system communication network resources. While it is obvious that ETC is aimed at the optimization of the control input, another practical issue concerning the control input also deserves special attention, namely the saturation phenomenon of the actuator. In practical situations, the amplitude and frequency of the controller output current and voltage are limited, and the motor output torque and rotational speed are limited. A large number of engineering practices have shown that ignoring the constraints of the saturation phenomenon will degrade the performance of the system and even lead to catastrophic consequences. One of the most famous examples is the crash of Plane YF-22 [38]. Therefore, the saturation treatment of the system is a issue worthy of great efforts, and a series of important results emerged. Among the current methods, the sector-bounded condition, low-gain feedback, and convex hull representation are the most popular methods. Da Silva et al. [39] proposed the sector-bounded condition, and the stability analysis of system is successfully transformed into the solution of linear matrix inequalities (LMIs) by introducing the sector inequality. Lin [40] proved that by solving the parametric algebraic Riccati equation (ARE), the low-gain feedback method can retain the control input within the saturation threshold, i.e., the system does not exhibit the saturation. As a novel result, Hu [41] introduced the convex hull theory to the treatment of saturation terms. Since then, a large number of meaningful research results on saturation control emerged in academia; see [42–50].

It is widely acknowledged that research studies of event-triggered consensus control protocols for MASs with actuator saturation are more challenging compared to the ones for a single system. The difficulties of research studies mainly come from the following aspects.

(i) The distributed control protocol and ETM contain complex information coupling, i.e., the state information of individual agents and their neighbors in the communication network.

(ii) Distinct from the low-gain feedback method, the sector-bounded condition and convex hull representation focus on semi-global stabilization, so there is an effective vector space domain called the domain of attraction (DOA), and the processing of saturation terms is reasonable only in this domain. Due to the introduction of the ETM, which makes the control input more complex, the estimation of DOA will be more complicated and difficult than the situation without ETM.

(iii) Difficulty in ruling out Zeno phenomenon, which means an infinite number of triggered events for a limited period of time: When it comes to saturated systems under ETC, some existing studies use the sector-bounded condition method to simplify the estimation problem of the DOA [51,52], whlie few studies focus on convex hull representation, which is a less conservative approach. The estimation problem of the DOA will be analyzed in detail below.

On the basis of the review papers [23,53–55], we review the event-triggered consensus problem of saturated MASs in recent years. Based on the different methods dealing with saturation, the design problem of feasible event-triggered consensus control protocols for MAS subjects relative to actuator saturation is analyzed. The structure of this paper is organized as follows. Section 2 provides the description of common MASs actuator saturation and the introduction of the working mechanisms of ETC. Based on three methods dealing with saturation, i.e., sector-bounded condition, low-gain feedback, and convex hull representation, this paper reviews the design problems of control strategies with reference values in recent years and analyzes their respective advantages and disadvantages in Section 3. Section 4 reviews one simulation example and its comparative experiments to specify performance evaluation indicators. Section 5 summarizes challenging topics about related fields in the future. The conclusion of this paper is provided in Section 6.

Throughout the paper, the following symbols will be used. $I[1,N]$ represents the set of consecutive integers $\{1,2,\cdots,N\}$, and $sign(\cdot)$ means the symbol function. $\otimes$ is the Kronecker product. For matrix A, $A>0$ (≥ 0) represents a semi-positive definite matrix, and $A<0$ (≤ 0) represents a semi-negative definite matrix.

2. System Description and Preliminaries

In this section, we first provide a description of the common model of MAS that is subject to actuator saturation, along with the distributed control protocol. Next, a general framework of the distributed ETM is proposed.

2.1. Multi-Agent Systems with Actuator Saturation

In order to summarize the existing theoretical results in a unified manner, we provide the following system description based on [56]. Two types of MASs are provided: leaderless one and leader–follower one, respectively.

$$\dot{x}_i(t) = Ax_i(t) + B\mathscr{U}_{sat}(u_i(t)), i \in I[1,N]. \tag{1}$$

$$\begin{cases} \dot{x}_0(t) = Ax_0(t), \\ \dot{x}_i(t) = Ax_i(t) + B\mathscr{U}_{sat}(u_i(t)), i \in I[1,N]. \end{cases} \tag{2}$$

MAS (1) represents leaderless one, and (2) represents leader–follower one. N represents the number of agents in MAS (2), and the nunber of agents in the MAS (1) is $N+1$ because of the existence of one leader agent. $x_0(t)$ indicates the state of the leader agent, $x_i(t) \in R^n$ represents the state of the follower agent, and $u_i(t) \in R^m$ is the control input of the ith agent. $A \in R^{n\times n}$ and $B \in R^{n\times m}$ are given matrices related to the systems. It is assumed that matrices (A,B) are stabilizable. The saturation function $\mathscr{U}_{sat}(u_i(t)) \in R^m$ is described by $\mathscr{U}_{sat}(u_i(t)) = [\mathscr{U}_{sat}(u_{i1}(t)),\cdots,\mathscr{U}_{sat}(u_{im}(t))]^{\mathrm{T}}$, where $\mathscr{U}_{sat}(u_{ip}(t)) = sign(u_{ip})\min\{|u_{ip}(t)|,u_0\}, p \in I[1,m]$, and u_0 is the saturation threshold.

The communication network of the MASs can be represented by a directed or undirected graph $\mathscr{G} = (\mathscr{V}, \mathscr{E}, \mathscr{A})$, where $\mathscr{V} = \{v_1, v_2, \cdots, v_N\}$ stands for the set of vertex, which represents N agents (for example, v_1 stands for the 1th agent in the MAS), and $\mathscr{E} \subseteq \mathscr{V} \times \mathscr{V}$ represents the set of edges. In graph $\mathscr{G}$, edge $\varepsilon_{ij} = (v_i, v_j)$ represents the fact that agent v_j can receive information from agent v_i. Therefore, vertex v_l is called a neighbor of agent v_i and $\mathcal{N}_i = \{v_j \in \mathscr{V} : \varepsilon_{ij} \in \mathscr{E}\}$ is called the neighbourhood of the agent v_i. The weighting adjacency matrix is defined by $\mathscr{A} = [a_{ij}] \in R^{N \times N}$, which represents the existence and strength of inter-agent communications. Thus, it is defined that $a_{ij} > 0$ if $\varepsilon_{ij} \in \mathscr{E}$ and $a_{ij} = 0$ if $\varepsilon_{ij} \notin \mathscr{E}$. We define a degree matrix $\mathcal{D} = diag[d_{ii}] \in R^{N \times N}$ with $d_{ii} = \sum_{j=1}^{N} a_{ij}$. Afterall, the Laplacian matrix $\mathcal{L}$ of the araph $\mathscr{G}$ is given by $\mathcal{L} = \mathcal{D} - \mathcal{A}$.

The above graph theory describes a leaderless MAS. When MAS has a leader, an additional matrix needs to be defined by $\mathcal{B} = diag[b_i] \in R^{N \times N}$, where $b_i > 0$ means that agent v_i is able to receive information from leader agent v_0; otherwise, $b_i = 0$. We define the Laplacian matrix of graph $\mathscr{G}$ with the leader agent by $\mathcal{H} = \mathcal{L} + \mathcal{B}$.

It is said that MAS (1) with actuator saturation has achieved consensus if all agents' states converge to the same value under control protocol $u_i(t)$ and initial conditions $x_i(0) \in \mathcal{X} \subset R^n$, i.e., $\lim_{t \to \infty} \|x_i(t) - x_j(t)\| = 0$. Set $\mathcal{X}$ as the DOA mentioned above. As for the leader–follower MAS (2), the consensus requires all follower agents to be consistent with the leader agent eventually, i.e., $\lim_{t \to \infty} \|x_i - x_0\| = 0$.

Compared with the common linear MASs studied before, the difference between MASs (1), (2) and the linear ones is that there exists the limitation of actuator saturation $\mathscr{U}_{sat}(u_i(t))$, which also makes the system nonlinear. Therefore, the lemmas required to deal with the saturation term are presented below.

Lemma 1 ([57]). *Define the dead zone function $\Phi(s) = \mathscr{U}_{sat}(s) - s$, where $s \in R^m$. Then, for any diagonal positive definite matrix $T \in R^{m \times m}$, the following inequality holds:*

$$\Phi^T(s)T(\Phi(s) + w) \leq 0,$$

if vectors s and w belong to the set $S(s, w, u_0) = \{s \in R^m, w \in R^m : \|s - w\|_\infty \leq u_0\}$.

Lemma 2 ([58]). *If all eigenvalues of matrix A are in the closed left-half s-plane, then for any $\epsilon \in (0, 1]$, there exists a unique matrix $P(\epsilon) > 0$ such that the following ARE is satisfied:*

$$A^T P(\epsilon) + P(\epsilon)A - P(\epsilon)BB^T P(\epsilon) + \epsilon I_N = \mathbf{0},$$

and $\lim_{\epsilon \to 0} = \mathbf{0}$.

Lemma 3 ([59]). *If there exist matrices $F, H \in R^{m \times n}$, then the saturation term can be represented by the following:*

$$\mathscr{U}_{sat}(Fx) \in co\{D_r Fx + D_r^- Hx\}, k \in I[1, 2^m],$$

where $x \in \mathscr{L}(H, u_0)$, $co\{\cdot\}$ is the convex hull of a set, and D_r is a diagonal matrix with diagonal elements being either 1 or 0, $D_r^- = I - D_r$. $\mathscr{L}(H, u_0) = \{x \in R^n : \|Hx\|_\infty \leq u_0\}$.

Lemmas 1–3 are the basis for using the three saturation-processing methods, i.e., sector-bounded condition, low-gain feedback, and convex hull representation, respectively. For details, please refer to the papers in Section 1, and they will also be analyzed in the next section.

In order to achieve the consensus of MAS, the following common control protocol is proposed [56]:

$$u_i(t) = K \sum_{j \in \mathcal{N}_i} a_{ij}(x_i(t) - x_j(t)) + b_i(x_i(t) - x_0(t)), \tag{3}$$

where K is the gain matrix to be designed. Moreover, we have $b_i \neq 0$ if MASs have a leader agent, and $b_i = 0$ otherwise.

In the application of the control protocol (3), the control protocol needs to continuously acquire all required agents' states, which will cause a large consumption of communication resources. To solve this problem, ETC is proposed and widely used because it can effectively save communication resources since it avoids continuous updates of the controller. Next, we introduce its basic framework and mechanism.

Remark 1. *Different from the event-triggered consensus problem reviewed in [23], this paper considers the limitation of actuator saturation additionally, so the content of this paper can be regarded as a broader and general result. According to the commonly used saturation-processing methods mentioned above, this paper specifically discusses the processing methods for the event-triggered consensus problem in the presence of saturation phenomena.*

2.2. The Framework of Event-Triggered Mechanism

Figure 1 shows the basic working principle of ETC roughly, and it can be seen that the difference from traditional MASs is the introduction of event-triggered detectors (marked by dotted lines). As a key component of ETC, the detector is responsible for collecting measurement information from the sensor, and then it judges whether the triggered mechanism is violated according to the pre-designed ETM. If the triggered mechanism is violated, the trigger is switched on, allowing the information of the sensor to be transmitted to the controller of the agent i, along with the update of the information of the controller i. Moreover, the real-time information will be transmitted to the neighbors of agent i through the communication network. It should be noted that the communication between the sensor and the detector may be continuous, i.e., the ETM is continuously monitored for violations, which also causes a certain degree of waste of communication resources. Therefore, inspired by the principle of TTC, researchers propose to conduct the communication between sensors and detectors in time segments (see Section 3 for details).

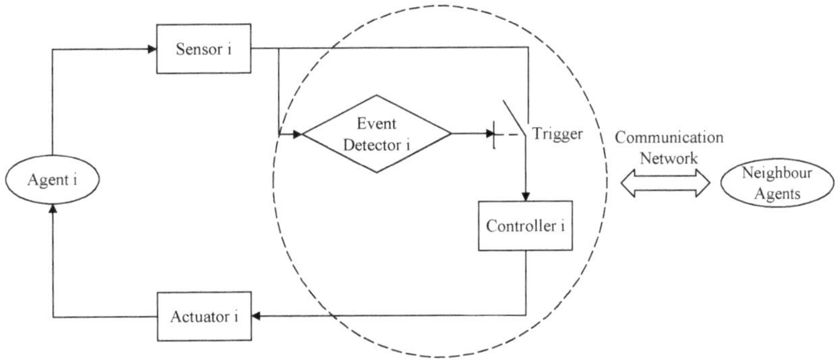

Figure 1. The framework of ETM.

In order to ensure that ETC can work effectively under MASs with actuator saturation, the following issues need to be considered:

(1) Design of ETM: When it comes to the event-triggered strategy, the design problem of ETM is of great importanance, which is related to the scheduling of communication resources and the update of the controller. However, the design of the ETM must also take into account the practical implementation, i.e., being practically executable. This leads to a contradiction between the mechanism's design and practical application. If the designed ETM is sophisticated, the controller is effective, but a sophisticated ETM may consume numerous communication and computing resources, becoming a real burden and vice versa. Therefore, the design of ETM is an art of achieving a balance between mechanism and reality. Thus far, the ETC widely used in the literature can be mainly divided into three types according to information utilization in the communication network: (i) centralized:

all agents' measurement information is required [60]; (ii) decentralized: only its own measurement information is required [61–63]; (iii) distributed: measurement information of itself and the neighbours is required [64–66]. Decentralized and distributed ETC are adopted by the mainstream because they cover less agents than centralized ones.

(2) Saturation Phenomenon: When the limitation of actuator saturation exists in systems, there are certain difficulties in designing the ETM. The first one is the estimation of DOA. When it comes to the sector-bounded condition and convex hull representation methods, the stability analysis of the system is constrained in a limited space, namely DOA. The DOA estimation problem has mature solutions in research studies without ETM. However, after ETM is introduced, the control input becomes more complicated, which adds difficulty in the estimation problem of DOA. Second, the saturation of the actuator means that there is a threshold for the control input, and it is also questionable whether the ETM can be successfully implemented within this limitation.

(3) Interval Between Events: Compared with traditional continuous control methods, the largest difference with respect to ETC is that it decides to update the controller according to whether the triggered mechanism is violated. It needs to ensure the elimination of the Zeno phenomenon; otherwise, it will degenerate into continuous control. However, this is not simple, and the difficulty comes from theoretical analyses and external information interferences. In addition, the interval between triggered events is often uncontrollable and may cause valuable information to be ignored generally.

The above issues deserve great attention when discussing event-triggered consensus in MASs with actuator saturation. Thus far, a majority of the literature studied control strategies in this field, which will be briefly reviewed in the next section.

3. Main Results

As discussed in Section 2, a key problem of the event-triggered consensus for MASs with actuator saturation lies in designing the control protocol, ETM, and handling the saturation terms. The mainstream saturation treatment methods include low-gain feedback, sector-bounded condition, and convex hull representation. In this section, we will review some interesting research results based on different approaches in dealing with saturation terms.

3.1. Low-Gain Feedback

The main idea of low-gain feedback is that, for any given bounded set $\mathcal{S}$ in the state space, there exists a linear feedback control that makes all system trajectories starting from $\mathcal{S}$ converge to the origin. The low-gain feedback method mainly uses a family of parameterized gain matrices $P(\epsilon)$ to design linear feedback controllers by solving the ARE (Lemma 2). The ϵ in matrix $P(\epsilon)$ is called a low-gain parameter. As low-gain parameter ϵ tends to zero, $P(\epsilon)$ also tends to the zero matrix. Achieving semi-global stabilization with low-gain feedback means that for any given bounded set $\mathcal{S}$, no matter how large it is, a low-gain parameter value ϵ always exists such that all control signals on $\mathcal{S}$ are within the saturation threshold. That is, the low-gain feedback can make the saturated system maintain the behavior of the unsaturated system in set $\mathcal{S}$. The advantage of the low-gain feedback compared with other methods lies in the fact that it can discard the saturation constraint in theoretical analysis, which reduces the complexity for the system's analysis and design, especially when ETM is introduced. Therefore, it has received extensive attention from researchers in the studies of event-triggered consensus for saturated MASs.

In [67], the following ARE is adopted,

$$A^T P(\epsilon) + P(\epsilon)A - \beta P(\epsilon)BB^T P(\epsilon) + \epsilon I_N = \mathbf{0}. \tag{4}$$

With the help of ARE (4), the actuator does not exhibit saturation, i.e., $\mathscr{U}_{sat}(u_i(t)) = u_i(t)$. For $t \in [t_k^i, t_{k+1}^i)$, the control protocol and ETM are given by the following:

$$\Sigma_1 = \begin{cases} \text{ETM}: & \|f_i(t)\| \le \vartheta \|q_i(t)\|, \\ \text{Protocol}: & u_i(t) = K\sum_{j\in\mathcal{N}_i} a_{ij}(x_j(t_k^i) - x_i(t_k^i)) \\ & +b_i(x_0(t_k^i) - x_i(t_k^i)),\ t \in [t_k^i, t_{k+1}^i), \end{cases} \tag{5}$$

where t_k^i is the kth-triggered instant of agent i, $\xi_i(t) = x_i(t) - x_0(t)$, $q_i(t) = \sum_{j\in\mathcal{N}_i} a_{ij}(\xi_j(t) -\xi_i(t)) - b_i\xi_i(t)$, $f_i(t) = q_i(t_k^i) - q_i(t)$; ϑ is the triggered threshold; $K = B^T P(\epsilon)$ is the control gain matrix to be designed. If the trigger function in (5) is violated, this means that the event is triggered. It can be explained that the difference between measurement error $f_i(t)$ in the system at the current instant and combined measurement $q_i(t)$ exceeds the threshold, so the controller needs to update the acquisition of the system state to stabilize the system. Furthermore, the proposed ETM has demonstrated that it can reduce the update frequency of the controller effectively and avoid the Zeno phenomenon successfully.

Although control strategy Σ_1 has considerable advantages, it also has certain disadvantages, which are given as follows.

(D1) Continuous Monitoring on ETM: According to the definition of $q_i(t)$ in ETM (5), it can be seen that the control strategy needs to continuously monitor the state of agent i itself and its neighbors, which will lead to the substantial consumption of network communication resources and is inconsistent with the intention of ETC.

(D2) Excessive Sampling: From the control protocol in strategy (5), it is found that the information required for the control input of a single agent needs to be updated under the same clock sequence. That is to say that the information collection of the neighbor agent needs to be implemented in its own clock sequence, along with its neighbors' clock additionally, which increases the consumption burden on a single agent and the entire MAS.

(D3) System Limitations: If the low-gain feedback method is adopted in the studies of saturated systems, the ARE of the system needs to be addressed first. From the AREs discussed above, it can be found that the involved systems are the simple linear systems only with additional saturation constraints. Therefore, this method may have some limitations, when the system studied has more complex characteristics, such as the presence of external nonlinear disturbances, unfixed communication topologies, and perturbed internal parameters of the systems.

In order to overcome the above disadvantages, researchers have made great efforts into improving (5). Considering the disadvantage of D1, [67] proposed a self-triggered ETM on the basis of (5) as follows:

$$\Sigma_2 = \begin{cases} \text{ETM}: & \|e_i(t)\| \le \tilde{g}(t_{k_i}^i, t_{k_j}^j), \\ \text{Protocol}: & u_i(t) = K\sum_{j\in\mathcal{N}_i} a_{ij}(x_j(t_{k_j}^j) - x_i(t_{k_i}^i)) \\ & +b_i(x_0(t_{k_i}^i) - x_i(t_{k_i}^i)),\ t \in [t_{k_i}^i, t_{k_i+1}^i), \end{cases} \tag{6}$$

where $e_i(t) = \xi_i(t_{k_i}^i) - \xi_i(t)$ is the measurement error, $\alpha_i > 0$ is the triggered threshold, and function $\tilde{g}(t_{k_i}^i, t_{k_j}^j)$ is defined as follows:

$$\tilde{g}(t_{k_i}^i, t_{k_j}^j) = \alpha_i \left\| \sum_{j\in\mathcal{N}_i} a_{ij}(\xi_j(t_{k_j}^j) - \xi_i(t_{k_i}^i)) + b_i\xi_i(t_{k_i}^i) \right\|.$$

Compared with (5), the most meaningful change in (6) is that it overcomes disadvantage D1. Specifically, the ETM and protocol in control strategy (6) only need to sample its information at the triggered instant ($t_{k_i}^i$ and $t_{k_j}^j$) of the desired agents. Instead of continuous sampling in (5), it can reduce the utilization of communication resources, in line with the intention of the ETC. However, this control strategy still has its limitation. For a certain agent, its control update depends on both its own triggered instant and the neighbors' triggered instant, while the control update only depends on the agent's own triggered instant in (5).

Therefore, as the number of neighbors of the agent increases, its control update interval may become shorter and shorter, even leading to the Zeno phenomenon [65,68].

It is worth mentioning that when discussing event-triggered strategies, the existing literature often studies the case where information sampling and actuator update are implemented synchronously [65,69,70], i.e., there does not exist time delay between this two actions. However, this type of delay phenomenon is widespread in the field of practical engineering, which is called update delay. ETC is sensitive to time delays, and ignoring the delay may degrade the control quality or even destroy the stability of the system. Therefore, it is of great practical significance to consider this time delay when designing ETC strategies. Inspired by the topics discussed above, Wang et al. [58] proposed one fully distributed ETC scheme with the consideration of update delays, and the control strategy is given as follows:

$$A^T P(\epsilon) + P(\epsilon)A - \frac{4}{(N-1)N} P(\epsilon)BB^T P(\epsilon) + \epsilon I_N = \mathbf{0}. \tag{7}$$

$$\Sigma_3 = \begin{cases} \text{ETM}: \\ 2\|M_i(t)\| - \frac{1}{2}\|\omega_i(t)\| - \gamma e^{-\theta t} \leq 0, t \in [r_k^i, r_{k+1}^i) \\ 2\|m_i(t)\| - \frac{1}{2}\|\omega_i(t)\| - \gamma e^{-\theta t} \leq 0, t \in [t_k^i, t_{k+1}^i) \\ \text{Protocol}: \\ u_i(t) = K\sum_{j\in\mathcal{N}_i} a_{ij}(x_j(t_{k-1}^i) - x_i(t_{k-1}^i)), \\ t \in [t_k^i, r_k^i) \\ u_i(t) = K\sum_{j\in\mathcal{N}_i} a_{ij}(x_j(t_k^i) - x_i(t_k^i)), \\ t \in [r_k^i, t_{k+1}^i), \end{cases} \tag{8}$$

where N is the number of agents, t_k^i and r_k^i represent updating sequences and sampling sequences, respectively. Define $E_i(t) = x_i(t) - x_i(t_{k-1}^i)$, $E_i^j(t) = x_j(t) - x_j(t_{k-1}^i)$, $\omega_i(t) = \sum_{j\in\mathcal{N}_i} a_{ij}K(x_i(t) - x_j(t))$, $M_i(t) = \omega_i(t) - \omega_i(t_{k-1}^i)$, and $M_i(t) = \omega_i(t) - \omega_i(t_k^i)$. By solving the ARE (7), the gain matrix is obtained by $K = B^T P(\epsilon)$.

Note that the control protocol in (8) is different from that in (6). The control input $u_i(t)$ in (8) only updates according to its corresponding triggered instant sequence for a certain agent i and does not depend the sequence of other agents, which will greatly reduce the consumption of resources in the network of MASs. That is to say that the aforementioned shortcoming D2 is overcome. However, it also has limitations with respect to D1. The ETM in (8) requires continuous information sampling of agent i and its neighbors, which may cause the burden of communication. Moreover, meticulous differentiation of the time sequencecs may lead to shorter triggered intervals, reducing the quality of control performance.

The ETMs discussed above have one thing in common, that is, the coefficients of their triggered functions are all fixed constants. In [71], a dynamic strategy is proposed as follows:

$$\Sigma_4 = \begin{cases} \text{ETM}: & g_i(t) \leq \mu_i\theta_i(t), \\ \text{Protocol}: & u_i(t) = K\sum_{j\in\mathcal{N}_i} c_{ij}(t)(\bar{x}_i(t) - \bar{x}_j(t)), \\ & t \in [t_k^i, t_{k+1}^i). \end{cases} \tag{9}$$

Define $\bar{x}_i(t) = e^{A(t-t_k^i)}x(t_k^i)$ as the measurement of $x_i(t)$ between the triggered interval, and the measurement error is defined by $e_i(t) = \bar{x}_i(t) - x_i(t)$. Furthermore, the parameters and functions of this dynamic control strategy can be described by the following equations:

$$\dot{c}_{ij}(t) = a_{ij}(\bar{x}_i(t) - \bar{x}_j(t))^T\Gamma(\bar{x}_i(t) - \bar{x}_j(t)),$$
$$\dot{\theta}_i(t) = -\pi_i\theta_i(t) - \omega_i g_i(t),\ \theta_i(0) = 0,$$
$$g_i(t) = \sum_{j\in\mathcal{N}_i}(1 + \beta c_{ij}(t))a_{ij}e_i^T(t)\Gamma e_i(t)$$
$$-\frac{1}{4}\sum_{j\in\mathcal{N}_i} a_{ij}(\bar{x}_i(t) - \bar{x}_j(t))^T\Gamma(\bar{x}_i(t) - \bar{x}_j(t)),$$

where $\pi_i > 0$, $\mu_i > 0$, $\omega_i > 0$, and β satisfy $\omega_i\beta > 1$.

The following linear matrix inequality (LMI) is introduced.

$$\begin{aligned} &AP^{-1}(\epsilon) + P^{-1}(\epsilon)A^T + (\rho + \epsilon)P^{-1}(\epsilon)P^{-1}(\epsilon) \\ &+\rho AA^T - BB^T < 0. \end{aligned} \tag{10}$$

By solving the LMI (10), feedback matrices are obtained by $K = B^TP(\epsilon), \Gamma = P(\epsilon)BB^TP(\epsilon)$.

Compared with the control gain matrix scheme (5), (6), and (8) based on solving ARE above, this control strategy provides greater flexibility because it only needs to solve matrix inequalities (10) instead of AREs (4) and (7). The design of gain matrix dose not rely on the solution of a parametric ARE. Another difference is that the parameters in control strategy (10) are dynamically changed rather than fixed ones aforementioned. The dynamic ETM adopted by strategy (10) can ensure that the interval time between triggered events is longer than that one of the fixed-parameter ETM, which is beneficial for the saving of communication resources.

In addition to the low-gain feedback-based event-triggered consensus studies discussed above, there are many interesting results that have not been discussed in detail. Xu [72] studied the bipartite consensus problem for high-order MAS subject to actuator saturation. The centralized and distributed event-triggering strategies for saturated MASs are both presented in [73]. Thus, it is concluded that the distributed control strategy can effectively reduce the number of triggered instants and the update frequency of the system, which saved the utilization of communication resources.

Remark 2. *The low-gain feedback method has significant advantages in the analysis of the system and the design of the control protocol because it can reduce saturation constraints. However, it depends on the solution of ARE, so it has higher requirements on studied systems; that is to say that the analyzed system needs to be relatively simple. If the system has more complex characteristics, such as external nonlinear interference, or the communication topology of the system is time-varying, the low-gain feedback method may fail. Therefore, other saturation processing methods will be introduced next, which can be applied to more complex systems.*

3.2. Sector-Bounded Condition

The sector-bounded condition is the most widely and frequently used saturation treatment method in the study of event-triggered consensus problems for MASs with actuator saturation. The system with saturation limition is difficult to analyze by using common Lyapunov stability theory, so the saturation term needs to be dealt with in advance. The sector-bounded condition provides an effective solution for transforming the stability analysis of the system into solvable LMIs. Its main idea is to convert the saturation term $\mathscr{U}_{sat}(u_i(t))$ into a dead-band function $\Phi(\mathscr{U}_{sat}(u_i(t))) = \mathscr{U}_{sat}(u_i(t)) - u_i(t)$ so that sector inequalities can be used (see Lemma 1). Unlike the low-gain feedback method discussed above, this method can be used in a wider range of systems because it does not avoid the saturation term, but processes it directly. In addition, unlike the low-gain feedback method, the sector-bounded condition and convex hull representation have their applicable range, namely DOA. The estimation of DOA is a common problem in the use of these two types of methods. The method widely adopted by researchers is to use the level set of the Lyapunov function to estimate the range of DOA. Compared with the convex hull representation, the sector-bounded condition has unique advantages in DOA estimation.

Because of the higher flexibility of the sector inequality, it is widely used in event-triggered related research studies.

Unlike the systems studied above, Yin [52] investigated the MAS that is additionally accompanied by a nonlinear term, and the system can be described as follows:

$$\begin{cases} \dot{x}_0(t) = Ax_0(t) + f(x_0(t)), \\ \dot{x}_i(t) = Ax_i(t) + B\mathscr{U}_{sat}(u_i(t)) + f(x_i(t)), i \in I[1,N], \end{cases} \tag{11}$$

where $f(x_i(t))$ represents the nonlinear function that satisfies the following Lipschitz condition.

Definition 1 ([74]). *The nonlinear function $f(\cdot)$: $R^n \to R^n$ satisfies the Lipschitz condition if there exists $l \in R^+$ such that*

$$\|f(x) - f(y)\| \le l\|x - y\|, x, y \in R^n,$$

and l is the Lipschitz constant.

As a novel research achievement, [52] proposed an adaptive dynamic ETM as follows:

$$\Sigma_5 = \begin{cases} \text{ETM}: & e_i^T(t)\Omega_i e_i(t) \le \mu_i(t) y_i^T(t)\Omega_i y_i(t), \\ \text{Protocol}: & u_i(t) = -K\sum_{j\in N_i} a_{ij}(\tilde{x}_i(t) - \tilde{x}_j(t)) \\ & +b_i(\tilde{x}_i(t) - x_0(t)), \ t \in [t_k^i, t_{k+1}^i), \end{cases} \tag{12}$$

where $\tilde{x}_i(t) = x(t_k^i)$ is the detection value of agent i for $t \in [t_k^i, t_{k+1}^i)$. $y_i(t) = x_i(t) - x_0(t)$, $\tilde{y}_i(t) = \tilde{x}_i(t) - x_0(t)$ if agent i can receive information from the leader. Otherwise, $y_i(t) = x_i(t) - x_{j_i}(t)$, $\tilde{y}_i(t) = \tilde{x}_i(t) - \tilde{x}_{j_i}(t)$, j_i is any neighbor of agent i, $e_i(t) = y_i(t) - \tilde{y}_i(t)$, Ω_i is the undetermined coefficient matrix, and $\mu_i(t)$ is determined by the following differential equation.

$$\dot{\mu}_i(t) = -d_i\mu_i^2(t)e_i^T(t)\Omega_i e_i(t).$$

Different from the ETMs discussed above, this triggered mechanism uses dynamic parameters instead of fixed ones. Unlike the ETMs proposed in [58,70], the triggered parameters $\mu_i(t)$ will dynamically adjust as the system's state changes instead of being fixed, which gives the triggered mechanism more flexibility. In addition, most triggered functions above take the form of multiplying a vector norm and a constant coefficient, and the constant coefficient is generally preset. This mechanism adopts the form of multiplying a vector and a coefficient matrix Ω_i, and the coefficient matrix Ω_i is designed together with the control protocol, which increases the flexibility of the control protocol design and expands the solvable range.

After completing the design of the control protocol and the ETM, it is necessary to consider the problem of DOA range estimation. The author of [52] provides a typical demonstration of DOA estimation.

Define $\delta_i(t) = x_i(t) - x_0(t)$, and the following variables, $\delta(t) = [\delta_1^T(t), \delta_2^T(t), \cdots, \delta_N^T(t)]^T$, $e(t) = [e_1^T(t), e_2^T(t), \cdots, e_N^T(t)]^T$, $u(t) = [u_1^T(t), u_2^T(t), \cdots, u_N^T(t)]^T$.

The Lyapunov function is defined as follows:

$$V(t) = \delta^T(t)(I_N \otimes P)\delta(t),$$

along with the level set $\mathcal{E}(P, \eta) \triangleq \{\delta(t) \in R^{Nn} : V(t) \le \eta\}$.

The control input u can be rewritten as follows:

$$u(t) = -(\mathcal{H} \otimes K)(\delta(t) + e(t)),$$

where $\mathcal{H}$ is the Laplace matrix mentioned in Section 2, and K is the gain matrix to be designed. We set $w(t) = u(t) + G\delta(t)$, and G is a suitable dimensional matrix; we obtain the following set:

$$\varphi(G, u_0) = \left\{\delta(t) \in R^{Nn} : \left|G_{(j)}\delta(t)\right| \leq u_0\right\},$$

where $G_{(j)}$ represents the jth row of matrix G. Lemma 1 ensures that if $\delta(t)$ belongs to set $\varphi(G, u_0)$, then the following sector inequality holds.

$$\Phi^T(u(t))T(\Phi(u(t)) + u(t) + G\delta(t)) \leq 0.$$

Set $\varphi(G, u_0)$ is the required DOA, but it is difficult to directly measure $\varphi(G, u_0)$, so the level set of Lyapunov function $\mathcal{E}(P, \eta)$ is used for indirect estimation. As [52] stated, if the following inequality is satisfied, then it can be proved that $\mathcal{E}(P, \eta)$ is enclosed in $\varphi(G, u_0)$.

$$\begin{bmatrix} I_N \otimes P & G_{(j)}^T \\ G_{(j)} & \dfrac{u_0^2}{\eta} \end{bmatrix} \geq 0.$$

Thus far, the issues related to ETC based on the sector-bounded condition have been fully considered, including the design of ETM, the control protocol, and the estimation of DOA, which are also three issues that must be considered in the research studies based on this method. An interesting point can be found from the above discussion, the construction of $w(t)$ has flexibility, and the researchers design $w(t) = u(t) + G\delta(t)$. The introduction of $G\delta(t)$ enables the range estimation of DOA to be concatenated with $\mathcal{E}(P, \eta)$. So even in the context of ETC, where the input is more complex, the utilization of sector-bounded conditions is not affected. This is different from the convex hull representation, which will be shown in the next subsection.

When studying the consensus problem of saturated MASs by a sector-bounded condition, the systems studied can be more complicated than those of low-gain feedback methods. The above discussion focused on MASs with nonlinear disturbances, while Dai [75] studied the event-triggered consensus problem of a class of saturated MASs with Markovian switching topologies. A novel ETM was adopted in [75], and the feature of which is that the inspection of events is not continuous but depends on a time-interval. The novel control strategy can be described as follows:

$$\Sigma_6 = \begin{cases} \text{ETM}: & e_i^T(t_k^i + lh)\Omega_i e_i(t_k^i + lh) \\ & \leq \delta_i z_i^T(t_k^i + lh)\Omega_i z_i(t_k^i + lh), \\ \text{Protocol}: & u_i(t) = -K\sum_{j\in\mathcal{N}_i} a_{ij}(x_i(t_k^i) - x_j(t_{k'}^j) \\ & +v_i(t_k^i) - v_j(t_{k'}^j)),\ t \in [t_k^i h, t_{k+1}^i h), \end{cases} \tag{13}$$

where h is the sampling period, t_k^i is the kth sequence at the sampling instant of agent i, $t_k^i + lh$ represents the current sampling instant, δ_i is the triggered threshold, Ω_i is consistent with the one in (12), and $t_{k'}^j = max\left\{t : t \in \left\{t_k^j, k = 0, 1, \cdots\right\}, t \leq t_k^i + lh\right\}$.

The relevant variables are defined as follows.

$$e_i^T(t_k^i + lh) = [e_i^{x^T}(t_k^i + lh), e_i^{v^T}(t_k^i + lh)],\ z_i^T(t_k^i + lh) = [z_i^{x^T}(t_k^i + lh), z_i^{v^T}(t_k^i + lh)],$$
$$e_i^x(t_k^i + lh) = x_i(t_k^i) - x_i(t_k^i + lh),\ e_i^v(t_k^i + lh) = v_i(t_k^i) - v_i(t_k^i + lh),$$
$$z_i^x(t_k^i + lh) = \sum_{j\in\mathcal{N}_i} a_{ij}(x_i(t_k^i) - x_j(t_{k'}^j)),\ z_i^v(t_k^i + lh) = \sum_{j\in\mathcal{N}_i} a_{ij}(v_i(t_k^i) - v_j(t_{k'}^j)).$$

Different from the event-based triggered mechanisms ($\Sigma_1 - \Sigma_5$) discussed above, this type of mechanism is a type of sampled-data-based ETM. It is worth mentioning that this type of triggered mechanism only judges the violation of ETM at the sampling interval, and the sampling interval is h, which leads to an interesting conclusion that this type of mechanism can naturally avoid the Zeno phenomenon. A number of improved ETMs have been proposed above for the disadvantage D1, but these improvements have limitations, and such a sampled-data-based ETM overcomes this disadvantage and completely avoids the continuous monitoring of the ETM. Therefore, there is no doubt that this type of triggered mechanism can save communication resources and computing costs effectively.

Although this type of sampled-data-based ETM has its advantages, such as avoiding continuous monitoring of ETM and ruling out the Zeno phenomenon, it still has certain limitations. Firstly, the existence of sampling interval h greatly reduces the update frequency of the controller, but the long interval may lead to ignoring useful information, especially when the system has large oscillations. Secondly, existing research studies on this type of triggered mechanism assume that all agents follow the same clock sequence, so when the scale of MAS is quite large, this type of mechanism may be difficult to implement practically.

On the basis of (12), [52] proposed an adaptive sampled-data-based ETM as follows.

$$\Sigma_7 = \begin{cases} \text{ETM}: \\ \alpha_i^T(t_k^i h + l_i h)\Omega_i \alpha_i(t_k^i h + l_i h) \\ \leq \mu_i(t) y_i^T(t_k^i h + l_i h)\Omega_i y_i(t_k^i h + l_i h), \\ \text{Protocol}: \\ u_i(t) = -K\sum_{j\in\mathcal{N}_i} a_{ij}(x_i(t_k^i h) - x_j(t_{k_j}^j h) \\ +b_i(x_i(t_k^i h) - x_0(mh))),\ t \in [t_k^i h, t_{k+1}^i h). \end{cases} \tag{14}$$

The adaptive coefficient $\mu_i(t)$ is determined by the following:

$$\begin{aligned} &\mu_i(t_k^i h + l_i h) - \mu_i(t_k^i h + l_i h - h) = -d_i h \mu_i(t_k^i h + l_i h) \\ &\mu_i(t_k^i h + l_i h - h)\alpha_i^T(t_k^i h + l_i h)\Omega_i \alpha_i(t_k^i h + l_i h), \end{aligned}$$

where $\alpha_i(t_k^i h + l_i h) = x_i(t_k^i h + l_i h) - x_i(t_k^i h)$, $y_i(t_k^i h + l_i h) = x_i(t_k^i h + l_i h) - x_0(t_k^i h + l_i h)$, if agent i can receive information from the leader. Otherwise, $y_i(t_k^i h + l_i h) = x_i(t_k^i h + l_i h) - x_{j_i}(t_{k_{j_i}}^{j_i} h + l_i h)$, and j_i is any neighbor of agent i. Define $l_i h = mh - t_k^i h$, m is an integer satisfying $t_k^i \leq m < t_{k+1}^i$, and $k_{j_i} = \arg \min_{p\in Z^+ : t_p^{j_i h} \leq h} \left\{ t - t_p^{j_i} h \right\}$.

Compared with the ETM in (13), this control strategy uses dynamically changing parameters instead of preset fixed ones, which avoids some difficulties in choosing suitable initial values. Each agent has its specific triggered clock sequence $l_i h$ and a not uniformly fixed one lh in (13). This overcomes the second limitation mentioned above and provides favorable conditions for implementation in the context of large-scale MASs. In addition, it can be seen that the adaptive parameter $\mu_i(t)$ in (12) depends on the differential equation, while that in (14) depends on the difference equation, which is more conducive to the implementation and operation.

Recently, the event-triggered consensus problem of saturated MASs based on the sector-bounded condition received extensive attention from the academic community [76–78]. Based on this flexible saturation-processing method, researchers are no longer limited by the limitations of simple systems with low-gain feedback and turn to more complex systems. The event-triggered consensus problem for one type of second-order MAS subject to actuator saturation and input time delay was investgated in [76], and Ref. [77] focused on the bipartite-tracking consensus problem of nonlinear MASs with cooperative–competitive interactions. Furthermore, [78] dealt with the leaderless consensus problem for saturated MASs with a directed communication topology.

Remark 3. *It is worth noting that most of the systems studied in the above mentioned references are linear systems or simple Lipschitz nonlinear systems, and there is still an open topic to study the event-triggered consensus problem for more general nonlinear systems, such as one-side Lipschitz or* *incremental quadratic constraints. The proper treatment of nonlinear systems is a challenge in this field; thus, the research in this direction is worthy of future efforts.*

3.3. Convex Hull Representation

Compared with the two saturation processing methods introduced above, the convex hull representation method is less studied. However, the convex hull representation method is less conservative than the sector-bounded condition since it introduces a convex hull to analyze the saturation term and it is not necessary to introduce additional sector inequality conditions in the analysis. The convex hull representation method is the least conservative in terms of the design of the control protocol. As stated in Lemma 3, the convex hull representation transforms the saturation term into a linear superposition by introducing auxiliary matrices H. The utilization of the convex hull representation method to study systems with actuator saturation has been welcomed by more and more researchers, but in the context of event-triggered controls, such studies are still scarce. The main reason is that, similarly to the sector-bounded condition, the convex hull representation method also needs to provide an estimation of DOA. Distinct from the flexible selection object of the former method, the DOA estimation of the convex hull representation method is directly related to auxiliary matrix H. Although researchers have given many mature methods for estimating the DOA of the convex hull representation method, the complexity of the control input creates a huge challenge with respect to the estimation problem of DOA when the ETC is introduced.

As an outstanding achievement, [79] presented a output–feedback control strategy, and the MAS can be described as follows:

$$\begin{cases} \dot{x}_0(t) = Ax_0(t) + f(x_0(t)), \\ y_0(t) = Cx_0(t), \\ \dot{x}_i(t) = Ax_i(t) + B\mathscr{U}_{sat}(u_i(t)) + f(x_i(t)), \\ y_i(t) = Cx_i(t), i \in I[1, N], \end{cases} \tag{15}$$

where $y_i(t)$ is the measurement output of agent i, and the rest of the parameters are the same as MAS (11). The output feedback control strategy is given as follows.

$$\Sigma_8 = \begin{cases} \text{ETM}: \\ t^i_{k+1} = t^i_k + \max\{\tau^i_k, c_i\}, \\ \tau^i_k = \min_t\{t - t^i_k : \|\tilde{\delta}_i(t)\| \geq \gamma\|X_i(t)\|\}, \\ \text{Protocol}: \\ u_i(t) = Kz_i(t^i_k),\ t \in [t^i_k, t^i_{k+1}) \\ \dot{z}_i(t) = (A+G)z_i(t) + \bar{G}e_i(t^i_k) + B\mathscr{U}_{sat}(Kz_i(t^i_k)). \end{cases} \tag{16}$$

Define consensus error $\tilde{x}_i(t) = x_i(t) - x_0(t)$, current output consensus error $e_i(t) = \sum_{j \in \mathcal{N}_i} a_{ij}(y_j(t) - y_i(t)) + b_i(y_0(t) - y_i(t))$, measurement error $s_i(t) = e_i(t^i_k) - e_i(t)$, measurement error of $z_i(t)$ as $w_i(t) = z_i(t^i_k) - z_i(t)$, $r_i(t) = \tilde{x}_i(t) - z_i(t)$, and matrix G and $\bar{G}$ are matrices that satisfy certain properties (see Lemma 2 in [79]). Let $\delta_i(t) = [z^T_i(t), r^T_i(t)]^T$, $\tilde{\delta}^T_i = [s^T_i(t), w^T_i(t)]^T$, $X^T_i(t) = [e^T_i(t), z^T_i(t)]^T$, $\delta(t) = [\delta^T_1(t), \delta^T_1(t), \cdots, \delta^T_1(t)]^T$, $\tilde{x} = [\tilde{x}^T_1(t), \tilde{x}^T_2(t), \cdots, \tilde{x}^T_N(t)]^T$.

There are some advantages about strategy (16). Firstly, agent i samples information $z_i(t)$ and $e_i(t)$ to update the control protocol in (16) only at triggered instant t^i_k. Second, the next trigger instant t^i_{k+1} depends on triggered variable $X_i(t)$, which consists of $e_i(t^i_k)$ and $z_i(t^i_k)$. This can avoid continuous communication between neighbors in the MAS network and save communication resources. Moreover, unlike the event-based ETMs above, the event interval of this ETM not only relies on whether the event is triggered but also takes c_i as the lower limit; that is, if the interval between two events is less than c_i, the event will not be triggered even if the ETM is satisfied. Therefore, the triggered interval of the ETM is at least greater than c_i, which can effectively avoid the Zeno phenomenon. Finally, in the actual background, the state of the agent may not be fully acquired, so the control protocol using output-feedback instead of state-feedback can effectively avoid the difficulty of state acquisition and save information sampling consumption.

As discussed in (12), after completing the design of ETM and control protocol, the convex hull representation method also needs to deal with the estimation problem of DOA. Define the set $\mathscr{L}(H, u_0) = \{\delta(t) \in R^{2Nn} : |(l_1 \otimes h_m)\delta(t)| \leq Nu_0\}$, where h_m denotes the mth row of the auxiliary matrix H and $l_1 = (1, 0, 1, 0, \cdots, 1, 0) \in R^{2Nn}$. It is well known that the premise of using the above convex hull representation method is to satisfy condition $|h_m z_i(t_k^i)| \leq u_0$, and it is ensured by $|h_m z_i(t)| \leq u_0$. Taking the context of MASs into consideration, premise $|h_m z_i(t)| \leq u_0$ can be expressed as $|(l_1 \otimes h_m)\delta(t)| \leq Nu_0$. Therefore, the overall premise of using the convex hull representation method to solve the design of the ETM and control protocol is to provide an estimation for set $\mathscr{L}(H, u_0)$. Similarly to the DOA discussion of the sector-bounded condition above, the direct solution of set $\mathscr{L}(H, u_0)$ has computational difficulties, so the indirect estimation method using the level set of the Lyapunov function is adopted. The Lyapunov function is chosen as follows:

$$V(t) = \delta^T(t)P\delta(t),$$

and the level set is $\mathcal{E}(P, \eta) = \{\delta(t) \in R^{Nn} : \delta^T(t)P\delta(t) \leq \eta\}$. It is worth noting that vector $\delta(t)$ corresponding to level set $\mathcal{E}(P, \eta)$ is a composite vector composed of $z_i(t)$ and $r_i(t)$, and it is difficult to describe DOA in detail. Therefore, a subset $\Omega(Q, \varrho) = \{\tilde{x}(t) \in R^{Nn} : \tilde{x}^T(t)Q\tilde{x}(t) \leq \varrho\}$ of level set $\mathcal{E}(P, \eta)$ is defined, and it can be seen that vector $\tilde{x}(t)$ of subset $\Omega(Q, \varrho)$ has a specific meaning, that is, the state difference between the leader agent and the follower agent. Using the result in [80], the optimal estimation of DOA can be obtained by solving the formulated problem (see Theorem 1 in [79]). The outstanding contribution of [79] is that it not only provides a method for estimating DOA but also gives an optimization problem on this basis, that is, maximizing the estimation of DOA, which is not presented in previous results.

In recent years, some researchers also studied the event-triggered consensus problem for MASs with actuator saturation using the convex hull representation method, and some interesting results have been proposed. The problem of event-triggered stabilization for positive systems subject to actuator saturation was investgated in [47]. However, the studied system was limited to a single system, and conclusions were not generalized to MASs. Moreover, a self-triggered consensus control strategy for nonlinear MASs with sensor saturation was proposed in [81].

Remark 4. *The convex hull representation method can effectively reduce the conservatism when dealing with saturated terms, but this method also has its drawbacks. First, as discussed above, this method is more cumbersome than the sector-bounded condition in terms of estimating DOA, which is more popular among researchers, especially in the context of ETC. Second, it can be seen from Lemma 3 that using the convex hull representation method to design the control protocol will increase the computation burden of LMI, which is directly related to input dimension m. In detail, the computational complexity of the convex hull representation method is* 2^m*, so the computational burden grows exponentially, which also suggests that the method may not be suitable for systems with large inputs. Therefore, this method is rarely adopted in the study of event-triggered consensus for MASs with actuator saturation. The existence of few related studies shows that this is a topic that requires further exploration.*

Remark 5. *In Section 3, we review some representative studies about the event-triggered consensus for satarated MASs in detail. In order to show the advantages and disadvantages of each research result more intuitively, we provide Table 1 to facilitate readers' better understanding. In Table 1, the important feature of the control strategies reviewed in this paper is listed. As observed from the table, although research studies have been conducted extensively on the event-triggered consensus problem for MASs with actuator saturation, there are still some important issues worthy of consideration in the future.*

Table 1. Advantages and disadvantages of control strategies.

Strategies	Methods	Advantages & Disadvantages				
		D1	D2	D3	Analysis of Zeno Phenomenon	Estimation of DOA
Σ_1	Low-gain feedback	×	✓	×	Complicated	Not needed
Σ_2		✓	×	×	Complicated	Not needed
Σ_3		×	✓	×	Complicated	Not needed
Σ_4		✓	✓	×	Complicated	Not needed
Σ_5	Sector-bounded condition	×	×	✓	Simple	Simple
Σ_6		✓	×	✓	Not needed	Simple
Σ_7		✓	×	✓	Not needed	Simple
Σ_8	Convex hull representation	×	✓	✓	Complicated	Complicated

If the strategy can overcome the disadvantage, it is marked by ✓; otherwise, it is marked by ×.

4. Simulation

In this section, we will review one simulation example and its comparative experiments in [52] to specify the performance evaluation indicators that should be paid attention to in the event-triggered consensus problem for saturated MASs. It mainly includes performance indicators related to ETC, such as the number of triggered instants and average interval time between events.

Consider the MAS (11) with four followers and one leader, and the agents are determined by a vertical taking-off and landing (VTOL) aircraft model in [52], where the following is the case.

$$A = \begin{bmatrix} -0.0366 & 0.0271 & 0.0188 & -0.4555 \\ 0.0482 & -1.01 & 0.0024 & -4.0208 \\ 0.1002 & 0.3681 & -0.707 & 1.420 \\ 0 & 0 & 1 & 0 \end{bmatrix}, B = \begin{bmatrix} 0.4422 & 0.1761 \\ 3.5446 & -7.5922 \\ -5.52 & 4.49 \\ 0 & 0 \end{bmatrix},$$

$$f(x_i(t)) = [0 \ 0 \ 0 \ -0.1sin(x_{i3}(t))]^T.$$

The meaning of the state variable is as follows: x_{i1}—horizontal velocity; x_{i2}—vertical velocity; x_{i3}—pitch rate; x_{i4}—pitch angle.

Figure 2 shows the communication topology graph between agents, and the numbers represent the agents labeled 0–5. On the basis of control strategies Σ_5 (12) and Σ_7 (14), ETMs are designed by event-based triggered mechanisms and sampled-data-based mechanisms, respectively. Effects of the control protocols are shown in the following figures.

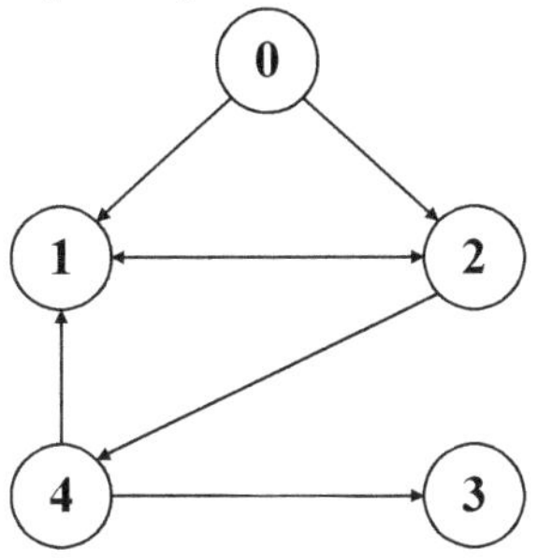

Figure 2. The communication topology graph.

The tracking errors and control input under Σ_5 are shown in Figures 3 and 4. It can be seen from the figure that the tracking error of the system finally tends to zero, indicating that the consensus of the MAS (11) is achieved. At the same time, the control input is different from the traditional continuous one, and the update of the control input is intermittent rather than continuous, which depends on the predefined ETM. Moreover, the tracking errors and control input under Σ_7 are shown in Figures 5 and 6.

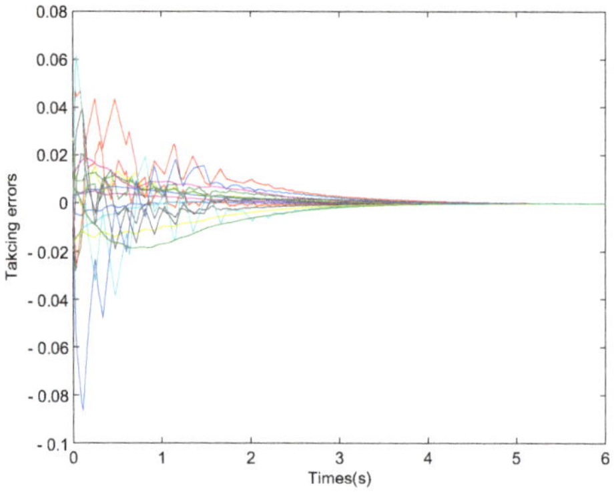

Figure 3. The tracking errors under Σ_5 .

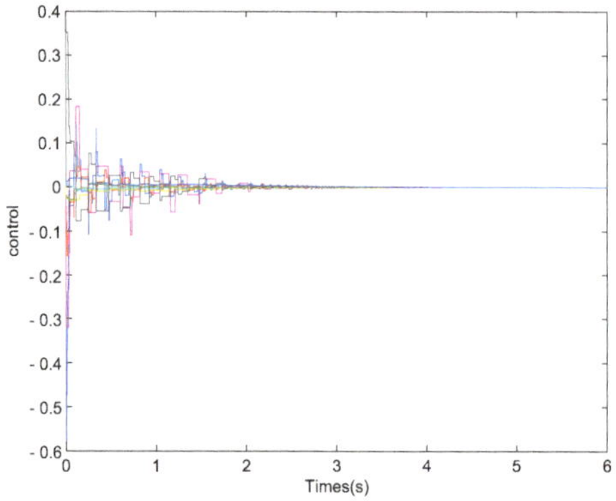

Figure 4. The control input under Σ_5 .

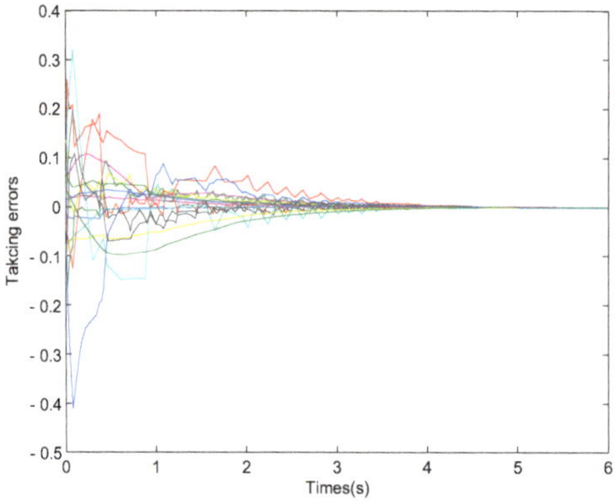

Figure 5. The tracking errors under Σ_7 .

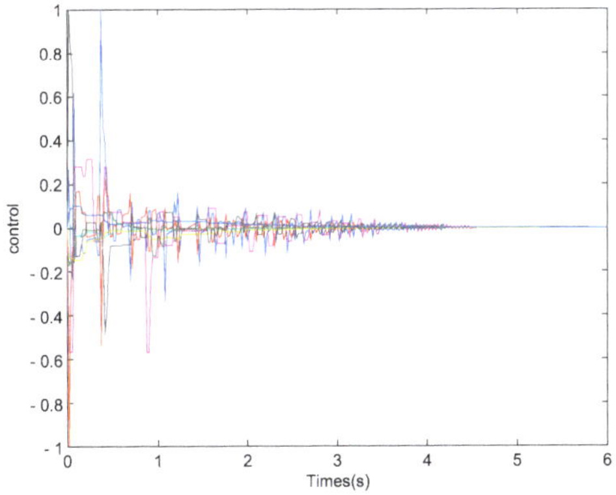

Figure 6. The control input under Σ_7 .

However, appropriate control performances often require the utilization of communication resources. It can be seen from Figures 4 and 6 that the control input is updated intermittently. In the context of ETC, the number of triggered instants and the average interval time between triggered events are important performance indicators to measure the ETM, so we will provide a quantitative experiment next.

According to the data in Tables 2 and 3, compared with Σ_5, control strategy Σ_7 can reduce the number of triggered instants by about 86.12% and prolong the average interval time between events by about 87.25%. The data prove that control strategy Σ_7 has significant advantages in saving communication resources. Compared with event-based triggered mechanisms, the important feature of sampled-data-based mechanisms is that it checks the ETM according to sampling period h. However, the selection of h is also sensitive. If the selection of h is large, the update of control input may be slow, leading to the failure of the ETM; if the selection of h is small, the update of control input will be frequent, and the significance of ETC will be lost, resulting in a huge waste of communication resources.

Table 2. The number of triggered instants.

	Agent 1	**Agent 2**	**Agent 3**	**Agent 4**	**Total**
Σ_5	1349	1388	920	1040	4697
Σ_7	199	197	124	132	652

Table 3. Average interval time between triggered events.

	Agent 1	**Agent 2**	**Agent 3**	**Agent 4**
Σ_5	4.4×10^{-3} s	4.3×10^{-3} s	6.5×10^{-3} s	5.8×10^{-3} s
Σ_7	0.0302 s	0.0305 s	0.0484 s	0.0455 s

5. Prospects for Future Research

A detailed review of event-triggered consensus has been provided in the previous section. Although some control problems have been studied in detail, there are still limitations on mechanistic studies and system limitations, which also brings potential room for improvements to existing research studies. Next, some challenging but meaningful topics will be raised.

(1) Diversified event-triggered mechanisms: Most triggered mechanisms involved in this paper are limited to two types of triggered mechanisms: event-based and sampled-data-based ETMs. In fact, with the development of resaerch studies on ETC, various novel ETMs have been proposed in academia, such as model-based schemes [82–85] and self-triggered sampling schemes [86–89]. Under the background that research studies on

actuator saturation have been developed in recent years, it is a topic worthy of researchers' efforts to study the issue of event-triggered consensus problem of MASs with actuator saturation by using novel ETMs.

(2) Complex conditions about the MASs: In existing studies, most studied systems are described by simple dynamical models in order to simplify the difficulty of theoretical analysis. However, in practice (robots, unmanned aerial vehicle, and complex industrial process), such simple dynamics cannot fully describe the characteristics of the system, and many important factors may be ignored. A notable example is stochastic processes. In practice, stochastic processes can manifest in many aspects, such as stochastic external noise, stochastic measurement errors, and stochastic communication topologies. These stochastic phenomena pose a huge challenge to the event-triggered consensus for MASs due to its uncertainty. To the best of the authors' knowledge, investigations on this issue under the premise of stochastic phenomena are still lacking.

(3) Optimal problems for the estimation of DOA: When the phenomenon of actuator saturation is involved in MASs, the estimation problem of the DOA is an unavoidable topic, especially when dealing with saturated items using the sector-bounded condition or convex hull representation methods. In the context of ETC, estimating the DOA of the MASs is a difficult task, and it is even more difficult to provide its optimization problem based on the estimation of the DOA, i.e., to maximize the estimation of the DOA. As pointed out in Table 3, most resaerch studies only consider the estimation problem of DOA and do not give a method to maximize the estimation, so this area is also an area worthy of future research.

(4) Event-triggered consensus for the MASs in finite time: Notably, most studies currently focus on the asymptotic consensus of MASs. However, in practical engineering, the convergence speed of the system is a key indicator to measure the control effect, and it is generally expected that the consensus of MASs can be achieved in a short and finite time [90]. However, this contradicts the mechanism of ETC. Since the purpose of ETC is to reduce the sampling of information and the frequency of the controller update and finally decrease the utilization of communication resources, but this will inevitably slow down the convergence speed of the system. So it will be a difficult but promising topic for designing a suitable control strategy, which can not only reduce the utilization of communication resources but also ensure a fast convergence effect.

6. Conclusions

This paper mainly reviews recent studeis on the issue of event-triggered consensus for MASs with actuator saturation, classifies them according to the saturation-processing methods used, and summarizes their advantages and disadvantages, as well as room for improvement. It is worth noting that ETC and actuator saturation are aimed at different aspects of the control input. ETC is intended to enable the control input to still meet the performance requirements at a lower cost, while the saturation phenomenon focuses on solving the practical limitation of the control input. The studies on event-triggered consensus for MASs with actuator saturation have brought out certain results, and it is interesting to witness more in the future.

7. Discussion

Recent studies on the issue of event-triggered consensus for MASs with actuator saturation are discussed in this paper, and our future research in this area will focus on novel ETMs and the optimal estimation of DOA.

Author Contributions: Conceptualization, J.X. and J.H.; methodology, J.X.; software, J.X.; validation, J.X. and J.H.; formal analysis, J.X.; investigation, J.X.; resources, J.X. and J.H.; data curation, J.X.; writing—original draft preparation, J.X.; writing—review and editing, J.H.; visualization, J.X.; supervision, J.X.; project administration, J.H.; funding acquisition, J.H. All authors have read and agreed to the published version of the manuscript.

Funding: This work is supported by the Natural Science Foundation of Jiangsu Province of China (BK2021-1309), the Open Fund for Jiangsu Key Laboratory of Advanced Manufacturing Technology (HGAMTL-2101), and the open project (No. Scip202207) of Key Laboratory of System Control and 589 Information Processing, Ministry of Education, China.

Institutional Review Board Statement: Not applicable.

Informed Consent Statement: Not applicable.

Data Availability Statement: Not applicable.

Conflicts of Interest: The authors declare no conflicts of interest.

Abbreviations

The following abbreviations are used in this manuscript:

MASs	Multi-agent systems;
TTC	Time triggered control;
ETC	Event-triggered control;
ETM	Event-triggered mechanism;
LMIs	Linear matrix inequalities;
ARE	Algebraic Riccati equation;
DOA	Domain of attraction.

References

1. Jolly, K.; Kumar, R.S.; Vijayakumar, R. Intelligent task planning and action selection of a mobile robot in a multi-agent system through a fuzzy neural network approach. *Eng. Appl. Artif. Intell.* **2010**, *23*, 923–933. [CrossRef]
2. Nazarova, A.V.; Zhai, M. Distributed Solution of Problems in Multi Agent Robotic Systems. In *Studies in Systems, Decision and Control*; Springer International Publishing: Berlin/Heidelberg, Germany, 2018; pp. 107–124. [CrossRef]
3. Freudenthaler, G.; Meurer, T. PDE-based multi-agent formation control using flatness and backstepping: Analysis, design and robot experiments. *Automatica* **2020**, *115*, 108897. [CrossRef]
4. Kapitonov, A.; Lonshakov, S.; Krupenkin, A.; Berman, I. Blockchain-based protocol of autonomous business activity for multi-agent systems consisting of UAVs. In Proceedings of the 2017 Workshop on Research, Education and Development of Unmanned Aerial Systems (RED-UAS), Linköping, Sweden, 3–5 October 2017; IEEE: Piscataway, NJ, USA , 2017. [CrossRef]
5. Silva, L.A.; Blas, H.S.S.; García, D.P.; Mendes, A.S.; González, G.V. An Architectural Multi-Agent System for a Pavement Monitoring System with Pothole Recognition in UAV Images. *Sensors* **2020**, *20*, 6205. [CrossRef] [PubMed]
6. Barriuso, A.; González, G.V.; Paz, J.D.; Lozano, Á.; Bajo, J. Combination of Multi-Agent Systems and Wireless Sensor Networks for the Monitoring of Cattle. *Sensors* **2018**, *18*, 108. [CrossRef]
7. Ge, X.; Han, Q.L.; Wang, Z. A Threshold-Parameter-Dependent Approach to Designing Distributed Event-Triggered H_∞ Consensus Filters Over Sensor Networks. *IEEE Trans. Cybern.* **2019**, *49*, 1148–1159. [CrossRef]
8. Cai, X.; Wang, J.; Zhong, S.; Shi, K.; Tang, Y. Fuzzy quantized sampled-data control for extended dissipative analysis of T–S fuzzy system and its application to WPGSs. *J. Frankl. Inst.* **2021**, *358*, 1350–1375. [CrossRef]
9. Pipattanasomporn, M.; Feroze, H.; Rahman, S. Multi-agent systems in a distributed smart grid: Design and implementation. In Proceedings of the 2009 IEEE/PES Power Systems Conference and Exposition, Seattle, WA, USA, 15–18 March 2009; IEEE: Piscataway, NJ, USA, 2009. [CrossRef]
10. Olfati-Saber, R.; Fax, J.A.; Murray, R.M. Consensus and Cooperation in Networked Multi-Agent Systems. *Proc. IEEE* **2007**, *95*, 215–233. [CrossRef]
11. Li, Z.; Duan, Z.; Chen, G.; Huang, L. Consensus of multiagent systems and synchronization of complex networks: A unified viewpoint. *IEEE Trans. Circuits Syst. I Regul. Pap.* **2009**, *57*, 213–224.
12. Shao, J.; Zheng, W.X.; Huang, T.Z.; Bishop, A.N. On Leader–Follower Consensus With Switching Topologies: An Analysis Inspired by Pigeon Hierarchies. *IEEE Trans. Autom. Control* **2018**, *63*, 3588–3593. [CrossRef]
13. Yang, Y.; Xu, H.; Yue, D. Observer-Based Distributed Secure Consensus Control of a Class of Linear Multi-Agent Systems Subject to Random Attacks. *IEEE Trans. Circuits Syst. I Regul. Pap.* **2019**, *66*, 3089–3099. [CrossRef]
14. Wei, X.; Yu, W.; Wang, H.; Yao, Y.; Mei, F. An Observer-Based Fixed-Time Consensus Control for Second-Order Multi-Agent Systems with Disturbances. *IEEE Trans. Circuits Syst. II Express Briefs* **2019**, *66*, 247–251. [CrossRef]
15. Zhang, J.; Zhang, H.; Sun, S.; Gao, Z. Leader-follower consensus control for linear multi-agent systems by fully distributed edge-event-triggered adaptive strategies. *Inf. Sci.* **2021**, *555*, 314–338. [CrossRef]
16. Yu, W.; Zheng, W.X.; Chen, G.; Ren, W.; Cao, J. Second-order consensus in multi-agent dynamical systems with sampled position data. *Automatica* **2011**, *47*, 1496–1503. [CrossRef]

17. Wen, G.; Duan, Z.; Yu, W.; Chen, G. Consensus of multi-agent systems with nonlinear dynamics and sampled-data information: A delayed-input approach. *Int. J. Robust Nonlinear Control* **2012**, *23*, 602–619. [CrossRef]
18. Ding, L.; Guo, G. Sampled-data leader-following consensus for nonlinear multi-agent systems with Markovian switching topologies and communication delay. *J. Frankl. Inst.* **2015**, *352*, 369–383. [CrossRef]
19. Ding, L.; Zheng, W.X. Consensus tracking in heterogeneous nonlinear multi-agent networks with asynchronous sampled-data communication. *Syst. Control Lett.* **2016**, *96*, 151–157. [CrossRef]
20. Tabuada, P. Event-Triggered Real-Time Scheduling of Stabilizing Control Tasks. *IEEE Trans. Autom. Control* **2007**, *52*, 1680–1685. [CrossRef]
21. Åarzén, K.E. A simple event-based PID controller. *IFAC Proc. Vol.* **1999**, *32*, 8687–8692. [CrossRef]
22. Peng, C.; Li, F. A survey on recent advances in event-triggered communication and control. *Inf. Sci.* **2018**, *457–458*, 113–125. [CrossRef]
23. Ding, L.; Han, Q.L.; Ge, X.; Zhang, X.M. An Overview of Recent Advances in Event-Triggered Consensus of Multiagent Systems. *IEEE Trans. Cybern.* **2018**, *48*, 1110–1123. [CrossRef]
24. Heemels, W.; Johansson, K.; Tabuada, P. An introduction to event-triggered and self-triggered control. In Proceedings of the 2012 IEEE 51st IEEE Conference on Decision and Control (CDC), Maui, HI, USA, 10–13 December 2012; IEEE: Piscataway, NJ, USA, 2012. [CrossRef]
25. Nguyen, K.H.; Kim, S.H. Event-Triggered Non-PDC Filter Design of Fuzzy Markovian Jump Systems under Mismatch Phenomena. *Mathematics* **2022**, *10*, 2917. [CrossRef]
26. Djordjevic, V.; Stojanovic, V.; Tao, H.; Song, X.; He, S.; Gao, W. Data-driven control of hydraulic servo actuator based on adaptive dynamic programming. *Discret. Contin. Dyn. Syst.-S* **2022**, *15*, 1633. [CrossRef]
27. Voortman, Q.; Efimov, D.; Pogromsky, A.Y.; Richard, J.P.; Nijmeijer, H. An event-triggered observation scheme for systems with perturbations and data rate constraints. *Automatica* **2022**, *145*, 110512. [CrossRef]
28. Song, X.; Sun, P.; Song, S.; Stojanovic, V. Event-driven NN adaptive fixed-time control for nonlinear systems with guaranteed performance. *J. Frankl. Inst.* **2022**, *359*, 4138–4159. [CrossRef]
29. Heemels, W.P.M.H.; Sandee, J.H.; Bosch, P.P.J.V.D. Analysis of event-driven controllers for linear systems. *Int. J. Control* **2008**, *81*, 571–590. [CrossRef]
30. Eqtami, A.; Dimarogonas, D.V.; Kyriakopoulos, K.J. Event-triggered control for discrete-time systems. In Proceedings of the 2010 American Control Conference, Baltimore, MA, USA, 30 June–2 July 2010; IEEE: Piscataway, NJ, USA, 2010. [CrossRef]
31. Heemels, W.P.M.H.; Donkers, M.C.F.; Teel, A.R. Periodic Event-Triggered Control for Linear Systems. *IEEE Trans. Autom. Control* **2013**, *58*, 847–861. [CrossRef]
32. Yue, D.; Tian, E.; Han, Q.L. A Delay System Method for Designing Event-Triggered Controllers of Networked Control Systems. *IEEE Trans. Autom. Control* **2013**, *58*, 475–481. [CrossRef]
33. Peng, C.; Han, Q.L.; Yue, D. To Transmit or Not to Transmit: A Discrete Event-Triggered Communication Scheme for Networked Takagi–Sugeno Fuzzy Systems. *IEEE Trans. Fuzzy Syst.* **2013**, *21*, 164–170. [CrossRef]
34. Zhang, D.; Han, Q.L.; Jia, X. Network-based output tracking control for T–S fuzzy systems using an event-triggered communication scheme. *Fuzzy Sets Syst.* **2015**, *273*, 26–48. [CrossRef]
35. Wen, S.; Yu, X.; Zeng, Z.; Wang, J. Event-Triggering Load Frequency Control for Multiarea Power Systems With Communication Delays. *IEEE Trans. Ind. Electron.* **2016**, *63*, 1308–1317. [CrossRef]
36. Zhang, X.M.; Han, Q.L. A Decentralized Event-Triggered Dissipative Control Scheme for Systems With Multiple Sensors to Sample the System Outputs. *IEEE Trans. Cybern.* **2016**, *46*, 2745–2757. [CrossRef] [PubMed]
37. Wen, S.; Zeng, Z.; Chen, M.Z.Q.; Huang, T. Synchronization of Switched Neural Networks With Communication Delays via the Event-Triggered Control. *IEEE Trans. Neural Netw. Learn. Syst.* **2017**, *28*, 2334–2343. [CrossRef] [PubMed]
38. Dornhein, M. Report pinpoints factors leading to YF-22 crash. *Aviat. Week Space Technol.* **1992**, *9*, 53–54.
39. Da Silva, J.G.; Tarbouriech, S. Antiwindup design with guaranteed regions of stability: An LMI-based approach. *IEEE Trans. Autom. Control* **2005**, *50*, 106–111. [CrossRef]
40. Lin, Z.; Saberi, A. Semi-global exponential stabilization of linear discrete-time systems subject to input saturation via linear feedbacks. In Proceedings of the 1994 American Control Conference-ACC '94, Baltimore, MD, USA, 29 June–1 July 1994; IEEE: Piscataway, NJ, USA, 1994. [CrossRef]
41. Hu, T.; Lin, Z.; Chen, B.M. Analysis and design for discrete-time linear systems subject to actuator saturation. *Syst. Control Lett.* **2002**, *45*, 97–112. [CrossRef]
42. Meng, Z.; Zhao, Z.; Lin, Z. On global leader-following consensus of identical linear dynamic systems subject to actuator saturation. *Syst. Control Lett.* **2013**, *62*, 132–142. [CrossRef]
43. Yang, T.; Meng, Z.; Dimarogonas, D.V.; Johansson, K.H. Global consensus for discrete-time multi-agent systems with input saturation constraints. *Automatica* **2014**, *50*, 499–506. [CrossRef]
44. Geng, H.; Chen, Z.; Liu, Z.; Zhang, Q. Consensus of a heterogeneous multi-agent system with input saturation. *Neurocomputing* **2015**, *166*, 382–388. [CrossRef]
45. Su, H.; Chen, M.Z.Q. Multi-agent containment control with input saturation on switching topologies. *IET Control Theory Appl.* **2015**, *9*, 399–409. [CrossRef]

46. Deng, C.; Yang, G.H. Consensus of Linear Multiagent Systems with Actuator Saturation and External Disturbances. *IEEE Trans. Circuits Syst. II Express Briefs* **2017**, *64*, 284–288. [CrossRef]
47. Yin, Y.; Lin, Z.; Liu, Y.; Teo, K.L. Event-triggered constrained control of positive systems with input saturation. *Int. J. Robust Nonlinear Control* **2018**, *28*, 3532–3542. [CrossRef]
48. Fu, J.; Wen, G.; Huang, T.; Duan, Z. Consensus of Multi-Agent Systems With Heterogeneous Input Saturation Levels. *IEEE Trans. Circuits Syst. II Express Briefs* **2019**, *66*, 1053–1057. [CrossRef]
49. Su, H.; Sun, Y.; Zeng, Z. Semiglobal Observer-Based Non-Negative Edge Consensus of Networked Systems With Actuator Saturation. *IEEE Trans. Cybern.* **2020**, *50*, 2827–2836. [CrossRef]
50. Lu, M.; Wu, J.; Zhan, X.; Han, T.; Yan, H. Consensus of second-order heterogeneous multi-agent systems with and without input saturation. *ISA Trans.* **2022**, *126*, 14–20. [CrossRef] [PubMed]
51. Zuo, Z.; Li, Y.; Wang, Y.; Li, H. Event-triggered control for switched systems in the presence of actuator saturation. *Int. J. Syst. Sci.* **2018**, *49*, 1478–1490. [CrossRef]
52. Yin, X.; Yue, D.; Hu, S. Adaptive periodic event-triggered consensus for multi-agent systems subject to input saturation. *Int. J. Control* **2015**, *89*, 653–667. [CrossRef]
53. Ge, X.; Yang, F.; Han, Q.L. Distributed networked control systems: A brief overview. *Inf. Sci.* **2017**, *380*, 117–131. [CrossRef]
54. Qin, J.; Ma, Q.; Shi, Y.; Wang, L. Recent Advances in Consensus of Multi-Agent Systems: A Brief Survey. *IEEE Trans. Ind. Electron.* **2017**, *64*, 4972–4983. [CrossRef]
55. Ge, X.; Han, Q.L.; Ding, D.; Zhang, X.M.; Ning, B. A survey on recent advances in distributed sampled-data cooperative control of multi-agent systems. *Neurocomputing* **2018**, *275*, 1684–1701. [CrossRef]
56. Ma, C.Q.; Zhang, J.F. Necessary and Sufficient Conditions for Consensusability of Linear Multi-Agent Systems. *IEEE Trans. Autom. Control* **2010**, *55*, 1263–1268. [CrossRef]
57. Tarbouriech, S.; Prieur, C.; Silva, J.G.D. Stability Analysis and Stabilization of Systems Presenting Nested Saturations. *IEEE Trans. Autom. Control* **2006**, *51*, 1364–1371. [CrossRef]
58. Wang, X.; Su, H.; Wang, X.; Chen, G. Fully Distributed Event-Triggered Semiglobal Consensus of Multi-agent Systems with Input Saturation. *IEEE Trans. Ind. Electron.* **2017**, *64*, 5055–5064. [CrossRef]
59. Hu, T.; Lin, Z. *Control Systems with Actuator Saturation*; Birkhauser: Boston, MA, USA, 2001. [CrossRef]
60. Dimarogonas, D.V.; Frazzoli, E.; Johansson, K.H. Distributed Event-Triggered Control for Multi-Agent Systems. *IEEE Trans. Autom. Control* **2012**, *57*, 1291–1297. [CrossRef]
61. Seyboth, G.S.; Dimarogonas, D.V.; Johansson, K.H. Event-based broadcasting for multi-agent average consensus. *Automatica* **2013**, *49*, 245–252. [CrossRef]
62. Garcia, E.; Cao, Y.; Yu, H.; Antsaklis, P.; Casbeer, D. Decentralised event-triggered cooperative control with limited communication. *Int. J. Control* **2013**, *86*, 1479–1488. [CrossRef]
63. Yang, D.; Ren, W.; Liu, X.; Chen, W. Decentralized event-triggered consensus for linear multi-agent systems under general directed graphs. *Automatica* **2016**, *69*, 242–249. [CrossRef]
64. Fan, Y.; Feng, G.; Wang, Y.; Song, C. Distributed event-triggered control of multi-agent systems with combinational measurements. *Automatica* **2013**, *49*, 671–675. [CrossRef]
65. Zhu, W.; Jiang, Z.P.; Feng, G. Event-based consensus of multi-agent systems with general linear models. *Automatica* **2014**, *50*, 552–558. [CrossRef]
66. Guo, G.; Ding, L.; Han, Q.L. A distributed event-triggered transmission strategy for sampled-data consensus of multi-agent systems. *Automatica* **2014**, *50*, 1489–1496. [CrossRef]
67. Zhou, B.; Liao, X.; Huang, T.; Li, H.; Chen, G. Event-Based Semiglobal Consensus of Homogenous Linear Multi-Agent Systems Subject to Input Saturation. *Asian J. Control* **2016**, *19*, 564–574. [CrossRef]
68. Li, H.; Liao, X.; Huang, T.; Zhu, W. Event-Triggering Sampling Based Leader-Following Consensus in Second-Order Multi-Agent Systems. *IEEE Trans. Autom. Control* **2015**, *60*, 1998–2003. [CrossRef]
69. Hu, W.; Liu, L.; Feng, G. Consensus of Linear Multi-Agent Systems by Distributed Event-Triggered Strategy. *IEEE Trans. Cybern.* **2016**, *46*, 148–157. [CrossRef] [PubMed]
70. Du, S.L.; Liu, T.; Ho, D.W.C. Dynamic Event-Triggered Control for Leader-Following Consensus of Multiagent Systems. *IEEE Trans. Syst. Man Cybern. Syst.* **2020**, *50*, 3243–3251. [CrossRef]
71. Zhao, G.; Wang, Z.; Fu, X. Fully Distributed Dynamic Event-Triggered Semiglobal Consensus of Multi-agent Uncertain Systems with Input Saturation via Low-gain Feedback. *Int. J. Control Autom. Syst.* **2021**, *19*, 1451–1460. [CrossRef]
72. Xu, Y.; Wang, J.; Zhang, Y.; Xu, Y. Event-triggered bipartite consensus for high-order multi-agent systems with input saturation. *Neurocomputing* **2020**, *379*, 284–295. [CrossRef]
73. Chen, S.; Jiang, H.; Yu, Z. Fully Distributed Event-triggered Semi-global Consensus of Multi-agent Systems with Input Saturation and Directed Topology. *Int. J. Control Autom. Syst.* **2019**, *17*, 3102–3112. [CrossRef]
74. Min, H.; Wang, S.; Sun, F.; Zhang, J. Robust consensus for networked mechanical systems with coupling time delay. *Int. J. Control Autom. Syst.* **2012**, *10*, 227–237. [CrossRef]
75. Dai, J.; Guo, G. Event-based consensus for second-order multi-agent systems with actuator saturation under fixed and Markovian switching topologies. *J. Frankl. Inst.* **2017**, *354*, 6098–6118. [CrossRef]

76. Gao, H.Y.; Hu, A.H. Event-triggered Pinning Bipartite Tracking Consensus of the Multi-agent System Subject to Input Saturation. *Int. J. Control Autom. Syst.* **2020**, *18*, 2195–2205. [CrossRef]
77. Wang, J.; Luo, X.; Yan, J.; Guan, X. Event-triggered consensus control for second-order multi-agent system subject to saturation and time delay. *J. Frankl. Inst.* **2021**, *358*, 4895–4916. [CrossRef]
78. Rehan, M.; Tufail, M.; Ahmed, S. Leaderless consensus control of nonlinear multi-agent systems under directed topologies subject to input saturation using adaptive event-triggered mechanism. *J. Frankl. Inst.* **2021**, *358*, 6217–6239. [CrossRef]
79. You, X.; Hua, C.; Guan, X. Event-Triggered Leader-Following Consensus for Nonlinear Multiagent Systems Subject to Actuator Saturation Using Dynamic Output Feedback Method. *IEEE Trans. Autom. Control* **2018**, *63*, 4391–4396. [CrossRef]
80. Hu, T.; Lin, Z.; Chen, B.M. An analysis and design method for linear systems subject to actuator saturation and disturbance. *Automatica* **2002**, *38*, 351–359. [CrossRef]
81. Chen, D.; Liu, X.; Yu, W.; Zhu, L.; Tang, Q. Neural-Network Based Adaptive Self-Triggered Consensus of Nonlinear Multi-Agent Systems With Sensor Saturation. *IEEE Trans. Netw. Sci. Eng.* **2021**, *8*, 1531–1541. [CrossRef]
82. Zhang, H.; Feng, G.; Yan, H.; Chen, Q. Observer-Based Output Feedback Event-Triggered Control for Consensus of Multi-Agent Systems. *IEEE Trans. Ind. Electron.* **2014**, *61*, 4885–4894. [CrossRef]
83. Yin, X.; Yue, D.; Hu, S.; Peng, C.; Xue, Y. Model-Based Event-Triggered Predictive Control for Networked Systems with Data Dropout. *SIAM J. Control Optim.* **2016**, *54*, 567–586. [CrossRef]
84. Xu, W.; Ho, D.W.C. Clustered Event-Triggered Consensus Analysis: An Impulsive Framework. *IEEE Trans. Ind. Electron.* **2016**, *63*, 7133–7143. [CrossRef]
85. Liu, X.; Du, C.; Lu, P.; Yang, D. Distributed event-triggered feedback consensus control with state-dependent threshold for general linear multi-agent systems. *Int. J. Robust Nonlinear Control* **2016**, *27*, 2589–2609. [CrossRef]
86. Wang, X.; Lemmon, M. Self-Triggered Feedback Control Systems With Finite-Gain $\mathcal{L}_2$ Stability. *IEEE Trans. Autom. Control* **2009**, *54*, 452–467. [CrossRef]
87. Wang, X.; Lemmon, M.D. Self-Triggering Under State-Independent Disturbances. *IEEE Trans. Autom. Control* **2010**, *55*, 1494–1500. [CrossRef]
88. Anta, A.; Tabuada, P. To Sample or not to Sample: Self-Triggered Control for Nonlinear Systems. *IEEE Trans. Autom. Control* **2010**, *55*, 2030–2042. [CrossRef]
89. Peng, C.; Han, Q.L. On Designing a Novel Self-Triggered Sampling Scheme for Networked Control Systems With Data Losses and Communication Delays. *IEEE Trans. Ind. Electron.* **2016**, *63*, 1239–1248. [CrossRef]
90. Lu, Q.; Han, Q.L.; Zhang, B.; Liu, D.; Liu, S. Cooperative Control of Mobile Sensor Networks for Environmental Monitoring: An Event-Triggered Finite-Time Control Scheme. *IEEE Trans. Cybern.* **2017**, *47*, 4134–4147. [CrossRef] [PubMed]

mathematics MDPI

Article

Another Case of Degenerated Discrete Chenciner Dynamic System and Economics

Sorin Lugojan [1], Loredana Ciurdariu [1,*] and Eugenia Grecu [2]

[1] Department of Mathematics, Politehnica University of Timisoara, 300006 Timisoara, Romania
[2] Department of Management, Politehnica University of Timisoara, 300006 Timisoara, Romania
* Correspondence: loredana.ciurdariu@upt.ro

Abstract: The non-degenerate Chenciner bifurcation of a discrete dynamical system is studied using a transformation of parameters which must be regular at the origin of the parameters (the condition CH.1 of the well-known treatise of Kuznetsov). The article studies a complementary case, where the transformation is no longer regular at the origin, representing a degeneration. Four different bifurcation diagrams appear in that degenerated case, compared to only two in the non-degenerated one. Degeneracy may cause volatility in economics systems modeled by discrete Chenciner dynamical systems.

Keywords: bifurcation; discrete-time systems; Chenciner; degeneracy

MSC: 37L10; 37G10

Citation: Lugojan, S.; Ciurdariu, L.; Grecu, E. Another Case of Degenerated Discrete Chenciner Dynamic System and Economics. *Mathematics* **2022**, *10*, 3782. https://doi.org/10.3390/math10203782

Academic Editors: Jun Huang and Yueyuan Zhang

Received: 29 August 2022
Accepted: 2 October 2022
Published: 13 October 2022

1. Introduction

Continuous and discrete-time dynamical systems can be used for modeling many applications in the surrounding world [1–3]. Discrete dynamical systems may appear in "practical applications when a phenomenon cannot be observed continuously in time" [4], but in certain moments of time [5]. Additionally, they can be obtained from dynamic systems with continuous time by discretizing time, that is, if we only take certain values for time [6] or as return maps that are return applications defined by the intersections of the system flows with certain "surfaces transversal to the flows" [4].

From a computational point of view, the use of dynamical systems with discrete time is more efficient in modeling because it can capture complex behaviors that cannot be easily captured otherwise [7–9]. Among the most "important topics in the qualitative theory" [10] of continuous and discrete dynamic systems is the analysis of bifurcations (see [11]).

One of the topics of interest in discrete dynamical systems is represented by the Chenciner bifurcation. Using the notations of the fundamental book of Kuznetsov, [12], page 405, a discrete Chenciner bifurcation happens when $r(0) = 1$, $Re(b_1(0)) = 0$ and $L_2(0) \neq 0$.

A parametric transformation $(\alpha_1, \alpha_2) \to (\beta_1, \beta_2)$ is needed in the regular case where the functions

$$\beta_1(\alpha) = \sum_{i+j=1}^{p} a_{ij}\alpha_1^i\alpha_2^j + O(|\alpha^{p+1}|)$$

$$\beta_2(\alpha) = \sum_{i+j=1}^{q} b_{ij}\alpha_1^i\alpha_2^j + O(|\alpha^{q+1}|),\ p, q \geq 1,$$

$$a_{10} = \frac{\partial\beta_1}{\partial\alpha_1}|_{\alpha=0}; a_{01} = \frac{\partial\beta_1}{\partial\alpha_2}|_{\alpha=0}; b_{10} = \frac{\partial\beta_2}{\partial\alpha_1}|_{\alpha=0}; b_{01} = \frac{\partial\beta_2}{\partial\alpha_2}|_{\alpha=0} \quad (1)$$

and so on, see [12], page 405. That transformation must be regular at the origin in order to have a non-degenerated Chenciner bifurcation.

The non-degenerate Chenciner bifurcation was firstly studied in the papers [6,13,14]. More recently this bifurcation appears in many papers from different areas of research, in "biology, physics, economy, informatics" [15] as well as multidisciplinary and applied sciences [12,16–30]. For example, in [31], the Chenciner bifurcation was observed when a potential mechanism from bifurcation analyses was used for studying the occurrence of modulated oscillations in synchronous machine nonlinear dynamics, being reported for the first time in power engineering for this bifurcation. Other authors have analyzed the normal forms to provide the parameter conditions for the Chenciner bifurcation [32] or the conditions to obtain a Chenciner bifurcation in macroeconomics [33].

Rational expectations are the foundation of modern finance. However, in principle, the efficient market hypothesis cannot help accurately predict future prices. There is ample empirical evidence that developments in financial time series, in the form of "stylized facts", cannot be explained by fundamentals alone, and markets appear to have specific internal dynamics. Among the so-called "stylized facts" is volatility clustering. It appears that if changes in asset prices are unpredictable, the magnitude of those changes is predictable; Thus, "large changes tend to be followed by large changes" [19] (either increasing or decreasing), while" small changes tend to be followed by small changes" [19]. That is why it is found that asset price fluctuations present "episodes of high volatility" [19] (with large price changes), which alternate irregularly with "episodes of low volatility" [19] (with small price changes).

In economics, in a series of empirical studies, the used model is useful only for a statistical description of the data [34]. However, these models cannot explain the clustering of volatility that is recorded in many financial time series. Typically, such models assume that volatility clustering is generated by factors external to the analyzed system.

Some structural explanations of volatility clustering are provided by "multi-agent systems" [19], where financial markets have been approached as "complex evolutionary systems" [19]. In such systems, two large categories of traders have been identified: fundamentalists (who state that prices are oriented toward the value of their fundamental rational expectations, generated by future dividends) and technical analysts (who, starting from the past prices, and based on some established models, try to project them in the future). Such systems show an irregular transition between low volatility situations (during which prices tend toward the fundamental price and then the market is dominated by fundamentalists) and high volatility situations (during which "prices move away from the fundamental price" [19] and then the market is dominated by technical analysts) [19]. In these conditions, the grouping of volatility can have endogenous explanations, that is, it could be caused and even amplified by the process of the heterogeneity of trading, but also by the interaction between agents, as well as by the phenomenon of adaptive learning.

The evolutionary model proposed by A. Gaunersdorfer, C.H. Hommes and F.O.O. Wagener presents the "coexistence of a stable state and a stable limit cycle" [19]. When such a system is subject to dynamic noise, there is an irregular switching between fundamental equilibrium fluctuations close to rational expectations (in which "the market is dominated by fundamentalists) and large-amplitude price fluctuations" [19] (in which the market is dominated by technical analysts). "The coexistence of a stable equilibrium state and a stable limit cycle " [19] is explained mathematically by means of the discrete Chenciner bifurcation. This is not caused by a particular specification of the model, but "is a generic feature for nonlinear systems with two or more parameters" [19].

The discrete degenerated Chenciner bifurcation is produced when the above mentioned regularity of the β transformation is not fulfilled. That results in a much more difficult scenario. A first type of such a degenerated Chenciner discrete dynamical was solved in [4]. Two other types of possible degeneration were studied in [10,15]. Each of those cases has a quite different method of solving. In the present article, we study another case of a possible degeneration. So, why bother with such particular cases, each having a specific kind of approach? An Edmund Hillary type of answer would be, "because they exist", and also one may see the complexity of nature's singularities reflected by mathematics.

In [4], the bifurcation diagrams were discovered in a general case, where the functions $\beta_1(\alpha)$ and $\beta_2(\alpha)$ both have linear terms different from zero that satisfy the degeneracy condition $a_{10}b_{01} - a_{01}b_{10} = 0$, or $a_{10} = a_{01} = 0$, see (1). In that case, 32 bifurcation diagrams were obtained. A parallel approach to that of [4] is studied in [35] by using another regular transformation of parameters, where the product $a_{10}a_{01}b_{10}b_{01} \neq 0$. In the article [15], the functions $\beta_1(\alpha)$ and $\beta_2(\alpha)$ have $a_{10} = 0; a_{01} = 0; b_{10} = 0; b_{01} = 0$, obtaining four bifurcation diagrams. Ref. [10] studied the case when $a_{20} = a_{11} = a_{02} = 0$ and $b_{10} \neq 0,\ b_{01} \neq 0$ or $a_{20} = a_{11} = a_{02} = 0$ and $b_{10} = 0,\ b_{01} = 0$, obtaining 18 different bifurcation diagrams. The stability of the fixed point O for $|\alpha|$ that is sufficiently small and, respectively, "the existence of closed invariant curves in the" [4] truncated normal form in all the cases was treated before [10,15,35].

A possible application of the degenerated Chenciner bifurcation was presented in [15], but one could analyze in all previous mentioned Chenciner papers what happens when degeneration occurs. For example, the volatility of the economics systems based on discrete Chenciner bifurcation may be interpreted as a variant of input data implying the degeneration of the bifurcation. One possible cause of that may be the presence of a noise, rendering a sequence of different degenerated and non-degenerated variants of the initial system in case the coefficients a_{ij} have small values.

The purpose of this article is to investigate the behavior of the dynamical system when $\beta_1(\alpha)$ or $\beta_2(\alpha)$ has a zero linear part $a_{10} = a_{01} = 0$ or $b_{10} = b_{01} = 0$, see (1), and the second function has at least a term of order one different from zero. This aspect has not been analyzed before. As it is not possible to choose new coordinates β_1, β_2, the idea is to use only the initial parameters (α_1, α_2). This leads to the modifications of the structure of the sets of points $B_{1,2}$ and C, thus obtaining concurrent lines at the origin, similar to the situation analyzed in other articles [10,15], but different from the cases studied in [4,35]. We want to specify how many bifurcation diagrams are obtained, many or few. The first case studied, when $\Delta_1 > 0$, is the most important and complex of the two and requires different methods of approach (the second is when $\Delta_1 < 0$).

The starting hypothesis in this study is that in the case of a degeneracy, a larger number of bifurcation diagrams is needed than in a non-degeneracy setting. The objective of this article is to verify the mentioned hypothesis in a degeneracy case that does not involve resonance.

The work is structured in six sections; after the Introduction (Section 1 and Appendices A and B), Section 2 presents the analysis of degenerate Chenciner bifurcation that "means the existence and stability of equilibrium points and invariant closed curves" [4] for this form of degeneracy, known as non-transversality, i.e., the "transformation of parameters is not regular at $(0,0)$" [35]. In Section 3, it is described the existence of bifurcations curves and their dynamics in the parametric plane (α_1, α_2) in Theorem 1. Section 4 shows the bifurcation diagrams for this type of degeneracy of Chenciner bifurcation when the smooth function $\beta_1(\alpha)$ is of order two. These bifurcation diagrams are different from the bifurcation diagrams from the non-degenerate framework. In Section 5, several numerical simulations using Matlab check the theoretical results from the previous section. Section 6 indicates the relevant discussions and conclusions of the paper.

2. Materials and Methods

Since Chenciner bifurcation happens for the discrete dynamical system, we consider

$$x_{n+1} = f(x_n, \alpha) \tag{2}$$

where $x_n \in \mathbb{R}^2$, $n \in \mathbb{N}$, $\alpha = (\alpha_1, \alpha_2) \in \mathbb{R}^2$ and f is a smooth function of class C^r with $r \geq 2$. In order to avoid indices, the Equation (2) is sometimes written in the form

$$x \longmapsto f(x, \alpha) \tag{3}$$

or $\tilde{x} = f(x, \alpha)$.

A bifurcation as in (A4) which satisfies $r(0) = 1$ and $Re(b_1(0)) = 0$ but $L_2(0) \neq 0$ is known as "the *Chenciner bifurcation* (or generalized Neimark–Sacker bifurcation)" [4]. It follows from $\beta_1(0) = 0$ that

$$L_2(0) = \frac{1}{2}\Big(Im^2(b_1(0)) + 2Re(b_2(0))\Big).$$

When the transformation of parameters

$$(\alpha_1, \alpha_2) \longmapsto (\beta_1(\alpha), \beta_2(\alpha)) \tag{4}$$

is regular at $(0,0)$, then the dynamics system of (A4) can be put in a simpler form. "This is the *non-degenerated* Chenciner bifurcation" [15] as it is studied in [12]. However, "the degenerate case when the change of parameters is not regular at $(0,0)$ is not any" [15] longer considered there. The purpose of the present article is to study an aspect of the degenerate Chenciner bifurcation. Since it is not possible to choose new coordinates β_1, β_2, the idea is to work only using the initial parameters (α_1, α_2).

3. Bifurcation Curves

Analysis of degenerated Chenciner bifurcation is performed in Appendix B and [4]. Since the smooth functions $\beta_{1,2}(\alpha)$ can be written as $\beta_1(\alpha) = a_{10}\alpha_1 + a_{01}\alpha_2 + \sum_{i+j\geq 2} a_{ij}\alpha_1^i\alpha_2^j$ and $\beta_2(\alpha) = b_{10}\alpha_1 + b_{01}\alpha_2 + \sum_{i+j\geq 2} b_{ij}\alpha_1^i\alpha_2^j$, the transformation (4) is not regular at $(0,0)$ and, thus, "the Chenciner bifurcation is *degenerate*" [4], if and only if $\frac{\partial\beta_1}{\partial\alpha_1}\frac{\partial\beta_2}{\partial\alpha_2}\Big|_{\alpha=0} - \frac{\partial\beta_1}{\partial\alpha_2}\frac{\partial\beta_2}{\partial\alpha_1}\Big|_{\alpha=0} = 0$, that is,

$$a_{10}b_{01} - a_{01}b_{10} = 0. \tag{5}$$

Remark 1. *In [4], we studied "the case when* (5) *is satisfied with non-zero terms" [15], that is* $a_{10}b_{01}a_{01}b_{10} \neq 0$. *In this work, we assume "that the linear part of* $\beta_1(\alpha)$ *nullifies, while* $\beta_2(\alpha)$ *has at least one linear term" [15]. Thus, "the degeneracy condition* (5) *remains valid while the functions* $\beta_{1,2}(\alpha)$ *become*

$$\beta_1(\alpha) = a\alpha_2^2 + b\alpha_1\alpha_2 + c\alpha_1^2 + \sum_{i+j=3}^{p_1} a_{ij}\alpha_1^i\alpha_2^j + O\Big(|\alpha|^{p_1+1}\Big) \tag{6}$$

and

$$\beta_2(\alpha) = p\alpha_1 + q\alpha_2 + \sum_{i+j=2}^{q_1} b_{ij}\alpha_1^i\alpha_2^j + O\Big(|\alpha|^{q_1+1}\Big) \tag{7}$$

for some $p_1 \geq 3$ and $q_1 \geq 2$, where $abcq \neq 0$" [15]. We denote by $a = a_{02}$, $b = a_{11}$ and $c = a_{20}$, respectively, $p = b_{10}$ and $q = b_{01}$.

Denote also by $B_{1,2}$ and C the following sets of points in $\mathbb{R}^2$

$$B_{1,2} = \Big\{(\alpha_1, \alpha_2) \in \mathbb{R}^2, \beta_{1,2}(\alpha) = 0, |\alpha| < \varepsilon\Big\} \tag{8}$$

and

$$C = \Big\{(\alpha_1, \alpha_2) \in \mathbb{R}^2, \Delta(\alpha) = 0, |\alpha| < \varepsilon\Big\} \tag{9}$$

for some $\varepsilon > 0$ that is sufficiently small. The expression $\Delta(\alpha) = \beta_2^2(\alpha) - 4\beta_1(\alpha)L_2(\alpha)$ becomes

$$\Delta(\alpha) = h\alpha_2^2(1 + O(|\alpha|)) + k\alpha_1\alpha_2(1 + O(|\alpha|)) + l\alpha_1^2(1 + O(|\alpha|)) \tag{10}$$

where $h = q^2 - 4aL_0$, $k = 2pq - 4bL_0$ and $l = p^2 - 4cL_0$. Assume $hkl \neq 0$. When $p = 0$ and $h \neq 0$, this condition is satisfied in general since $bcL_0 \neq 0$. Notice that

$$\Delta_2 = k^2 - 4hl = 16L_0^2\left(b^2 - 4ac\right) + 16L_0\left(ap^2 - bpq + cq^2\right). \tag{11}$$

In the following, we prove a theorem that was only stated in [15]. The structure of the set of points $B_{1,2}$ and C represents the main result in order to obtain the bifurcation diagrams; see also Remark 2. Recalling that $a = a_{02}, b = a_{11}, c = a_{20}, p = b_{10}, q = b_{01}, h = q^2 - 4aL_0, k = 2pq - 4bL_0, l = p^2 - 4cL_0$ and $\Delta_1 = b^2 - 4ac, \Delta_2 = k^2 - 4hl$, the following theorem is stated:

Theorem 1. *1. The set B_2 is a smooth curve of the form*

$$\alpha_2 = d_1\alpha_1 + d_2\alpha_1^2 + O\left(\alpha_1^3\right), \tag{12}$$

$d_1 = -\frac{p}{q}, d_2 = -\frac{1}{q}\left(b_{02} + d_1^2 b_{20} + d_1 b_{11}\right)$*, tangent to the line* $p\alpha_1 + q\alpha_2 = 0$.

2. *If $\Delta_1 = b^2 - 4ac > 0$, the set B_1 is a reunion of two smooth curves of the form*

$$\alpha_2 = e_{1,2}\alpha_1(1 + O(\alpha_1)), \tag{13}$$

where $e_1 = \frac{-b-\sqrt{\Delta_1}}{2a}$ *and* $e_2 = \frac{-b+\sqrt{\Delta_1}}{2a}$*. If $\Delta_1 < 0$, then $sign(\beta_1(\alpha)) = sign(a)$ for $|\alpha| < \varepsilon$.*

3. *If $\Delta_2 = k^2 - 4hl > 0$, the set C is a reunion of two smooth curves of the form*

$$\alpha_2 = m_{1,2}\alpha_1(1 + O(\alpha_1)), \tag{14}$$

where $m_1 = \frac{-k-\sqrt{\Delta_2}}{2h}$ *and* $m_2 = \frac{-k+\sqrt{\Delta_2}}{2h}$*. If $\Delta_2 < 0$, then $sign(\beta_1(\alpha)) = sign(h)$ for $|\alpha| < \varepsilon$.*

Proof. 1. Consider the function $\beta_2 : V_0 \subset \mathbb{R}^2 \to \mathbb{R}$ given by (7), where $V_0 = \left\{\alpha \in \mathbb{R}^2, |\alpha| < \varepsilon\right\}$ for $\varepsilon > 0$ sufficiently small. Then $\beta_2(0,0) = 0$ and $\frac{\partial \beta_2}{\partial \alpha_2}(0,0) = q \neq 0$. Thus, from the implicit function theorem (IFT) applied to β_2, there exists a unique curve $\alpha_2 = \alpha_2(\alpha_1)$, which satisfies $\beta_2(\alpha_1, \alpha_2(\alpha_1)) = 0$ for $|\alpha_1|$ that is small enough and can be written in the form (12). Notice that d_1 can be 0.

2. One further writes $\beta_1(\alpha)$ in the form

$\beta_1(\alpha) = a\alpha_2^2(1 + O(|\alpha|)) + b\alpha_1\alpha_2(1 + O(|\alpha|)) + c\alpha_1^2(1 + O(|\alpha|))$. Then $\beta_1(\alpha) = 0$ becomes

$$a\alpha_2^2 + b\alpha_1\alpha_2(1 + O(|\alpha|)) + c\alpha_1^2(1 + O(|\alpha|)) = 0. \tag{15}$$

Solving for α_2 in (15), one obtains $\alpha_2 = e_{1,2}\alpha_1(1 + O(|\alpha|))$, where $\Delta_1 = b^2 - 4ac$ and $e_{1,2} = \frac{-b\pm\sqrt{\Delta_1}}{2a}$, when $\Delta_1 > 0$. Denote further by

$$F(\alpha_1, \alpha_2) = \alpha_2 - e_{1,2}\alpha_1(1 + O(|\alpha|)),$$

where $F : V_0 \subset \mathbb{R}^2 \to \mathbb{R}$. Since $F(0,0) = 0$ and $\frac{\partial F}{\partial \alpha_2}(0,0) = 1 \neq 0$, the IFT yields the conclusion. When $\Delta_1 < 0$, it does not exist $\alpha \neq 0$ with $|\alpha| < \varepsilon$ such that $\beta_1(\alpha) = 0$. Thus, $\beta_1(\alpha)$ keeps a constant sign on V_0, which is given, for example, by $\beta_1(\alpha_2, 0) = a\alpha_2^2(1 + O(|\alpha|))$. This yields the conclusion. For 3, one proceeds similarly to 2. □

Theorem 1 was only stated in [15], but the proof is also given here because it is used in the present article. In this theorem, the structure of the sets of points $B_{1,2}$ and C is established, i.e., what kind of curves appear in the three situations from points 1,2 and 3; Theorem 1 provides the necessary theoretical basis for drawing bifurcation diagrams.

4. Bifurcation Diagrams

Assume $\beta_{1,2}(\alpha)$ and $\Delta(\alpha)$ have nonzero coefficients in their lowest terms, that is, $abcq \neq 0$ and $hkl \neq 0$. Thus, "the three bifurcation curves are well-defined when $|\alpha|$ is sufficiently small " [4]. B_2 is a unique curve, while each of B_1 and "C is a reunion of two curves" [15].

Remark 2. *Figure A1 presents generic phase portraits "corresponding to different regions of the bifurcation diagrams, including the phase portraits on the bifurcation curves defined by $\Delta(\alpha) = 0$," [4] respectively, $\beta_1(\alpha) = 0$. We summarize in Table A1 the correspondence between Δ, $\beta_{1,2}$, L_0 and "the generic phase portraits, respectively, different regions from bifurcation diagrams. When $\beta_{1,2}(\alpha) = 0$, then $\alpha = 0$" [4].*

The sign of a 2-nd degree polynomial of two real variables is discussed below.

Let us consider a polynomial

$$\Delta(\alpha_1, \alpha_2) = a\alpha_2^2 + b\alpha_1\alpha_2 + c\alpha_1^2, \quad a, b, c \in \mathbb{R}_*.$$

Considering its associated one-variable-polynomial $\delta(m) = am^2 + bm + c$, the signs of $\Delta(\alpha_1, \alpha_2)$ and $\delta(m)$ are the same, for all the pairs (α_1, α_2), which are solutions of the equation, $\alpha_2 = m\alpha_1$.

We use the convention that

$$m_1 = \begin{cases} \frac{-b-\sqrt{\Delta_1}}{2a}, & \text{if} \quad a > 0 \\ \frac{-b+\sqrt{\Delta_1}}{2a}, & \text{if} \quad a < 0 \end{cases}$$

and the corresponding formula for m_2.

The sign of $\Delta(\alpha_1, \alpha_2)$ is shown in Figure 1a for $a > 0$, and in Figure 1b for $a < 0$, where $(d_i) : \alpha_2 = m_i\alpha_1$, for $i = 1, 2$.

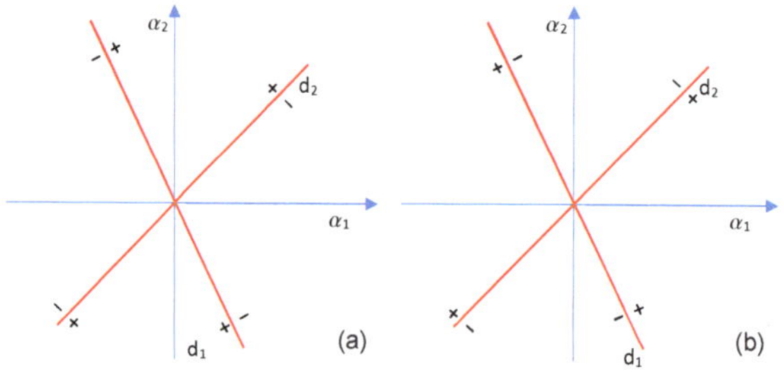

Figure 1. The sign of $\Delta(\alpha_1, \alpha_2)$ when **(a)** $a > 0$; **(b)** $a < 0$.

4.1. Bifurcation Diagrams When the First Discriminant Is Strictly Positive

Bifurcation diagrams for $\Delta_1 > 0$ are given in this subsection.

Firstly, we suppose that $\Delta_2 > 0$, and we consider the polynomials of $\mathbb{R}_*[T] : \beta_1(T) = aT^2 + bT + c, \quad \delta(T) = hT^2 + kT + l$, having the distinct real roots $e_1,\ e_2$, respectively, $m_1,\ m_2$.

There will be considered the following cases of root ordering:

I $\quad : e_1 < e_2 < m_1 < m_2$,
II $\quad : e_1 < m_1 < e_2 < m_2$,
III $\quad : e_1 < m_1 < m_2 < e_2$,
IV $\quad : m_1 < e_1 < e_2 < m_2$,
V $\quad : m_1 < e_1 < m_2 < e_2$.

There is only one more case, $m_1 < m_2 < e_1 < e_2$, which will not be taken into account, since it is a rotated case of I.

That ordering will be applied to the associated polynomials of $\beta_1(\alpha_1, \alpha_2)$, $\Delta(\alpha_1, \alpha_2)$, that is $\beta_1(T)$, $\delta(T)$; see Section 4.

Theorem 2. *The polynomials $\beta_1(T)$ and $\delta(T)$ have the following properties:*

1. $\delta(e_1) + \delta(e_2) > 0$
2. $\beta_1(m_1) \cdot \beta_1(m_2) \geq 0$

Proof. 1. $\delta(e_1) + \delta(e_2) = he_1^2 + ke_1 + l + he_2^2 + ke_2 + l$ by Viete relations

$\delta(e_1) + \delta(e_2) = \frac{1}{a^2}(b^2h - 2ach - abk + 2a^2 l)$, and by using the relations (11), $\delta(e_1) + \delta(e_2) = \frac{1}{a^2q^2}[2a^2(\frac{p}{q})^2 - 2ab\frac{p}{q} + b^2 - 2ac]$, which is positive since the polynomial $P(T) = 2a^2T^2 - 2abT + b^2 - 2ac > 0,\ (\forall)T \in \mathbb{R}$.

Indeed, the reduced discriminant of P is $\Delta' = -a^2\Delta_1 < 0$.

2. $\beta_1(m_1) \cdot \beta_1(m_2) = a^2m_1^2m_2^2 + abm_1m_2(m_1 + m_2) + ac(m_1^2 + m_2^2) + b^2m_1m_2 + bc(m_1 + m_2) + c^2$ by using Viete relations $\beta_1(m_1) \cdot \beta_1(m_2) = \frac{1}{h^2}(a^2l^2 - abkl + ack^2 - 2achl + b^2hl - bchk + c^2h^2)$.

Using (11), one concludes that

$$\beta_1(m_1)\beta_1(m_2) = \frac{1}{h^2q^4}\left[a^2\left(\frac{p}{q}\right)^4 - 2ab\left(\frac{p}{q}\right)^3 + (2ac + b^2)\left(\frac{p}{q}\right)^2 - 2bc\frac{p}{q} + c^2\right] =$$

$$= \frac{a^2}{q^4h^2}\left(\frac{p}{q} - \frac{b+\sqrt{\Delta_1}}{2a}\right)^2 \cdot \left(\frac{p}{q} - \frac{b-\sqrt{\Delta_1}}{2a}\right)^2. \quad \square$$

Corollary 1. *The cases II and V do not fulfill condition (2) of Theorem 2, and therefore they are eliminated.*

Considering the possible sub-cases of I, III, and IV, depending on the signs of a, h, one remarks that the numbers of sub-cases is halved by condition (1) of Theorem 2.

The left sub-cases are graphically represented in Figures 2–4.

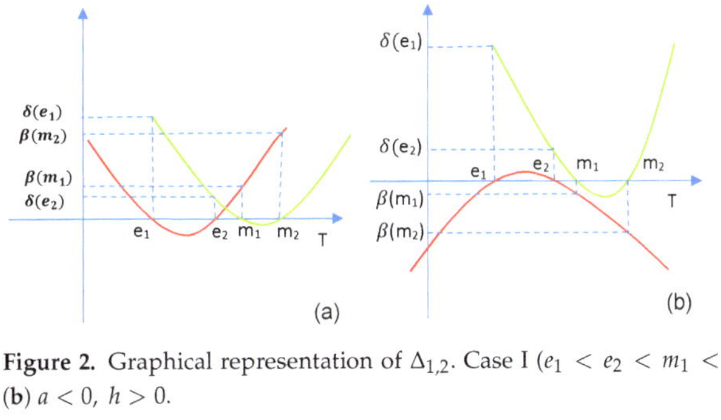

Figure 2. Graphical representation of $\Delta_{1,2}$. Case I ($e_1 < e_2 < m_1 < m_2$) when (**a**) $a > 0,\ h > 0$; (**b**) $a < 0,\ h > 0$.

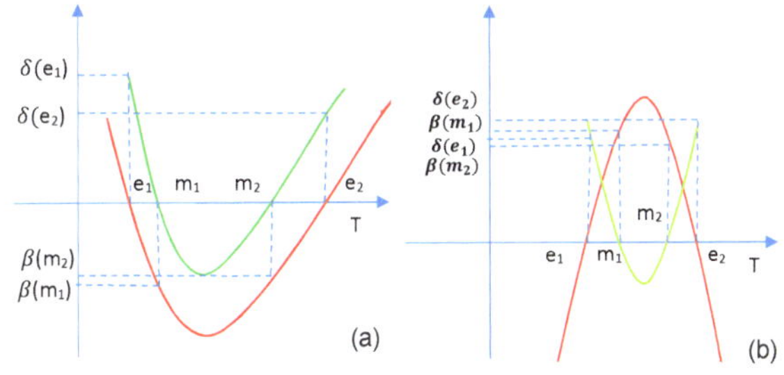

Figure 3. Graphical representation of $\Delta_{1,2}$. Case III ($e_1 < m_1 < m_2 < e_2$): (**a**) $a > 0,\ h > 0$; (**b**) $a < 0,\ h > 0$.

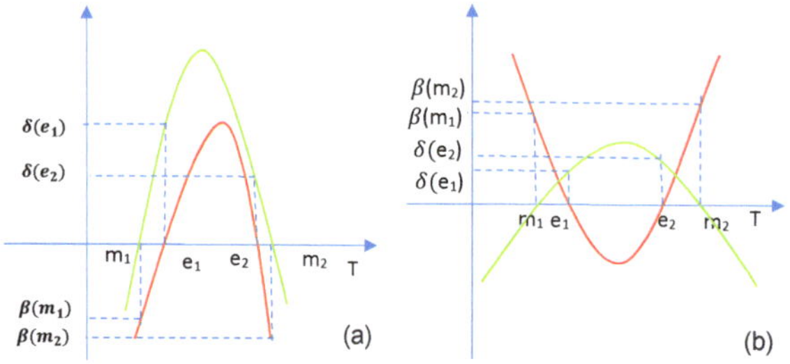

Figure 4. Graphical representation of $\Delta_{1,2}$. Case IV ($m_1 < e_1 < e_2 < m_2$): (**a**) $a < 0,\ h < 0$; (**b**) $a > 0,\ h < 0$.

Theoretically, for any of the previous sub-cases, one must consider two possibilities, depending on the sign of the L_0. However, the following theorem assigns a determined sign for any case.

Theorem 3. *The sign of $\beta(m_1) + \beta(m_2)$ equals that one of L_0.*

Proof. We calculate $\beta(m_1) + \beta(m_2) = a(m_1^2 + m_2^2) + b(m_1 + m_2) + 2c =$
$= \frac{1}{h^2}(ak^2 - ahl - bhk + 2ch^2)$.
By using the relation (11): $\beta(m_1) + \beta(m_2) = \frac{1}{4hL_0}(2h^2p^2 - 2hlq^2 - 2hkpq + k^2q^2)$.
Hence, the sign of $\beta(m_1) + \beta(m_2)$ is that of the expression in T:
$L_0(2h^2T^2 - 2hkT + k^2 - 2hl)$.
The reduced discriminant of the last parenthesis is $\Delta' = -4h^2\Delta_2 < 0$. Therefore, $sign(\beta(m_1) + \beta(m_2)) = sign(L_0)$. □

Corollary 2. *By the previous theorem, one may specify the sign of L_0 in the following cases:*

1. *I a, III b, IV b have $L_0 > 0$,*
2. *I b, III a, IV a have $L_0 < 0$.*

We may further reduce the sub-cases by the following theorems:

Theorem 4. *Denoting $M = ap^2 - bpq + cq^2$, $N = hp^2 - kpq + lq^2$, it results that $N = -4L_0M$.*

Proof. $N = hp^2 - kpq + lq^2$ equals, by (11), $(q^2 - 4aL_0)p^2 - (2pq - 4bL_0)pq + (p^2 - 4cL_0)q^2 = -4L_0M$. □

Corollary 3. *In cases I a, III b, and IV b, M and N have different signs, and for the rest of the sub-cases, they have the same sign.*

Theorem 5. *The sum $M + N$ has no definite sign.*

Proof. $M + N = (a + h)p^2 - (b + k)pq + (c + l)q^2$, and by (11), $M + N = (1 - 4L_0)(ap^2 - bpq + cq^2)$. L_0 is fixed, so $1 - 4L_0$ has a fixed sign. The second parenthesis has no fixed sign for all $p, q \in \mathbb{R}$, since $\Delta_1 > 0$. □

Corollary 4. *If M, N have the same sign, then $M + N$ has a definite sign for all p, $q \in \mathbb{R}$. If M, N do not have the same sign, then $M + N$ do not have a definite sign for all p, $q \in \mathbb{R}$. Hence, by Theorem 5 and Corollary 3, we may eliminate the cases I b, III a, and IV a. The remaining cases are I a, III b, and IV b.*

By Corollary 4, the cases for the graphical representation of the lines B_1, B_2, C are as follows:

I a1 : $a > 0,\ h > 0,\ L_0 > 0,\ M < 0,\ N > 0.$
I a2 : $a > 0,\ h > 0,\ L_0 > 0, M > 0,\ N < 0.$
III b1: $a < 0,\ h > 0,\ L_0 > 0,\ M < 0,\ N > 0.$
III b2: $a < 0,\ h > 0,\ L_0 > 0,\ M > 0,\ N < 0.$
IV b1: $a > 0,\ h < 0,\ L_0 > 0,\ M > 0,\ N < 0.$
IV b2: $a > 0,\ h < o,\ L_0 > 0,\ M < 0,\ N > 0.$

The bifurcation diagrams of cases I a1, III b1, and IV b2 are the same, represented in Figure 5a, and the bifurcation diagrams of cases I a2, III b2, and IV b1 are the same represented in Figure 5b.

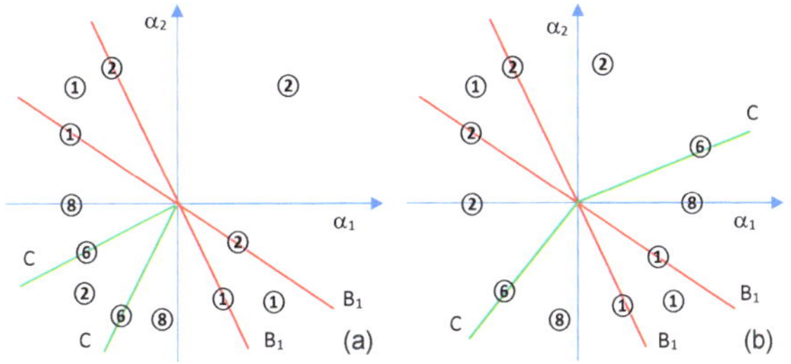

Figure 5. Bifurcation diagrams when $\Delta_1 > 0$ and (**a**) when case I a1, III b1, or IV b2 holds; (**b**) when case I a2, III b2 or IV b1 holds. The numbers represent the corresponding phase portraits.

Remark 3. *The case $\Delta_1 > 0,\ \Delta_2 < 0$ is solved by Theorem 1, Section 3 since if $\Delta_2 < 0$, then $sign\beta_1(\alpha) = sign(h)$ for all $|\alpha| < \varepsilon$. That is, the single straight line which remains is B_1, and this case is trivial.*

4.2. Bifurcation Diagrams When the First Discriminant Is Strictly Negative

Bifurcation diagrams for $\Delta_1 < 0$ are given in this subsection.

Remark 4. *If $\Delta_1 < 0$ and $aL_0 < 0$ then $\Delta(\alpha) > 0$. We will show that the single bifurcation curve is $\beta_2(\alpha) = 0$ in this case.*

We observe that $h = q^2 - 4aL_0$ and by $aL_0 < 0$ we have that $h > 0$. Taking into account Theorem 1, (3), we have that $sign(\beta_1(\alpha)) = sign(a)$ and $sign(\Delta(\alpha)) = sign(h)$.

There are more two trivial bifurcation diagrams which are not taken into account due to their triviality:

Remark 5. *(a) If $\Delta_1 < 0,\ a > 0$ and $L_0 < 0$ then the bifurcation diagrams contain only region 3.*
(b) If $\Delta_1 < 0,\ a < 0$ and $L_0 > 0$, then the bifurcation diagrams contain only region 1.

Proof of Remark 4. Using $aL_0 < 0$ results in $h = q^2 - 4aL_0 > 0$. Taking into account that $\Delta_1 < 0$, we obtain $\Delta_1 h < 0$. Because

$$\Delta_3 = b^2q^2 - 4aL_0\Delta_1 - 4acq^2 = q^2(b^2 - 4ac) - 4aL_0\Delta_1 = \Delta_1(q^2 - 4aL_0) = \Delta_1 h$$

it follows that $\Delta_3 < 0$. However,

$$\Delta_2 = k^2 - 4hl = 16L_0^2(b^2 - 4ac) + 16L_0(ap^2 - bpq + cq^2) =$$

$$= 16L_0[l_0(b^2 - 4ac) + ap^2 - bpq + cq^2] = 16L_0[L_0\Delta_1 + ap^2 - bpq + cq^2] =$$

$$= 16L_0(ap^2 - bpq + cq^2 + L_0\Delta_1)$$

and by $aL_0 < 0$, we have that $\Delta_2 < 0$.

Using Theorem 1, (3) and that $\Delta_2 < 0,\ h > 0$, we have $sign(\Delta(\alpha)) = sign(h) > 0$. □

Case 4.2.1 When $\Delta_1 < 0$, $aL_0 > 0$ and $h > 0$.

We see that $\Delta_3 = \Delta_1 h < 0$, and from $aL_0 > 0$, it follows that $\Delta_2 > 0$. Thus, the equation $\Delta(\alpha) = 0$ has two real distinct roots, $m_{1,2} = \frac{-k \pm \sqrt{\Delta_2}}{2h}$. We notice that $m_1 < m_2$.

We consider the expression $P = (m_1 + \frac{p}{q})(m_2 + \frac{p}{q})$.

By calculus, we obtain $P = \frac{1}{h}\left(\frac{p^2}{q^2}h - k\frac{p}{q} + l\right)$. We replace further in the previous expression h by $q^2 - 4aL_0$, k by $2pq - 4bL_0$, and l by $p^2 - 4cL_0$, and we have $P = -\frac{4L_0}{hq^2}(ap^2 - pqb + cq^2)$.

Now using that $aL_0 > 0,\ h > 0$ and $\Delta_1 < 0$, we will find that $P < 0$. In this situation we have only the following two systems:

$$\begin{cases} m_1 + \frac{p}{q} > 0 \\ m_2 + \frac{p}{q} < 0 \end{cases} \text{ or } \begin{cases} m_1 + \frac{p}{q} < 0 \\ m_2 + \frac{p}{q} > 0 \end{cases}.$$

By solving these systems, we find only the solution $m_1 < -\frac{p}{q} < m_2$ because $m_1 < m_2$.

In the previous case, two sub-cases arise:

Remark 6. *(a) If $a > 0,\ L_0 > 0$ and $-\frac{p}{q} \in (m_1, m_2)$, then the following bifurcation diagram appears in Figure 6a:*

(b) If $a < 0,\ L_0 < 0$ and $-\frac{p}{q} \in (m_1, m_2)$, then the following bifurcation diagram appears in Figure 6b:

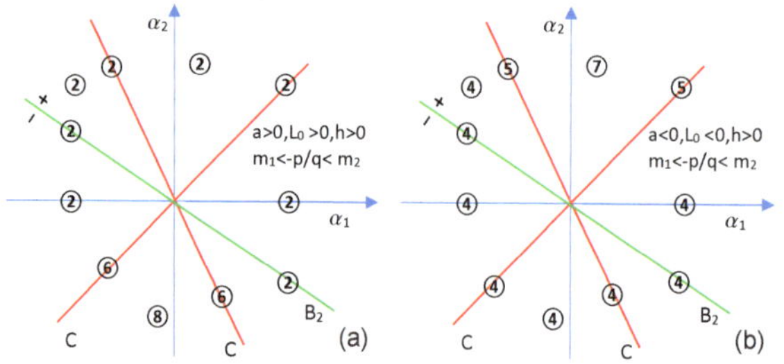

Figure 6. Bifurcation diagrams corresponding to the case: (**a**) $a > 0,\ L_0 > 0,\ -\frac{p}{q} \in (m_1, m_2)$; (**b**) $a < 0,\ L_0 < 0,\ -\frac{p}{q} \in (m_1, m_2)$.

Case 4.2.2 If $\Delta_1 < 0,\ aL_0 > 0,$ and $h < 0$, then $\Delta_3 > 0$ and the equation $\Delta_2 = 0$ has two real and distinct roots:

$$p_{1,2} = \frac{bq \pm \sqrt{\Delta_3}}{2a} = \frac{bq \pm \sqrt{h\Delta_1}}{2a}.$$

Taking into account that $h < 0,\ aL_0 > 0$ and $\Delta_1 < 0$. We obtain this time that $P > 0$. Now we compute also the sum, S, thus

$S = m_1 + m_2 + 2\frac{p}{q} = -\frac{k}{h} + \frac{2p}{q} = \frac{4L_0(bq - 2ap)}{qh}$.

From here, two cases arise.

When

$$\begin{cases} P > 0 \\ S > 0 \end{cases} \text{ or } \begin{cases} P > 0 \\ S < 0. \end{cases}$$

first sub-case, $\begin{cases} P > 0 \\ S > 0 \end{cases}$, is equivalent to

$L_0\left(b - 2a\frac{p}{q}\right) < 0$, and in this point we also have two possibilities:

(a) $\begin{cases} L_0 > 0 \\ b - 2a\frac{p}{q} < 0 \end{cases}$ or (b) $\begin{cases} L_0 < 0 \\ b - 2a\frac{p}{q} > 0. \end{cases}$

In case (a), from $L_0 > 0$ we obtain $a > 0$ and then $\frac{b}{2a} < \frac{p}{q}$.
In case (b), from $L_0 < 0$ we have $a < 0$ and further $\frac{b}{2a} < \frac{p}{q}$.
This means that

$$\begin{cases} m_1 + \frac{p}{q} > 0 \\ m_2 + \frac{p}{q} > 0 \end{cases}$$

and using that $m_2 < m_1$, we obtain $-\frac{p}{q} < m_2 < m_1$.

Now, the second sub-case becomes $L_0\left(b - 2a\frac{p}{q}\right) > 0$, and in this point, we also have two possibilities:

(a) $\begin{cases} L_0 > 0 \\ b - 2a\frac{p}{q} > 0 \end{cases}$ or (b) $\begin{cases} L_0 < 0 \\ b - 2a\frac{p}{q} < 0. \end{cases}$

In case (a), from $L_0 > 0$ we obtain $a > 0$ and then $\frac{b}{2a} > \frac{p}{q}$.
In case (b), from $L_0 < 0$ we have $a < 0$ and further $\frac{b}{2a} > \frac{p}{q}$.

Therefore, now we have instead, $\begin{cases} m_1 + \frac{p}{q} < 0 \\ m_2 + \frac{p}{q} < 0 \end{cases}$ and using that $m_2 < m_1$, we obtain $-\frac{p}{q} > m_1 > m_2$.

Here, it does not appear to be the case that $m_2 < -\frac{p}{q} < m_1$.

Case 4.2.2 I If $p \in (p_1, p_2)$, then $\Delta_2 < 0$ and from here, using that $h < 0$, we obtain $\Delta(\alpha) < 0$.

There are other two more trivial bifurcation diagrams which were not taken into account due to their triviality.

Remark 7. *(a)* *If $a > 0$, $L_0 > 0$ and $p \in (p_1, p_2)$ then the bifurcation diagram contain only the region 2 in the whole plane of coordinates, $\alpha_1 O \alpha_2$.*
Using that $sign(\beta(\alpha)) = sign(a) = +$, $sign(\Delta(\alpha)) = -$, $L_0 > 0$ and taking into account that $\beta(\alpha)$ can have any sign, we see in Table A1 that for this configuration of signs will appear only the region 2.

(b) *If $a < 0$, $L_0 < 0$, $p \in (p_1, p_2)$ then the bifurcation diagram will contain only region 4 in the whole plane of coordinate $\alpha_1 O \alpha_2$.*
By the same reason, using that $sign(\beta_1(\alpha)) = sign(a) = -$, $sign(\Delta(\alpha)) = -$, $L_0 < 0$ and taking into account that $\beta(\alpha)$ can have any sign, we see in Table A1 that for this configuration of signs, it will appear only in region 4.

4.2.2 II If $p \in (-\infty, p_1) \cup (p_2, \infty)$, then $\Delta_2 > 0$. However, $h < 0$ and therefore $\Delta(\alpha)$ has two distinct real roots m_1 and m_2.

$$sign(\Delta(\alpha)) = \begin{cases} +, & m \in (m_1, m_2); \\ -, & m \in (-\infty, m_1) \cup (m_2, \infty). \end{cases}$$

Because $-\frac{p}{q}$ is not between m_2 and m_1, we see that $sign(\Delta(\alpha)) = -$.

Remark 8. *In this case, the bifurcation diagrams are as in the case 4.2.1, Figure 6a,b, and only the conditions are different, not the dispersion of the regions.*

5. Numerical Simulations

In order to numerically illustrate "the existence of closed invariant curves" [4] in some of the studied cases, the Matlab software was used. In the particular case when the two-dimensional map is given in polar coordinates by

$$\rho_{n+1} = \rho_n + \beta_1(\alpha)\rho_n + \beta_2(\alpha)\rho_n^3 - \rho_n^5 \text{ and } \varphi_{n+1} = \varphi_n + \theta_0,$$

$|\alpha|$ being sufficiently small and $L_0 = -1,\ \theta_0 = 0.2$, we choose

$$\beta_1(\alpha) = \alpha_1^2 - \alpha_1\alpha_2 + \alpha_2^2,\ \beta_2(\alpha) = -\alpha_1 - 3\alpha_2,\ \alpha_1 = 0.1,\ \alpha_2 = -0.1.$$

Figure 7a,b shows the phase portraits 3 and 1 obtained when the conditions of Remark 5a,b are satisfied, respectively. In Figure 7a, the magenta orbit starting from $(\rho_1, \varphi_1) = (0.7, 0)$ approximates the invariant closed curve (invariant circle) from Theorem 1 [4], being obtained for $N = 400$ steps starting from the outside of the circle. The blue orbit starts from $(\rho_2, \varphi_2) = (0.06, 0)$ and it is also obtained for $N = 400$, which approximates the invariant circle starting from the inside and staying inside the circle. The red orbit starts in $(\rho_3, \varphi_3) = (0.4, 0)$, approximates the invariant circle from the inside, and is obtained for $N = 400$ steps. The green orbit starts from $(\rho_4, \varphi_4) = (0.59, 0)$, from the outside of the invariant circle and approximates it. This is how the phase 3 portrait appears here, the conditions in Remark 5a, Case 4.2 being satisfied ($\Delta_1 = -3 < 0,\ a = 1 > 0,\ L_0 = -1 < 0$). For the invariant circle, the radius is $\rho_n = \sqrt{y_2} = 0.547$; in our case, having $\beta_1(\alpha) = 0.03,\ \beta_2(\alpha) = 0{,}2 > 0,\ \Delta(\alpha) > 0,\ L_0 = -1$, we are also in the conditions of Theorem 1 (2) (b) [4]. We consider the particular case where the two-dimensional map is given in polar coordinates by

$$\rho_{n+1} = \rho_n + \beta_1(\alpha)\rho_n + \beta_2(\alpha)\rho_n^3 + \rho_n^5 \text{ and } \varphi_{n+1} = \varphi_n + \theta_0,$$

$|\alpha|$ being sufficiently small, $\theta_0 = 0.1,\ \alpha_1 = 0.1,\ \alpha_2 = -0.1$,

$$\beta_1(\alpha) = -\alpha_1^2 - \alpha_1\alpha_2 - \alpha_2^2,\ \beta_2(\alpha) = -\alpha_1 - 3\alpha_2,\ L_0 = 1,\ .$$

It is observed that $L_0 > 0,\ a = -1 < 0,\ \Delta_1 < 0$, so the conditions in Remark 5b are satisfied, and then the bifurcation diagram contains only phase portrait 1 (corresponding to region 1). We choose 3 orbits starting from the points $(\rho_1, \varphi_1) = (0.2035223739, 0)$, $(\rho_2, \varphi_2) = (0.181, 0)$ and $(\rho_3, \varphi_3) = (0.187, 0)$ of the colors magenta, red and blue, respectively, and which have $N = 850$, $N = 4000$ and $N = 4000$ steps, respectively; see Figure 7b. The magenta orbit moves away from the invariant circle and may escape to infinity, while the red orbit tends toward the origin $(0, 0)$, and the blue orbit likewise tends toward the origin. In addition, the radius of the invariant circle will be $\rho_n = \sqrt{y_1} = 0.2035$ (the conditions of Theorem 1 (2), (a) being satisfied) ($L_0 > 0,\ \beta_1(\alpha) = -0.01 < 0,\ \beta_2(\alpha) = 0.2 > 0$).

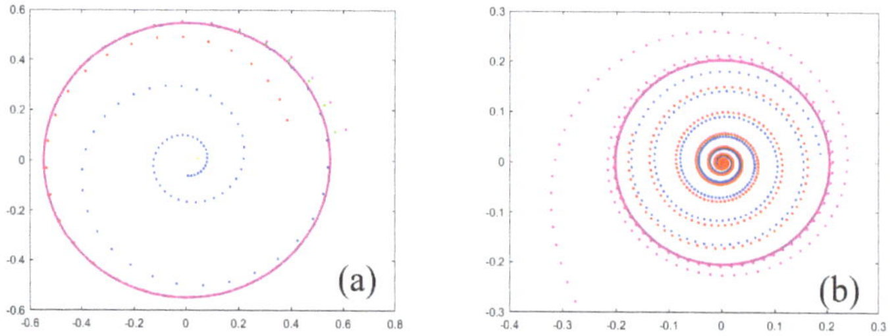

Figure 7. Numerical simulation for the map (A7) and (A8) with: (**a**) $\beta_1(\alpha) = \alpha_1^2 - \alpha_1\alpha_2 + \alpha_2^2$, $\beta_2(\alpha) = -\alpha_1 - 3\alpha_2$ and $L_0 = -1$, $\alpha_1 = 0.1$, $\alpha_2 = -0.1$; (**b**) $\beta_1(\alpha) = -\alpha_1^2 - \alpha_1\alpha_2 - \alpha_2^2$, $\beta_2(\alpha) = -\alpha_1 - 3\alpha_2$ and $L_0 = 1$, $\alpha_1 = 0.1$, $\alpha_2 = -0.1$.

For Figure 6a, we wanted to check on a particular case where the appearance of regions 2, 6, and 8 corresponds to phase portraits 2, 6, and 8. We consider the map given in polar coordinates by

$$\rho_{n+1} = \rho_n + \beta_1(\alpha)\rho_n + \beta_2(\alpha)\rho_n^3 + \rho_n^5 \text{ and } \varphi_{n+1} = \varphi_n + \theta_0,$$

$|\alpha|$ being small enough $L_0 = 1$, $\theta_0 = 0.1$. We took $\beta_1(\alpha) = \alpha_1^2 + \alpha_1\alpha_2 + \alpha_2^2$, $\beta_2(\alpha) = \alpha_1 + 3\alpha_2$, $\alpha_1 = 0.1$, $\alpha_2 = -0.1$ and we notice that the conditions are checked ($a > 0$, $L_0 > 0$, $h > 0$, $-p/q \in (m_1, m_2)$, $m_1 = -1$, $m_2 = 0.6$ and $\frac{\alpha_1}{\alpha_2} = m_1$) to be on one of the straight lines that form the curve (C) in Figure 6a, the point (α_1, α_2) being in quadrant IV, so it is region 6. For the orbits of blue, red, magenta and yellow colors from Figure 8a, starting at points $(\rho_1, \varphi_1) = (0.023, 0)$, $(\rho_2, \varphi_2) = (0.35, 0)$, $(\rho_3, \varphi_3) = (0.38, 0)$ and $(\rho_4, \varphi_4) = (0.078, 0)$, respectively, we consider $N = 4000$, $N = 141$, $N = 54$ and $N = 4000$ steps, respectively. It can be seen that the blue orbit approximates the invariant circle, the red orbit tends to infinity (if we increase the number of steps to $N = 142$ and $N = 58$ for the red and magenta curves, we obtain Figure 8b), the magenta orbit, like the red one, tends at infinity moving away from the invariant circle, and the yellow orbit, like the blue one, approximates (tends to) the invariant circle. This proves that we have phase portrait 6, so region 6 (as in the figure) is in accordance with the theoretical results. More than that, $\rho_n = \sqrt{y_1} = 0.3162$, is the radius of the invariant circle, and because $\Delta(\alpha) = 0$, the equation $y^2 - 0.2y + 0.01 = 0$ has a double root.

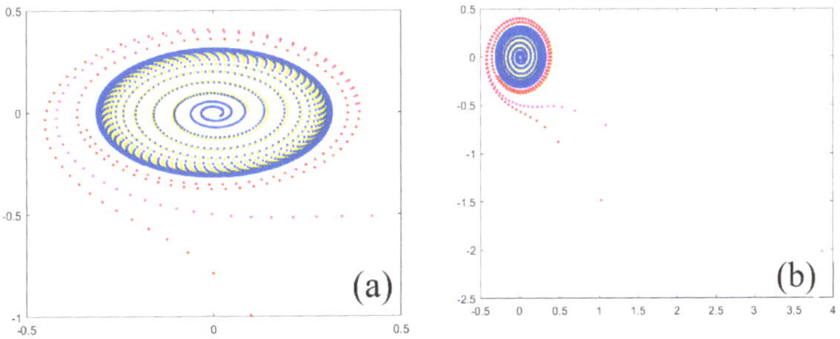

Figure 8. Numerical simulation for the map (A7) and (A8) with (**a**) $\beta_1(\alpha) = \alpha_1^2 + \alpha_1\alpha_2 + \alpha_2^2$, $\beta_2(\alpha) = \alpha_1 + 3\alpha_2$ and $L_0 = 1$, $\alpha_1 = 0.1$, $\alpha_2 = -0.1$; (**b**) like (**a**), but the step number is increased, and $N = 142$ and $N = 58$ for red orbit and magenta orbit, respectively.

However, with $\alpha_1 = 0.1$ and $\alpha_2 = 0.1$, the point (α_1, α_2) is in quadrant I, $\frac{\alpha_1}{\alpha_2}$ will be different from m_1 and m_2, and in Figure 6a, region 2 will appear. For the orbits of blue, green and brown colors starting from points $(\rho_1, \varphi_1) = (0.087, 0)$, $(\rho_2, \varphi_2) = (0.06, 0)$ and $(\rho_3, \varphi_3) = (0.04, 0)$, respectively, the numbers of steps are considered $N = 35$, $N = 45$ and $N = 60$, respectively. The 3 orbits tend to infinity corresponding to phase 2 portrait (region 2); see Figure 9a. If we take $N = 39$, $N = 50$ and $N = 64$ instead of the previous 3 values, we obtain Figure 9b, and it is observed that the last values increase a lot. Then, choosing $\alpha_1 = -0.01$, $\alpha_2 = -0.5$, the pair (α_1, α_2) is in quadrant III, and $\frac{\alpha_1}{\alpha_2}$ will be different from m_1 and m_2.

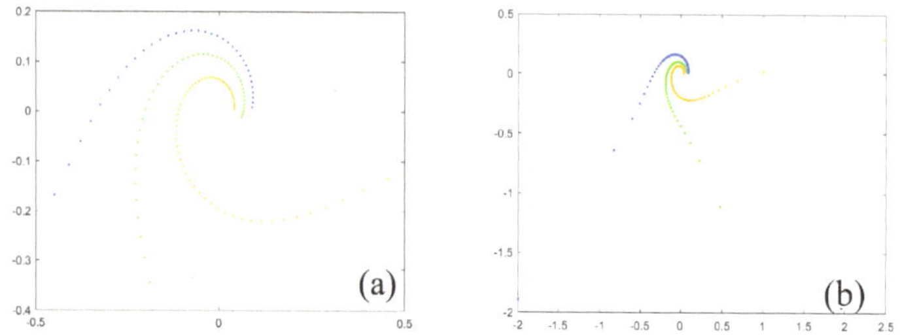

Figure 9. Numerical simulation for the map (A7) and (A8) with: (**a**) $\beta_1(\alpha) = \alpha_1^2 + \alpha_1\alpha_2 + \alpha_2^2$, $\beta_2(\alpha) = \alpha_1 + 3\alpha_2$ and $L_0 = 1$, $\alpha_1 = 0.1$, $\alpha_2 = 0.1$; (**b**) like to (**a**) but the step number is increased to $N = 39$, $N = 50$ and $N = 64$, respectively, for blue orbit, green orbit and brown orbit.

The six orbits start in Figure 10 from din $(\rho_1, \varphi_1) = (0.087, 0)$, $(\rho_2, \varphi_2) = (0.78, 0)$, $(\rho_3, \varphi_3) = (0.35, 0)$, $(\rho_4, \varphi_4) = (1.14724966464545445, 0)$, $(\rho_5, \varphi_5) = (0.023, 0)$ and $(\rho_6, \varphi_6) = (1.127, 0)$ having the colors yellow, magenta, red, green, blue and cherry, respectively, with steps $N = 400$, $N = 54$, $N = 141$, $N = 15$, $N = 400$ and $N = 400$, respectively. The cyan-colored orbit is the outer invariant circle. The cherry and magenta orbits approximate the inner invariant circle from the outside, and the blue, yellow and red orbits approximate the inner invariant circle from the inside. The green orbit moves away from the outer invariant circle tending to infinity, thus observing that the orbits move away from the outer circle and tend toward the inner invariant circle. We thus have the portrait of phase 8, region 8. The radii of the two invariant circles are known from Theorem 1 [4].

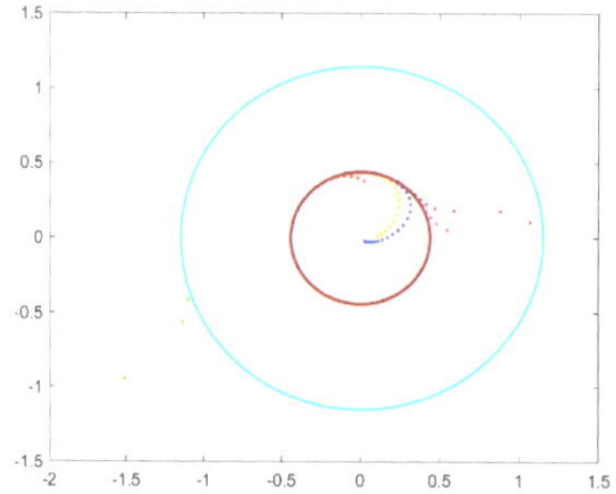

Figure 10. Numerical simulation for the map (A7) and (A8) with $\beta_1(\alpha) = \alpha_1^2 + \alpha_1\alpha_2 + \alpha_2^2$, $\beta_2(\alpha) = \alpha_1 + 3\alpha_2$ and $L_0 = 1$, $\alpha_1 = -0.01$, $\alpha_2 = -0.5$.

6. Discussions and Conclusions

6.1. Discussions

In this study, the truncated normal form of the Chenciner bifurcation was analyzed in a degeneracy case, where the degeneracy condition is given by $a_{10}b_{01} - a_{01}b_{10} = 0$ and $a_{10} = a_{01} = 0$ or $b_{10} = b_{01} = 0$, as an answer to the problem open in [4,35].

In this article, all eight regions corresponding to the eight phase portraits (see Figure A1) appear in the bifurcation diagrams, unlike [15] or [10], where all of these are not present. In [15], only regions 1–4 appear in the bifurcation diagrams. If in a previous study [15] only two alternating regions appeared, in this article, more alternating regions (4 and 3 regions, respectively) appear in the bifurcation diagrams. This situation indicates a more complex structure of bifurcation diagrams. By modifying the structure of the sets of points $B_{1,2}$ and C, concurrent lines at the origin are obtained in the bifurcation diagrams, as in some recent studies [10,15], and different from other previous works [4,35]. When $\Delta_1 > 0$ (Section 4.1) the analysis of the six cases obtained leads to the first two diagrams in Figure 5a,b. When $\Delta_1 < 0$ (Section 4.2) Figure 6 presents the last two nontrivial bifurcation diagrams. However, in this last case, there are additionally four trivial situations when the bifurcation diagrams contain only one region in the whole plane (see Remarks 5a,b and 7a,b) and therefore do not require the creation of an additional representation.

The obtained theoretical results could be verified by means of the Matlab program, which allowed the realization of several representative simulations.

The Chenciner bifurcation in this case acts similar to an "organizing center" of dynamic behavior, generating "global dynamic phenomena such as the creation or disappearance of stable limit cycles" [19]. Near a Chenciner bifurcation point, "there is an open region in the parameter space where a stable equilibrium state and a stable limit cycle coexist" [19].

6.2. Conclusions

The advantage of using Chenciner degenerate bifurcation for modeling economics volatility versus chaotic behavior is that the transition to chaos amplifies itself and requires several iterations, but the volatility may be transitory. The case studied in this article has the advantage that it leads to the reduction of the large number of bifurcation diagrams that appeared in [4,10]. Thus, the hypothesis that was made is confirmed: if the degeneracy is not so large, we have a small number of bifurcation diagrams. The limitations of the present procedure is that it is applicable to degenerated cases, which seldom represent cases that have importance in special situations. Moreover, the more restrictive method leading to a new parameter change as in [35] is not necessary for this study. The results obtained for "the truncated normal form give an approximate description of the complicated bifurcation structure, near a generic Chenciner bifurcation" [4]. As in the case of the Neumark–Sacker bifurcation and in the case of the degenerate Chenciner bifurcations, it is observed that the normal form thus obtained captures "only the appearance of a closed invariant curve but does not describe the structure of the orbit on this curve" [12]. The article completes the studies started in another reference material on the degenerate Chenciner bifurcation [4] and not addressed in other cases of degeneracy [10,15]. In the mentioned articles, the functions β_1 and β_2 do not contain any terms of the first degree [15], one of the two functions does not contain terms of the first or second degree, and the other may or may not contain terms of the first degree [10].

A number of four different bifurcation diagrams were obtained instead of "two as in the non-degenerate Chenciner case" [15]. The first two bifurcation diagrams were obtained in Case 4.1 when $\Delta_1 > 0$, and the last two bifurcation diagrams were generated in Case 4.2 when $\Delta_1 < 0$. Several subcases that appeared (discussed) in Case 4.1 could be removed.

So, the conclusion is that eight different bifurcation diagrams were recorded, four of them being trivial.

In the case studied now, the linear part of β_1 cancels, and β_2 has at least one linear term. Compared to the mentioned articles [4,10], much fewer bifurcation diagrams appear. Thus, eight bifurcations diagrams result (if we also consider the four trivial ones from Remark 5 and Remark 7), and only four non-trivial ones are recorded, which are different from those previously highlighted [15].

The obtained results "can be used in bifurcation theory" [15] as a field of dynamic systems, but could also be exploited in other fields of activity, where the evolution of some processes and phenomena is in the form of discrete dynamic systems (economy, biology, ecology, medicine and computers).

Author Contributions: Conceptualization, S.L. and L.C.; methodology, S.L. and L.C.; formal analysis, S.L., L.C., and E.G.; investigation, S.L., L.C., and E.G.; resources, L.C. and E.G.; data curation, L.C. and E.G.; writing—original draft preparation, L.C.; writing—review and editing, L.C.; visualization, S.L., L.C., and E.G.; supervision, S.L., L.C., and E.G.; project administration, S.L., L.C., and E.G.; funding acquisition, L.C. and E.G. All authors have read and agreed to the published version of the manuscript.

Funding: This research received no external funding.

Institutional Review Board Statement: Not applicable.

Informed Consent Statement: Not applicable.

Data Availability Statement: Not applicable.

Acknowledgments: This research was partially supported by the Horizon 2020-2017-RISE-777911 project.

Conflicts of Interest: The authors declare no conflict of interest.

Appendix A. Chenciner Bifurcation

The discrete-time system 2 may be written in complex coordinates as

$$z \to z.\delta(\alpha) + g(z, \overline{z}, \alpha), \tag{A1}$$

where δ and g are smooth functions of their arguments, given by

$$\delta(\alpha) = e^{i\theta(\alpha)}.r(\alpha) \quad \text{and} \qquad g(z, \overline{z}, \alpha) = \sum_{i+l\geq 2} \frac{g_{il}(\alpha) z^i \overline{z}^l}{i!l!},$$

also $r(0) = 1,\ \theta(0) = \theta_0$ and g_{il} are smooth functions with complex values.

Following the same steps as in [12], it will turn (A1) into

$$w \to w.\left(e^{i\theta(\alpha)}.r(\alpha) + w\overline{w}.a_1(\alpha) + w^2\overline{w}^2.a_2(\alpha)\right) + O(|w|^6) =$$

$$= w.e^{i\theta(\alpha)}\left(r(\alpha) + w\overline{w}.b_1(\alpha) + w^2\overline{w}^2.b_2(\alpha)\right) + O(|w|^6), \tag{A2}$$

where $b_j(\alpha) = e^{-i\theta(\alpha)}.a_j(\alpha),\ j = 1, 2.$

It should be noted that the following smoothly reversible complex coordinate change was used:

$$z = w + \sum_{2\leq i+l\leq 5} \frac{h_{il}(\alpha) w^i \overline{w}^l}{i!l!}, \tag{A3}$$

with $h_{21}(\alpha) = h_{32}(\alpha) = 0$.

If $\beta_1(\alpha),\ \beta_2(\alpha)$ denote $r(\alpha) - 1$ and $Re(b_1(\alpha))$, respectively, and polar coordinates are used, then relation (A2) will be

$$\begin{cases} \rho_{n+1} = & (1 + \beta_1(\alpha) + \beta_2(\alpha)\rho_n^2 + L_2(\alpha)\rho_n^4)\rho_n + \rho_n O(\rho_n^6) \\ \varphi_{n+1} = & \varphi_n + \theta(\alpha) + \left(\frac{Im(b_1(\alpha))}{1+\beta_1(\alpha)} + O(\rho_n, \alpha)\right)\rho_n^2 \end{cases}, \tag{A4}$$

It is called **Chenciner bifurcation**, a state of system (A4) that satisfies the conditions $r(0) = 1,\ Re(b_1(0)) = 0$ and $L_2(0) \neq 0$.

Out of $\beta_1(0) = 0$, it results that

$$L_2(0) = \frac{Im^2(b_1(0)) + 2.Re(b_2(0))}{2}.$$

When the mapping

$$(\alpha_1, \alpha_2) \to (\beta_1(\alpha), \beta_2(\alpha)) \tag{A5}$$

is regular in $(0, 0)$, then the functions β_1 and β_2 become the new parameters of the system (A4). This is the **non-degenerate Chenciner bifurcation**.

It is known from [12], relation (13) page 4, that

$$\begin{cases} \beta_1(\alpha) = \sum_{i+l=1}^{p} a_{il}\alpha_1^i\alpha_2^l + O(|\alpha|^{p+1}) \\ \beta_2(\alpha) = \sum_{i+l=1}^{q} b_{il}\alpha_1^i\alpha_2^l + O(|\alpha|^{q+1}) \end{cases}, \tag{A6}$$

for $p \geq 1,\ q \geq 1$ and $a_{10} = \frac{\partial\beta_1}{\partial\alpha_1}|_{\alpha=0},\ a_{01} = \frac{\partial\beta_1}{\partial\alpha_2}|_{\alpha=0},\ b_{10} = \frac{\partial\beta_2}{\partial\alpha_1}|_{\alpha=0},\ b_{01} = \frac{\partial\beta_2}{\partial\alpha_2}|_{\alpha=0}$ and so on.

If the transformation (A5) is not regular in $(0,0)$, the Chenciner bifurcation is **degenerate**, i.e., if and only if $a_{10}.b_{01} - a_{01}.b_{10} = 0$.

Next, the higher-order terms of the ρ-map (of the application) (A4) will be eliminated, obtaining the truncated form

$$\rho_{n+1} = (1 + \beta_1(\alpha) + \rho_n^2 \beta_2(\alpha) + \rho_n^4.L_2(\alpha)).\rho_n. \tag{A7}$$

The φ-map application of system (A4) describes a rotation of an angle depending on α and ρ, and can be approximated by its truncated form

$$\varphi_{n+1} = \varphi_n + \theta(\alpha). \tag{A8}$$

It will be assumed that $0 < \theta(0) = \theta_0 < \pi$, and the system analyzed in this paper is (A7) and (A8). This system is also called the **truncated normal form of the system** (A2).

Appendix B. Degenerate Chenciner Bifurcation

Equation (A7) defines a one-dimensional dynamic system, which is independent of equation (A8)(φ-map) and will be studied separately. The system (A7) (ρ-map) has the fixed point $\rho = 0$, for any α which corresponds to the fixed point $O(0,0)$ in the normal forms (A7) and (A8). Each positive and non-zero fixed point of the ρ-map (8) corresponds to a closed invariant curve in the system, (A7) and (A8). We specify that we denote by $O(|\alpha|^n)$, $n \geq 1$ a series with real coefficients c_{ij} having the form, $\sum_{i+j \geq n} c_{ij} \alpha_1^i \alpha_2^j$. It can be easily shown that $sign(L_2(\alpha)) = sign(L_0)$ for $|\alpha| = \sqrt{\alpha_1^2 + \alpha_2^2}$ that is chosen to be small enough, bearing in mind that $L_2(\alpha)$ can be chosen as $L_2(\alpha) = (1 + O(|\alpha|)).L_0$ and $L_0 \neq 0$.

The following theorem describes the stability of the point O for $|\alpha|$ that is small enough, and it was demonstrated in [4].

Theorem A1. *"The fixed point O is(linearly) stable if $\beta_1(\alpha) < 0$ and unstable if $\beta_1(\alpha) > 0$,for any value of α with $|\alpha|$ small enough. On the bifurcation curve $\beta_1(\alpha) = 0$, O is (nonlinear) stable if $\beta_2(\alpha) < 0$ and unstable if $\beta_2(\alpha) > 0$, when $|\alpha|$ is small enough. When $\alpha = 0$, O is (non-linearly) stable if $L_0 < 0$ and unstable if $L_0 > 0$." [4]*

The fixed points of (A7) are the solutions of the equation $L_2(\alpha).y^2 + \beta_2(\alpha).y + \beta_1(\alpha) = 0$ where the variable $y = \rho_n^2$. The discriminant of the equation will be denoted by $\Delta(\alpha) = \beta_2^2(\alpha) - 4.L_2(\alpha).\beta_1(\alpha)$, and the roots will be $y_1 = \frac{\sqrt{\Delta(\alpha)} - \beta_2(\alpha)}{2.L_2(\alpha)}$ and $y_2 = -\frac{\sqrt{\Delta(\alpha)} + \beta_2(\alpha)}{2.L_2(\alpha)}$ "when they exist as real number" [4]. The following theorem studies the existence of closed invariant curves in the truncated normal form (A7) and (A8) and is given in [4].

Theorem A2.
1. *" When $\Delta(\alpha) < 0$ for all $|\alpha|$ sufficiently small, the system* (A7) *and* (A8) *has no invariant circles.*
2. *When $\Delta(\alpha) > 0$ for all $|\alpha|$ sufficiently small, the system* (A7) *and* (A8) *has*
   (a) One invariant unstable circle $\rho_n = \sqrt{y_1}$ if $L_0 > 0$ and $\beta_1(\alpha) < 0$;
   (b) One invariant stable circle $\rho_n = \sqrt{y_2}$ if $L_0 < 0$ and $\beta_1(\alpha) > 0$;
   (c) Two invariant circles, $\rho_n = \sqrt{y_1}$ unstable and $\rho_n = \sqrt{y_2}$ stable, if $L_0 > 0$, $\beta_1(\alpha) > 0$, $\beta_2(\alpha) < 0$ or $L_0 < 0$, $\beta_1(\alpha) < 0$, $\beta_2(\alpha) > 0$; in addition, $y_1 < y_2$ if $L_0 < 0$ and $y_2 < y_1$ if $L_0 > 0$;
   (d) No invariant circles if $L_0 > 0$, $\beta_1(\alpha) > 0$, $\beta_2(\alpha) > 0$ or $L_0 < 0$, $\beta_1(\alpha) < 0$, $\beta_2(\alpha) < 0$.
3. *On the bifurcation curve $\Delta(\alpha) = 0$, the system* (A7) *and* (A8) *has one invariant unstable circle $\rho_n = \sqrt{y_1}$ for all $L_0 \neq 0$. Moreover, if $L_0 < 0$, the invariant circle is stable from the exterior and unstable from the interior, and vice versa if $L_0 > 0$.*

4. *When* $\beta_1(\alpha) = 0$*, the system* (A7) *and* (A8) *has one invariant circle* $\rho_n = \sqrt{-\frac{\beta_2(\alpha)}{L_0}}$ *whenever* $L_0\beta_2(\alpha) < 0$*. It is stable if* $L_0 < 0$ *and* $\beta_2(\alpha) > 0$*, respectively, unstable if* $L_0 > 0$ *and* $\beta_2(\alpha) < 0$ *" [4,15,35].*

Table A1. Correspondence between Δ, $\beta_{1,2}$, L_0 and the generic phase portraits [4].

$\Delta(\alpha)$	L_0	$\beta_1(\alpha)$	$\beta_2(\alpha)$	Region
+	+	+	+	2
+	−	−	−	4
+	+	−	±, 0	1
+	−	+	±, 0	3
+	−	−	+	7
+	+	+	−	8
−	+	+	±, 0	2
−	−	−	±, 0	4
0	+	+	+	2
0	−	−	−	4
0	−	−	+	5
0	+	+	−	6
0	+	0	0	2
0	−	0	0	4
+	−	0	+	3
+	−	0	−	4
+	+	0	−	1
+	+	0	+	2

Corresponding to the studies we previously carried out [4,15], the following phase portraits are highlighted below. In this case, the phase portraits for the curves of bifurcation when $\Delta(\alpha) = 0$ are shown in Figure A1.

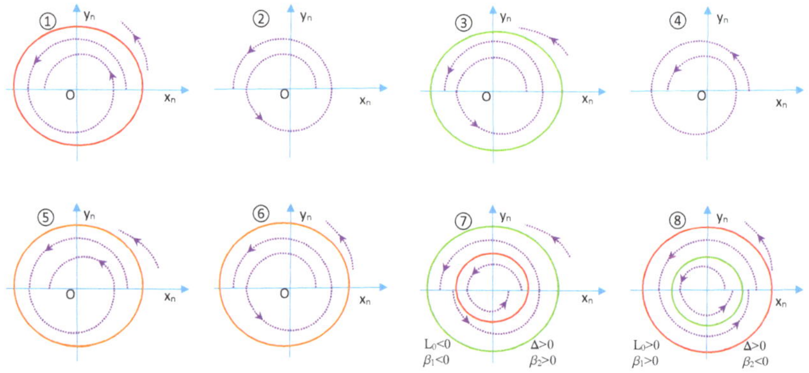

Figure A1. Generic portraits phase when $\theta_0 > 0$. The numbers represent the phase portraits [4].

The red invariant circles are unstable, the green invariant circle are stable, and the blue curves represent arbitrary orbits in Figure A1.

References

1. Muller, J.; Kuttler, C. *Methods and Models in Mathematical Biology*; Springer: Berlin/Heidelberg, Germany, 2015.
2. Moza, G.; Grecu, E.; Tirtirau, L. Analysis of a nonlinear financial model. *Carpathian J. Math.* **2022**, *38*, 477–487. [CrossRef]
3. Li, X.; Hou, X.R.; Yang, J.; Luo, M. Stability and Stabilization of 2D Linear Discrete Systems with Fractional Orders Based on the Discrimination System of Polynomials. *Mathematics* **2022**, *10*, 1862.
4. Tigan, G.; Lugojan, S.; Ciurdariu, L. Analysis of Degenerate Chenciner Bifurcation. *Int. J. Bifurcation Chaos* **2020**, *30*, 2050245. [CrossRef]

5. Biswas, M.; Bairagi, N. On the dynamic consistency of a two-species competitive discrete system with toxicity. *J. Comput. Appl. Math.* **2020**, *363*, 145155. [CrossRef]
6. Chenciner, A. Bifurcations de points fixes elliptiques. II. Orbites periodiques et ensembles de Cantor invariants. *Invent. Math.* **1985**, *80*, 81–106. [CrossRef]
7. Floudas, C.A.; Lin, X. Continuous-time versus diccrete-time approaches for scheduling of chemical processes: A review. *Comput. Chem. Eng.* **2004**, *28*, 2109–2129. [CrossRef]
8. Khanin, K.; Kocic, S. Hausdorff dimension of invariant measure of circle diffeomorphisms with a break point. *Ergod. Th. Dyn. Syst.* **2019**, *39*, 1331–1339. [CrossRef]
9. Llibre, J.; Sirvent, V.F. On Lefschetz periodic point free self-maps. *J. Fixed Point Th. Appl.* **2018**, *20*, 38. [CrossRef]
10. Lugojan, S.; Ciurdariu, L.; Grecu, E. Chenciner Bifurcation Presenting a Further Degree of Degeneration. *Mathematics* **2022**, *10*, 1603.
11. Wiggins, S. *Introduction to Applied Nonlinear Dynamical Systems and Chaos*; Springer Science and Business Media: Berlin/Heidelberg, Germany, 2003; Volume 2.
12. Kuznetsov, Y.A. *Elements of Applied Bifurcation Theory*, 2rd ed.; Springer: New York, NY, USA, 1998.
13. Chenciner, A.; Gasull, A.; Llibre, J. Une description complete du portrait de phase d'un modele d'elimination resonante. *C. R. Acad. Sci. Paris Ser. I Math.* **1987**, *305*, 623–626.
14. Chenciner, A. Bifurcations de points fixes elliptiques. III. Orbites periodiques de petites periodes. *Inst. Hautes Etudes Sci. Publ. Math.* **1988**, *66*, 5–91.
15. Lugojan S, Ciurdariu, L.; Grecu, E. New Elements of Analysis of a Degenerate Chenciner Bifurcation. *Symmetry* **2022**, *14*, 77. [CrossRef]
16. Shilnikov, L.P.; Shilnikov, A.L.; Turaev, D.V.; Chua, L.O. *Methods of Qualitative Theory in Non-linear Dynamics*; Part 2; World Scientific: Singapore, 2001.
17. Alidousti, J.; Eskandari, Z.; Avazzadeh, Z. Generic and symmetric bifurcations analysis of a three dimensional economic model. *Chaos Solitons Fractals* **2020**, *140*, 110251.
18. Hajnova, V.; Pribylova, L. Two-parameter bifurcations in LPA model. *J. Math. Biol.* **2017**, *75*, 1235–1251. [PubMed]
19. Gaunersdorfer, A.; Hommes, C.H.; Wagener, F.O.O. Bifurcation routes to volatility clustering under evolutionary learning. *J. Econ. Behav. Organ.* **2008**, *67*, 27–47.
20. Beso, E.; Kalabusic, S.; Pilav, E. Stability of a certain class of a host-parasitoid models with a spatial refuge effect. *J. Biol. Dyn.* **2020**, *14*, 1–31. [CrossRef]
21. Revel, G.; Alonso, D.M.; Moiola, J.L. A Degenerate 2:3 Resonant Hopf-Hopf Bifurcations as Organizing Center of the Dynamics: Numerical Semiglobal Results. *Siam J. Appl. Dyn. Syst.* **2015**, *14*, 1130–1164.
22. Alidousti, J.; Eskandari, Z.; Asadipour, M. Codimension two bifurcations of discrete Bonhoeffer-van der Pool oscillator model. *Soft Comput.* **2021**, *25*, 5261–5276. [CrossRef]
23. Pandey, V.; Singh, S. Bifurcations emerging from a double Hopf bifurcation for a BWR. *Prog. Nucl. Energy* **2019**, *117*, 103049. [CrossRef]
24. Gyllenberg, M.; Jiang, J.F.; Yan, P. On the dynamics of multi-species Ricker models admitting a carrying simplex. *J. Differ. Equ. Appl.* **2019**, *25*, 1489–1530. [CrossRef]
25. Gyllenberg, M.; Jiang, J.F.; Yan, P. On the classification of generalized competitive Atkinson-Allen models via the dynamics on the boundary of the carrying simplex. *Discret. Contin. Dyn. Syst.* **2018**, *38*, 615–650. [CrossRef]
26. Chow, S.-N., Li, C.; Wang, D. *Normal Forms and Bifurcations of Planar Vector Fields*; Cambridge University Press: Cambridge, UK, 1994.
27. Gils, S.A.; Horozov, E. Uniqueness of Limit Cycles in Planar Vector Fields Which Leave the Axes Invariant. *Contemp. Math.* **1986**, *56*, 117–129.
28. Guckenheimer, J. Multiple bifurcation problems of codimension two. *SIAM J. Math. Anal.* **1984**, *15*, 1–49. [CrossRef]
29. Lorenz, H.M. Nonlinear dynamical economics and chaotic motion. In *Lecture Notes in Economics and Mathematical System*; Springer: Berlin/Heidelberg, Germany, 1989; Volume 334.
30. Silva, V.B.; Vieira, J.P.; Leonel, E.D. A new application of the normal form description to a N dimensional dynamical systems attending the conditions of a Hopf bifurcation. *J. Vib. Syst. Dyn.* **2018**, *2*, 249–256. [CrossRef]
31. Wu, D.; Vorobev, P.; Turitsyn, K. Modulated Oscillations of Synchronous Machine Nonlinear Dynamics With Saturation. *IEEE Trans. Power Syst.* **2020**, *35*, 2915–2925. [CrossRef]
32. Deng, S.F. Bifurcations of a Bouncing Ball Dynamical System. *Int. J. Bifurcation Chaos* **2019**, *29*, 1950191. [CrossRef]
33. Zhong, J.Y.; Deng, S.F. Two codimension-two bifurcations of a second-order difference equation from macroeconomics. *Discret. Contin. Dyn.-Syst.-Ser.* **2018**, *23*, 1581–1600. [CrossRef]
34. Barros, M.F.; Ortega, F. An optimal equilibrium for a reformulated Samuelson economic discrete time system. *Econ. Struct.* **2019**, *8*, 29. [CrossRef]
35. Tigan, G.; Brandibur, O.; Kokovics, E.; Vesa, L.F. Degenerate Chenciner Bifurcation Revisited. *Int. J. Bifurcation Chaos* **2021**, *10*, 2150160. [CrossRef]

mathematics

Article

Stability and Stabilization of 2D Linear Discrete Systems with Fractional Orders Based on the Discrimination System of Polynomials

Xiaoxue Li [1], Xiaorong Hou [1,*], Jing Yang [1] and Min Luo [2]

1 School of Automation Engineering, University of Electronic Science and Technology of China, Xiyuan Ave, Chengdu 611731, China; lixx@uestc.edu.cn (X.L.); j.yang@uestc.edu.cn (J.Y.)
2 School of Electrical Engineering and Information, Southwest Petroleum University, Xindu Road, Chengdu 611731, China; luomin@swpu.edu.cn
* Correspondence: houxr@uestc.edu.cn

Abstract: This paper considers the stability and stabilization of two-dimensional (2D) fractional-order systems described by state-space model based on the discrimination system of polynomials. Necessary and sufficient conditions of stability and stabilization are established. We change the criterion for checking the stability of linear discrete-time 2D fractional-order systems into an easy checking criterion whether some polynomials are positive. We use the discrimination system of polynomials to check the new conditions. For the stabilization problem, we get a stable gain matrix region. The unstable system with the gain parameters of the stable gain matrix region is stable. We give the method of stability analysis and stabilization for the general 2D fractional-order system. An example shows the validity of the proposed stability and stabilization methods.

Keywords: 2D fractional-order systems; stability; stabilization; the discrimination system for polynomials

MSC: 93D09; 93D15

Citation: Li, X.; Hou, X.; Yang, J.; Luo, M. Stability and Stabilization of 2D Linear Discrete Systems with Fractional Orders Based on the Discrimination System of Polynomials. *Mathematics* **2022**, *10*, 1862. https://doi.org/10.3390/math10111862

Academic Editors: Jun Huang and Yueyuan Zhang

Received: 11 April 2022
Accepted: 24 May 2022
Published: 29 May 2022

1. Introduction

In recent years, 2D systems have increasingly attracted attention and become more important in the theory and practice field with broad applications, such as for river pollution models, photogrammetric data, batch processing, iterative learning control, and multi-dimensional digital filtering [1]. This paper focuses on the two models introduced by Roesser in [2] and Fornasini and Marchesini in [3,4], which are popular with people for their concise description. Although there are fruitful methods concerning stability and stabilization of 2D systems which have been proposed, they mostly focus on integer-order systems. Integer-order 2D systems can't accurately describe many practical systems, such as circuit components, electro-magnetic systems, heat transfer processes, or viscoelastic systems. On the contrary, they are well characterized by the fractional-order 2D systems [5,6]. Many results have been given about the stability analysis of fractional-order systems in [7–11]. Specifically, the methods based on Lyapunov functions were derived in [7–9] for analyzing stability of fractional one-dimensional (1D) systems. Yang and Hou in [10,11] studied the fractional-order systems with perturbation via cylindrical algebraic decomposition method. These methods for studying fractional systems mostly focus on 1D systems.

Kaczorek in [12] firstly proposed the concept of fractional-order 2D discrete systems. For 2D fractional systems, some results have been investigated in [13–18]. Specifically, for the fractional-orders continuous 2D systems represented by the FM (Fornasini–Marchesini) second model, the general solution formula was obtained based on 2D Laplace transform in [13]. Ref. [14] proposed an asymptotic stability criterion of fractional 2D non-linear continuous-time system based on they Lyapunov function method. In [15], the concept, the practical stability of positive fractional 2D linear systems, was proposed by Tadeusz Kaczorek. The 2D fractional system described by FM first type was derived. However, the dimensions of input and output vectors increase when the system variable decreases. Tadeusz Kaczorek in [16] showed the result of the asymptotic stability for positive fractional 2D linear system. Refs. [17,18] studied the stabilization issue of 2D fractional systems, they proposed the practical stability of the positive fractional 2D system. Specifically, in [17], Tadeusz Kaczorek introduced a class of fractional 2D system presented by Roesser model. The sufficient criteria of the positivity and stabilization were established. Laila Dami et al. studied the issues of positivity stabilization for the uncertain 2D fractional discrete-time systems in [18]. These papers have well studied the stability and stabilization problem for the positive fractional 2D linear systems. The problem of the stability and stabilization of the general 2D linear discrete systems with fractional order is still an open problem to be solved. Fractional-order 2D systems have attracted increasing interest, due to the fact that many real-world physical systems are well characterized by fractional-order 2D systems.

In this paper, we introduce the issues of stability and stabilization for 2D linear discrete systems with fractional orders. Firstly, based on the existing method, the fractional-order 2D system is transformed into an integral-order system. Secondly, based on the Hurwitz Theorem, we equivalently convert the existing stability condition into a new easily checked condition. Then, we use the discriminant theory of polynomials in [19] to solve the represented condition. We extend the stability results in this paper to the problem of stabilization. The key contributions related to this paper are shown as follows:

(1) We change checking whether the fractional 2D linear discrete system is stable into checking whether the polynomials are positive based on Hurwitz Theorems. Thus, the processing of stability analysis is changed into a mathematical problem whether some polynomials are definitely positive, which can be easily checked. It simplifies the existing methods based on Lyapunov functions in [7–9] and has low complexity.

(2) Based on the results proposed by Kaczorek in [15–18], we give a more general method of stability and stabilization for 2D linear fractional-order discrete systems not only for the positive systems.

(3) For the stabilization, because the condition is necessary and sufficient, we can get a complete solution of gain matrixes called the stable gain matrix region of the considered unstable system. The unstable system with the gain parameters of the stable gain matrix region is stabilizable.

The organization is as follows: Section 2 shows problem formulation and fractional 2D system representation. In Section 3, we give the results of stability analysis and an algorithm for obtaining the stable gain matrix region. In Section 4, an example is given to analyze the stability and get the stable gain matrix region to show the effectiveness of the proposed methods. Section 5 shows the conclusions.

Notations. $\mathbb{C}$ and $\mathbb{R}$ stand for the set of complex numbers and real numbers, respectively. The symbols $Re(x)$ is the real part of x. $\mathbb{C}^+ \triangleq \{x \in \mathbb{C} : Re(x) > 0\}$. I and 0 stand for identity matrix and zero block of appropriate sizes, respectively. j denotes an imaginary unit. $det(\Phi)$ denotes determinant of a matrix Φ. $\mathbb{D} \triangleq \{z \in \mathbb{C} \mid |z| \leq 1\}$,$\mathbb{P} \triangleq \{z \in \mathbb{C} \mid |z| = 1\}$, $\bar{\mathbb{U}} \triangleq \{(z_1, z_2) : z_1, z_2 \in \mathbb{C} \mid |z_1| \leq 1, |z_2| \leq 1\}$. $f(\tau) = \kappa_n\tau^n + \kappa_{n-1}\tau^{n-1} + \ldots + \kappa_0(\kappa_0 > 0)$ is a real coefficient polynomial, where κ_i is real. The $n \times n$ Hurwitz matrix of $f(\tau)$ denotes

$$M_f = \begin{bmatrix} \kappa_{n-1} & \kappa_{n-3} & \kappa_{n-5} & \cdots & 0 \\ \kappa_n & \kappa_{n-2} & \kappa_{n-4} & \cdots & 0 \\ 0 & \kappa_{n-1} & \kappa_{n-3} & \cdots & 0 \\ 0 & \kappa_n & \kappa_{n-2} & \cdots & 0 \\ \ldots & \ldots & \ldots & \ldots & \ldots \\ 0 & 0 & 0 & \cdots & \kappa_0 \end{bmatrix},$$

$\triangle(f)_k, k = 1, 2\ldots, n$, represent the k_{th} principal minor determinant of M_f, respectively. $y(\tau)$ is a complex coefficient polynomial and satisfies $y(j\tau) = \varrho_n\tau^n + \varrho_{n-1}\tau^{n-1} + \ldots +$

$\varrho_0 + j(\kappa_n \tau^n + \kappa_{n-1}\tau^{n-1} + \ldots + \kappa_0)$, $\kappa_n \neq 0$, where κ_i and ϱ_i are real. The $2n \times 2n$ Hurwitz matrix of $y(\tau)$ denotes

$$M_y = \begin{bmatrix} \kappa_n & \kappa_{n-1} & \kappa_{n-2} & \cdots & 0 \\ \varrho_n & \varrho_{n-1} & \varrho_{n-2} & \cdots & 0 \\ 0 & \kappa_n & \kappa_{n-1} & \cdots & 0 \\ 0 & \varrho_n & \varrho_{n-1} & \cdots & 0 \\ \cdots & \cdots & \cdots & \cdots & \cdots \\ 0 & 0 & 0 & \cdots & \kappa_0 \\ 0 & 0 & 0 & \cdots & \varrho_0 \end{bmatrix},$$

$\triangle(y)_{2k}, k = 1, 2 \ldots, n$, represent the $2k_{th}$ principal minor determinant of M_y, respectively.

2. Problem Formulation and Fractional-Order 2D System Representation

The aim of this section is to get a fractional 2D linear discrete system represented by integral-order 2D model. We show the process obtained fractional 2D system represented by Roesser model. The transformation from Roesser model to FM second model is introduced for further discussing the stability and stabilization problems. Focus on the following fractional 2D linear system represented as the state-space equations.

$$\begin{bmatrix} \Delta_\alpha^h x_{i+1,j} \\ \Delta_\beta^v x_{i,j+1} \end{bmatrix} = \begin{bmatrix} A_{11} & A_{12} \\ A_{21} & A_{22} \end{bmatrix} \begin{bmatrix} x^h(i,j) \\ x^v(i,j) \end{bmatrix} + \begin{bmatrix} B_1 \\ B_2 \end{bmatrix} u_{ij}, \quad (1)$$

$$y_{ij} = \begin{bmatrix} C_1 & C_2 \end{bmatrix} \begin{bmatrix} x^h(i,j) \\ x^v(i,j) \end{bmatrix} + Du_{ij}, \quad i,j \in \mathbb{Z}_+, \quad (2)$$

where $x_{ij}^h \in \mathbb{R}^{n_1}, x_{ij}^v \in \mathbb{R}^{n_2}, u_{ij} \in \mathbb{R}^m, y_{ij} \in \mathbb{R}^p$ are horizontal state vector, vertical state vector, input vector and output vector at the point (i,j), respectively. And $A_{11} \in \mathbb{R}^{n_1 \times n_1}, A_{12} \in \mathbb{R}^{n_1 \times n_2}, A_{21} \in \mathbb{R}^{n_2 \times n_1}, A_{22} \in \mathbb{R}^{n_2 \times n_2}, B_1 \in \mathbb{R}^{n_1 \times m}, B_2 \in \mathbb{R}^{n_2 \times m}, C_1 \in \mathbb{R}^{p \times n_1}, C_2 \in \mathbb{R}^{p \times n_2}, D \in \mathbb{R}^{p \times m}, n = n_1 + n_2$.

The boundary conditions are defined by

$$x_{0j}^h, j \in \mathbb{Z}_+ \text{ and } x_{i0}^v, i \in \mathbb{Z}_+ \quad (3)$$

For further getting the Roesser model representing the 2D fractional system, we recall the definitions, horizontal and vertical fractional differences described by the 2D functions, and Lemma 1.

Definition 1 ([17]). *The $\alpha-$order horizontal fractional difference of a 2D function $x_{ij}, i,j \in \mathbb{Z}_+$ is defined by*

$$\Delta_\alpha^h x_{ij} = \sum_{k=0}^{i} c_\alpha(k) x_{i-k,j}, \quad (4)$$

where $\alpha \in \mathbb{R}, n-1 < \alpha < n \in \mathbb{N} = 1, 2, \ldots$ and

$$c_\alpha(k,l) = \begin{cases} 1 & for \quad k = 0 \\ (-1)^k \dfrac{k!}{\alpha(\alpha-1)\ldots(\alpha-k+1)} & k > 0 \end{cases} \quad (5)$$

Definition 2 ([17]). *The $\beta-$order vertical fractional difference of a 2D function $x_{ij}, i,j \in \mathbb{Z}_+$ is defined by*

$$\Delta_\beta^v x_{ij} = \sum_{l=0}^{j} c_\beta(k) x_{i,j-l}, \quad (6)$$

where $\beta \in \mathbb{R}, n-1 < \beta < n \in \mathbb{N} = 1, 2, \ldots$ *and*

$$c_\beta(l) = \begin{cases} 1 & for \quad l = 0 \\ (-1)^l \dfrac{l!}{\beta(\beta-1)\ldots(\beta-l+1)} & l > 0 \end{cases} \tag{7}$$

Remark 1. *We have mentioned that many real-world physical systems are well characterized by a fractional-order model. And the physical systems are generally continuous. We present a definition of fractional derivative and integral by Grünwald-Letnikov as follow:*

$${}^{GL}_{t_0}\mathcal{D}^\alpha_t f(t) \approx \frac{1}{h^\alpha} \sum_{j=0}^{[(t-t_0)/h]} \omega_j f(t - jh)$$

Forthe purpose of easy calculation, the continuous physical models are usually discretized. Definitions 1 and 2 can be obtained by determining the step size of the fractional order equation. The size of h determines the accuracy of the model. The smaller h is, the closer the model is to the real system. We can select h according to the requirement of precision in practical production. And the fractional order system with different step h can be converted into different integer order system. The system is built as a 2D fractional-order model then convert to a 2D integer-order model instead of directly building a 2D integer-order model, which has a more accurate result.

Lemma 1 ([17]). *If* $n-1 < \alpha < n \in \mathbb{N}(n-1 < \beta < n)$*, then*

$$\sum_{k=0}^{\infty} c_\alpha(k) = 0 \quad (resp. \sum_{k=0}^{\infty} c_\beta(k) = 0). \tag{8}$$

According to Definitions 1 and 2, system (1) can be rewritten as follows

$$\begin{aligned} \begin{bmatrix} x^h(i+1,j) \\ x^v(i,j+1) \end{bmatrix} = \begin{bmatrix} \overline{A}_{11} & A_{12} \\ A_{21} & \overline{A}_{22} \end{bmatrix} \begin{bmatrix} x^h(i,j) \\ x^v(i,j) \end{bmatrix} + \begin{bmatrix} B_1 \\ B_2 \end{bmatrix} u_{ij} \\ - \begin{bmatrix} \sum_{k=2}^{i+1} c_\alpha(k) x_{i-k+1,j} \\ \sum_{l=2}^{j+1} c_\beta(k) x_{i,j-l+1} \end{bmatrix}, \end{aligned} \tag{9}$$

where $n = n_1 + n_2$, $\overline{A}_{11} = A_{11} + \alpha I_{n_1}$ and $\overline{A}_{22} = A_{22} + \beta I_{n_2}$. 2D fractional system (1) has been rewritten as the integer-order 2D system with delays.

From Equations (5) and (7), we can get that $c_\alpha(0) = c_\beta(0) = 1$, $c_\alpha(1) = -\alpha$ and $c_\beta(1) = -\beta$. Based on these equations and Lemma 1 we have

$$\sum_{k=2}^{\infty} c_\alpha(k) = \alpha - 1 \; and \; \sum_{k=2}^{\infty} c_\beta(k) = \beta - 1 \tag{10}$$

We firstly analyze the stability of 2D fractional system. Let the input vector $u_{ij} = 0$. We consider the open-loop system

$$\begin{bmatrix} x^h(i+1,j) \\ x^v(i,j+1) \end{bmatrix} = \begin{bmatrix} \overline{A}_{11} & A_{12} \\ A_{21} & \overline{A}_{22} \end{bmatrix} \begin{bmatrix} x^h(i,j) \\ x^v(i,j) \end{bmatrix} - \begin{bmatrix} \sum_{k=2}^{i+1} c_\alpha(k) x^h_{i-k+1,j} \\ \sum_{k=2}^{j+1} c_\beta(l) x^v_{i,j-l+1} \end{bmatrix}. \tag{11}$$

From [17], the system (11) is asymptotically stable if and only if the following 2D system

$$\begin{bmatrix} x^h(i+1,j) \\ x^v(i,j+1) \end{bmatrix} = \left(\begin{bmatrix} \overline{A}_{11} & A_{12} \\ A_{21} & \overline{A}_{22} \end{bmatrix} - \sum_{k=2}^{\infty} \begin{bmatrix} I_{n_1} c_\alpha(k) & 0 \\ 0 & I_{n_2} c_\beta(k) \end{bmatrix} \right) \cdot \begin{bmatrix} x^h(i,j) \\ x^v(i,j) \end{bmatrix} \tag{12}$$

is asymptotically stable.

According to (10), $\overline{A}_{11} = A_{11} + I_{n_1}\alpha$ and $\overline{A}_{22} = A_{22} + I_{n_2}\beta$, the system (12) is represented as the following form

$$\begin{bmatrix} x^h(i+1,j) \\ x^v(i,j+1) \end{bmatrix} = \begin{bmatrix} A_{11} + I_{n_1} & A_{12} \\ A_{21} & A_{22} + I_{n_2} \end{bmatrix} \begin{bmatrix} x^h(i,j) \\ x^v(i,j) \end{bmatrix} \tag{13}$$

Remark 2. *For the purposes of analyzing stability of the fractional-order 2D model, we convert the fractional order model into the integer-order model.*

Due to the above discussion, if the 2D system (13) is asymptotically stable, the 2D fractional system (1) with $u_{ij} = 0$ is asymptotically stable. System (13) is a typical Roesser model. The considered stability issue of fractional-order 2D systems is changed into considering the stability of 2D integral-order Roesser model.

Let $x(i,j) = [x^{hT}(i,j)\ x^{vT}(i,j)]^T$ and the matrices

$$A_1 = \begin{bmatrix} A_{11} + I_{n_1} & A_{12} \\ 0 & 0 \end{bmatrix}, A_2 = \begin{bmatrix} 0 & 0 \\ A_{21} & A_{22} + I_{n_2} \end{bmatrix}$$

System (13) can be represented as the following FM second model:

$$x(i+1,j+1) = A_1 x(i,j+1) + A_2 x(i+1,j), \tag{14}$$

where $x(i+1,j+1)$ denotes the state vector at $(i+1,j+1)$.

Thus, we can analyze the stability of system (14) to know the stability of fractional-order 2D system (1).

For better presenting and understanding the following content in this paper, we give some definitions and recall a lemma and the lemma of Hurwitz stable of the real coefficient polynomial and the complex coefficient polynomial, respectively.

Definition 3. *The gain matrix K is called the stable gain matrix of the closed-loop 2D fractional-order system if the closed-loop 2D fractional-order system with the stable gain matrix is stabilizable.*

Definition 4. *The set of the stable gain matrixes of the closed-loop 2D fractional-order system is called the stable gain matrix region of the closed-loop 2D fractional-order system.*

Lemma 2 ([20]). *System (14) is asymptotically stable if and only if*

$$\begin{cases} H(z_1,0) \neq 0, z_1 \in \mathbb{D} \\ H(z_1,z_2) \neq 0, z_1 \in \mathbb{P}, z_2 \in \mathbb{D} \end{cases} \tag{15}$$

where $H(z_1,z_2) = det(I_n - z_1 A_1 - z_2 A_2)$.

Remark 3. *Condition (15) is not numerically tractable [21]. Next, Condition (15) is transformed into new conditions that can be easily implemented.*

Lemma 3 ([22]). *The necessary and sufficient condition for the roots' real part of real coefficient polynomial $f(\tau)$ to be negative is $\triangle(f)_k > 0, k = 1,2\ldots,n$.*

Lemma 4 ([22]). *If $\triangle(y)_{2k} \neq 0$, the necessary and sufficient condition for the roots' real part of complex coefficient polynomial $y(\tau)$ to be negative is $\triangle(y)_{2k} > 0, k = 1,2\ldots,n$.*

3. Stability and Stabilization Analysis

3.1. Stability Analysis

This subsection is to obtain new tractable conditions based on traditional condition (15). By linear fraction transformation, the stability conditions of 2D system is equivalent to the issue whether the polynomials are Hurwitz stable shown in Theorem 1. We use the criterion of Lemmas 3 and 4 to deal with the new derived conditions of Theorem 1. And we can get the tractable conditions of Theorem 2.

Theorem 1. *System (1) with* $u_{ij} = 0$ *is asymptotically stable if and only if these criteria are satisfied,*

(1) $H(-1,0) \neq 0, L_1(\gamma,0) \neq 0, \gamma \in \mathbb{C}^+$

(2) $H(-1,-1) \neq 0, L_2(-1,\gamma) \neq 0, \gamma \in \mathbb{C}^+$

(3) $W(s,-1) \neq 0, L_3(s,\gamma) \neq 0, s \in \mathbb{R}, \gamma \in \mathbb{C}^+$

where $W(s,z_2) = (1-js)^m H(\frac{1+js}{1-js}, z_2)$,
$L_1(\gamma,0) = (1+\gamma)^m H(\frac{1-\gamma}{1+\gamma}, 0)$,
$L_2(-1,\gamma) = (1+\gamma)^m H(-1, \frac{1-\gamma}{1+\gamma})$,
$L_3(s,\gamma) = (1+\gamma)^n W(s, \frac{1-\gamma}{1+\gamma})$.

Proof. Substitute $z_1 = \frac{1+js}{1-js}$ to $H(z_1,z_2)$ of condition (15). We can get

$$W(s,z_2) = (1-js)^m H(\frac{1+js}{1-js}, z_2), \tag{16}$$

where m stands for the degree of $H(z_1,z_2)$ in z_1. We can easily obtain that condition (15) of Lemma 2 is equivalent to

$$\begin{cases} H(z_1,0) \neq 0, z_1 \in \mathbb{D} \\ H(-1,z_2) \neq 0, z_2 \in \mathbb{D} \\ W(s,z_2) \neq 0, s \in \mathbb{R}, z_2 \in \mathbb{D} \end{cases} \tag{17}$$

Substitute $z_1 = \frac{1-\gamma}{1+\gamma}$ to $H(z_1,0)$. Substitute $z_2 = \frac{1-\gamma}{1+\gamma}$ to $H(-1,z_2)$ and $W(s,z_2)$, respectively. We can obtain

$$\begin{cases} L_1(\gamma,0) = (1+\gamma)^m H(\frac{1-\gamma}{1+\gamma}, 0), \\ L_2(-1,\gamma) = (1+\gamma)^m H(-1, \frac{1-\gamma}{1+\gamma}), \\ L_3(s,\gamma) = (1+\gamma)^n W(s, \frac{1-\gamma}{1+\gamma}) \end{cases} \tag{18}$$

By the above transformations, condition (17) is equivalent to

$$\begin{cases} H(-1,0) \neq 0, L_1(\gamma,0) \neq 0, \gamma \in \mathbb{C}^+ \\ H(-1,-1) \neq 0, L_2(-1,\gamma) \neq 0, \gamma \in \mathbb{C}^+ \\ H(s,-1) \neq 0, L_3(s,\gamma) \neq 0, s \in \mathbb{R}, \gamma \in \mathbb{C}^+ \end{cases} \tag{19}$$

Condition (15) is equivalently converted into Condition (19). The 2D system (14) is asymptotically stable if and only if the condition (19) is satisfied. The fractional 2D linear discrete system (1) with $u_{ij} = 0$ can be represented by integral-order 2D model (14) in Section 2. So system (1) with $u_{ij} = 0$ is asymptotically stable if and only if the Condition (19) is satisfied. The proof is complete. □

The conditions of Theorem 1 can be represented by the criterion of Lemmas 3 and 4. We show it as follows.

Theorem 2. *System (1) with* $u_{ij} = 0$ *is asymptotically stable if and only if these conditions are satisfied,*

(1) $H(-1,0) \neq 0, \triangle(L_1)_k > 0, k = 1,2\ldots,n,$

(2) $H(-1,-1) \neq 0, \triangle(L_2)_k > 0, k = 1,2\ldots,n,$

(3) $W(s,-1) \neq 0, \triangle(L_3)_{2k} > 0, k = 1,2\ldots,n, s \in \mathbb{R}.$

Proof. In Theorem 1, $L_1(\gamma, 0)$, $L_2(-1, \gamma)$ are real coefficient polynomial in γ. $L_1(\gamma, 0) \neq 0, \gamma \in \mathbb{C}^+$ and $L_2(-1, \gamma) \neq 0, \gamma \in \mathbb{C}^+$, the roots' real part of L_1 and L_2 are negative, are the criterion of Lemma 3, $\triangle(L_1)_k > 0$ of Hurwitz matrix M_{L_1} and $\triangle(L_2)_k > 0$ of Hurwitz matrix M_{L_2}.

$L_3(s, \gamma) \neq 0, s \in \mathbb{R}, \gamma \in \mathbb{C}^+$, the roots' real part of $L_3(s, \gamma)$ to be negative, is the criterion of Lemma 4, the $2k_{th}$ principal minor determinants $\triangle(L_3)_{2k} > 0$ of the Hurwitz matrix M_{L_3}. According to the above, we rewrite the conditions of Theorem 1 and show them in Theorem 2. The proof is complete. □

In this section, we firstly transform the traditional stability conditions in [20] into Theorem 1 by the linear fraction transformation. Then the obtained conditions of Theorem 1 are the criterions whether the polynomials are Hurwitz stable. These new criterions are represented as Theorem 2 by using the conditions of Lemmas 3 and 4 in [22]. In next subsection, we focus on the stabilization problem applying the similar process of the above proposed method of checking stability.

3.2. Stabilization

This section is to design a state feedback to stabilize the system and get the stable gain matrix region. Consider the system (1) with the following state-feedback

$$u_{ij} = [K_1 K_2] \begin{bmatrix} x^h(i,j) \\ x^v(i,j) \end{bmatrix}, \tag{20}$$

where $K = [K_1\ K_2] \in \mathbb{R}^{m \times n}, K_j \in \mathbb{R}^{m \times n_j}, j = 1, 2$ is a gain matrix.

A gain matrix K need to be solved to ensure that the closed-loop system is stabilizable via state feedback. Specifically, K need to be fond to ensure that the following system

$$\begin{aligned} \begin{bmatrix} x^h(i+1,j) \\ x^v(i,j+1) \end{bmatrix} &= \begin{bmatrix} \overline{A}_{11} + B_1K_1 & A_{12} + B_1K_2 \\ A_{21} + B_2K_1 & \overline{A}_{22} + B_2K_2 \end{bmatrix} \begin{bmatrix} x^h(i,j) \\ x^v(i,j) \end{bmatrix} \\ &- \begin{bmatrix} \sum_{k=2}^{i+1} c_\alpha(k) x^h_{i-k+1,j} \\ \sum_{k=2}^{j+1} c_\beta(l) x^v_{i,j-l+1} \end{bmatrix} \end{aligned} \tag{21}$$

is asymptotically stable.

Same as the operations of stability analysis, we can easily get the results that the system (21) is asymptotically stable if and only if the following 2D system

$$\begin{aligned} \begin{bmatrix} x^h(i+1,j) \\ x^v(i,j+1) \end{bmatrix} &= \left(\begin{bmatrix} \overline{A}_{11} + B_1K_1 & A_{12} + B_1K_2 \\ A_{21} + B_2K_1 & \overline{A}_{22} + B_2K_2 \end{bmatrix} \right. \\ &\left. - \sum_{k=2}^{\infty} \begin{bmatrix} I_{n_1} c_\alpha(k) & 0 \\ 0 & I_{n_2} c_\beta(k) \end{bmatrix} \right) \begin{bmatrix} x^h(i,j) \\ x^v(i,j) \end{bmatrix} \end{aligned} \tag{22}$$

is asymptotically stable.

According to the Equation (10), $\overline{A}_{11} = A_{11} + I_{n_1}\alpha$ and $\overline{A}_{22} = A_{22} + I_{n_2}\beta$, system (22) can be represented as follows:

$$\begin{bmatrix} x^h(i+1,j) \\ x^v(i,j+1) \end{bmatrix} = \begin{bmatrix} A_{11} + I_{n_1} + B_1K_1 & A_{12} + B_1K_2 \\ A_{21} + B_2K_1 & A_{22} + I_{n_2} + B_2K_2 \end{bmatrix} \cdot \begin{bmatrix} x^h(i,j) \\ x^v(i,j) \end{bmatrix} \tag{23}$$

Denote

$$\widetilde{A}_1 = \begin{bmatrix} A_{11} + I_{n_1} + B_1K_1 & A_{12} + B_1K_2 \\ 0 & 0 \end{bmatrix},$$

$$\widetilde{A}_2 = \begin{bmatrix} 0 & 0 \\ A_{21} + B_2K_1 & A_{22} + I_{n_2} + B_2K_2 \end{bmatrix}.$$

We have a new 2D system in form of FM second model with matrices $\widetilde{A}_1$ and $\widetilde{A}_2$ as follow:

$$x(i+1,j+1) = \widetilde{A}_1 x(i,j+1) + \widetilde{A}_2 x(i+1,j), \tag{24}$$

The proposed method of stability can be applied to consider the stabilization of system (1) with the state-feedback (20). We represent Theorem 2 as follows:

Proposition 1. *The closed-loop 2D fractional-order system (21) is stabilizable if and only if these conditions are satisfied,*

(1) $\widetilde{H}(-1,0) \neq 0, \Delta(\widetilde{L_1}(\gamma))_k > 0, k = 1,2\ldots,n,$

(2) $\widetilde{H}(-1,-1) \neq 0, \Delta(\widetilde{L_2}(\gamma))_k > 0, k = 1,2\ldots,n,$

(3) $\widetilde{W}(s,-1) \neq 0, \Delta(\widetilde{L_3}(j\gamma))_{2k} > 0, k = 1,2\ldots,n, s \in \mathbb{R}.$

where $\widetilde{H}(z_1,z_2) = det(I_n - z_1\widetilde{A_1} - z_2\widetilde{A}_2),$
$\widetilde{W}(s,z_2) = (1-js)^m \widetilde{H}(\frac{1+js}{1-js}, z_2),$
$\widetilde{L_1}(\gamma,0) = (1+\gamma)^m \widetilde{H}(\frac{1-\gamma}{1+\gamma}, 0),$
$\widetilde{L_2}(-1,\gamma) = (1+\gamma)^n \widetilde{H}(-1, \frac{1-\gamma}{1+\gamma}),$
$\widetilde{L_3}(s,\gamma) = (1+\gamma)^n \widetilde{H}(s, \frac{1-\gamma}{1+\gamma}), (z_1,z_2) \in \overline{\mathbb{U}}, s \in \mathbb{R}, z_2 \in \mathbb{D}, \gamma \in \mathbb{C}^+$, *m and n respectively stand for the degree of* $\widetilde{H}(z_1,z_2)$*in* z_1 *and* z_2.

Proof. The proof is same as Theorem 2. □

For obtaining the stable gain matrix region of the closed-loop 2D fractional-order system, we give the following Algorithm 1.

Algorithm 1 2DF Stabilization.

Input: The characteristic equation $\widetilde{H}(z_1,z_2)$ of a closed-loop 2D fractional-order system.
Output: The stable gain matrix region.
Step 1. Let the gain matrix as K.
Step 2. Calculate $\widetilde{H}(z_1,z_2)$ of the closed-loop 2D fractional-order system.
Step 3. Solve the inequalities of Proposition 1 based on the discrimination system of polynomials.
Step 3.1. Get the stable gain matrix region of K by solving $\widetilde{H}(-1,0) \neq 0, \triangle(\widetilde{L}_1)_k > 0, k = 1,2\ldots,n.$
Step 3.2. Get the stable gain matrix region of K by solving $\widetilde{H}(-1,-1) \neq 0, \triangle(\widetilde{L}_2)_k > 0, k = 1,2\ldots,n.$
Step 3.3. Calculate $\widetilde{L}_3(s,\gamma)$, then get $L_3(j\gamma) = \varrho_n\gamma^n + \varrho_{n-1}\gamma^{n-1} + \ldots + \varrho_0 + j(\kappa_n\gamma^n + \kappa_{n-1}\gamma^{n-1} + \ldots + \kappa_0)$, where ϱ_i and $\kappa_i, i = 0,\ldots n$ are real coefficient polynomials in s. Solve $\widetilde{W}(s,-1) \neq 0$, and $\triangle(L_3)_{2k} > 0, k = 1,2\ldots,n, s \in \mathbb{R}$ to the stable gain matrix region of K.
Step 4. From step 3, obtain the final results of the stable gain matrix region of K.

4. Example

In this section, we show a numerical example that the fractional-order 2D system has generality from [17] to show the efficiency of the methods of stability and stabilization in this paper. We focus on the 2D fractional system (1) with $\alpha = 0.4, \beta = 0.5$ and $\begin{pmatrix} A_{11} & A_{12} \\ A_{21} & A_{22} \end{pmatrix} = \begin{pmatrix} -1.1255 & 0.8 \\ 0.149 & 0.24 \end{pmatrix}, B_1 = -19, B_2 = -10$

We show the steps of analyzing stability and stabilization processes of fractional 2D systems as follows.

First, we analyze the stability of the considered system.

Step 1. Change the 2D fractional system (1) with $u_{ij} = 0$ into the 2D system (14). Based on the given parameters, we firstly get the polynomial

$$H(z_1, z_2) = 1 - \frac{31}{25}z_2 - \frac{1}{5}z_1 - \frac{49}{125}z_1 z_2$$

Step 2. Check whether the inequalities of Theorem 2 hold using Maple.
Step 2.1. We have $H(-1, 0) = \frac{6}{5} \neq 0$
Calculate $L_1(\gamma, 0)$, as follows

$$L_1(\gamma, 0) = \frac{6}{5}\gamma + \frac{4}{5}$$

We have $\triangle(L_1)_2 = \frac{24}{25} > 0$. This conditions of Theorem 2 is satisfied from system (1) with the given parameters.
Step 2.2. We have $H(-1, -1) = \frac{256}{125} \neq 0$. Calculate $L_2(-1, \gamma)$, as follows

$$L_2(-1, \gamma) = \frac{256}{125}\gamma + \frac{44}{125}$$

We have $\triangle(L_2)_2 = \frac{11{,}264}{15{,}625} > 0$.
Step 2.3. We have $W(s, -1) = -\frac{256}{125}js + \frac{304}{125} \neq 0$.
Calculate $L_3(s, \gamma)$, as follows

$$L_3(s, \gamma) = \frac{304}{125}\gamma - \frac{104}{125} + j(-\frac{256}{125}s\gamma - \frac{44}{125}s)$$

$$L_3(s, j\gamma) = \frac{256}{125}s\gamma - \frac{104}{125} + j(-\frac{44}{125}s + \frac{304}{125}\gamma)$$

We have $\triangle(L_3)_2 = 88s^2 - 247$. It's easy to know that $\triangle(L_3)_2$ isn't satisfied the condition $\Delta(L_3(j\gamma))_2 > 0, k = 1, 2 \ldots, n.$ of Theorem 2.
Step 3. This fractional-order 2D system is unstable.

Remark 4. *As shown in the example, we know this considered fractional-order 2D system is unstable. The result is in agreement with the literature [17]. While we needn't stabilise the considered system based on the precondition that the system is positive as in the other methods. We extend the existing valuable methods of fractional-order 2D systems in the control theory. The stability condition of the general fractional-order 2D systems instead of the positive fractional-order 2D systems is given.*

Now, we consider the stabilization of the system and obtain the stable gain matrix region according to as follows:
Step 1. Let $K = [k_1 \quad k_2]$.
Step 2.Based on the given parameters of the fractional 2D system, we firstly get the polynomial

$$\begin{aligned}\widetilde{H}(z_1, z_2) =& 1 - \frac{31}{25}z_2 + 10k_2 z_2 - \frac{1}{5}z_1 - \frac{49}{125}z_1 z_2 \\ &+ \frac{66}{5}k_2 z_1 z_2 + 19k_1 z_1 - \frac{389}{25}z_1 k_1 z_2\end{aligned}$$

Step 3. Solve the inequalities of Proposition 1 by Maple.
Step 3.1. Calculate the polynomial $\widetilde{L}_1(\gamma, 0)$, as follows

$$\widetilde{L}_1(\gamma, 0) = (\frac{6}{5} - 19k_1)\gamma + 19k_1 + \frac{4}{5} \neq 0, \gamma \in \mathbb{C}^+.$$

Then, from

$$\triangle(\widetilde{L}_1)_2 = (19k_1 - \frac{6}{5})(19k_1 + \frac{4}{5}) < 0,$$

we can get

$$-\frac{4}{95} < k_1 < \frac{6}{95}. \tag{25}$$

From

$$\widetilde{H}(-1,0) = \frac{6}{5} - 19k_1 \neq 0,$$

we can get

$$k_1 \neq \frac{6}{95}. \tag{26}$$

Step 3.2. Calculate the polynomial $\widetilde{L}_2(-1,\gamma)$, as follows

$$\begin{aligned}\widetilde{L}_2(-1,\gamma) =&(-\frac{864}{25}k_1 + \frac{16}{5}k_2 + \frac{256}{125})\gamma - (\frac{86}{25}k_1 - \frac{16}{5}k_2) \\ &+ \frac{44}{125} \neq 0, \gamma \in \mathbb{C}^+.\end{aligned}$$

Then, we can get

$$\triangle(\widetilde{L}_2)_2 = (\frac{864}{25}k_1 - \frac{16}{5}k_2 - \frac{256}{125})(\frac{86}{25}k_1 + \frac{16}{5}k_2 - \frac{44}{125}) > 0, \tag{27}$$

we can get

$$\widetilde{H}(-1,-1) = \frac{256}{125} + \frac{16}{5}k_2 - \frac{864}{25}k_1 \neq 0. \tag{28}$$

Step 3.3. We have

$$\begin{aligned}\widetilde{W}(s,-1) = &(\frac{864}{25}k_1 - \frac{16}{5}k_2 - \frac{256}{125})js + \frac{864}{25}k_1 - \frac{116}{5}k_2 \\ &+ \frac{304}{125} \neq 0, s \in \mathbb{R},\end{aligned}$$

From the above condition, we can get

$$\frac{864}{25}k_1 - \frac{16}{5}k_2 - \frac{256}{125} \neq 0. \tag{29}$$

We have

$$\begin{aligned}\widetilde{L_3}(s,j\gamma) =& -4320k_1s\gamma - 400k_2s\gamma - 256s\gamma - 430k_1 \\ &- 2900k_2 + 104 + j(-430k_1s - 4320k_1\gamma \\ &- 400k_2x + 2900k_2\gamma + 44s - 304\gamma) \\ &\neq 0, s \in \mathbb{R}, \gamma \in \mathbb{C}^+.\end{aligned}$$

Then we can obtain the inequalities $\triangle(\widetilde{L_3}(s,j\gamma))_{2k} > 0, k = 1,2, s \in \mathbb{R}$ to get uncertain parameters k_1 and k_2 of K as follows:

$$\begin{aligned}\triangle(\widetilde{L_3}(s,j\gamma))_2 =&232{,}200k_1^2s^2 + 194{,}500k_1k_2s^2 - 20{,}000k_2^2s^2 \\ &- 37{,}520k_1s^2 - 10{,}600k_2s^2 + 232{,}200k_1^2 \\ &+ 1{,}410{,}125k_1k2 - 1{,}051{,}250k_2^2 + 1408s^2 \\ &- 39{,}820k_1 + 147{,}900k_2 - 3952 > 0, s \in \mathbb{R}.\end{aligned}$$

In order for $\triangle(\widetilde{L_3}(s,j\gamma))_2 > 0, s \in \mathbb{R}$ to be established, we have

$$\begin{cases} (215k_1 + 200k_2 - 22)(270k_1 - 25k_2 - 16) > 0 \\ (215k_1 + 1450k_2 - 52)(1080k_1 - 725k_2 + 76) > 0 \end{cases} \tag{30}$$

Step 4. From (25)–(30), we can find that when

$$\begin{cases} 215k_1 + 200k_2 - 22 < 0 \\ 270k_1 - 25k_2 - 16 < 0 \\ 215k_1 + 1450k_2 - 52 > 0 \\ 1080k_1 - 725k_2 + 76 > 0 \end{cases} \quad (31)$$

all the conditions of Proposition 1 are established. Thus, the closed-loop system (21) with the gain parameters of the stable gain matrix region (31) is stable.

From the above discussion, through solving the conditions of Proposition 1 for getting K, we obtain the stable gain matrix region (31) which is shown as Figure 1:

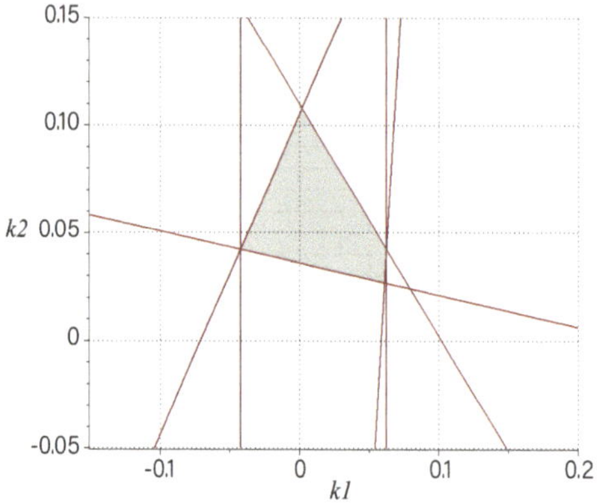

Figure 1. The stable gain matrix region.

For better showing the validity of the method in this paper, we give some simulations. Figures 2 and 3 show the state space responses of system (1). We can find that the state responses of the open-loop system (21) in Figures 2 and 3 is divergent and not stable.

According to the solution (31), let K = [0.01 0.08], and the state responses of the closed-loop system (21) are shown in Figures 4 and 5. As i and j get bigger, the state space response $x_1(i,j)$ and $x_2(i,j)$ go to 0. The system (1) is stable after stabilization. Similarly, let K = [0 0.05] from the region (31), and the state responses of the closed-loop system (21) are shown in Figures 6 and 7. The system (1) is stable after stabilization. It is noted that as long as the gain matrix K is selected in the stable gain matrix region shown in Figure 1, the closed-loop system (21) is stable.

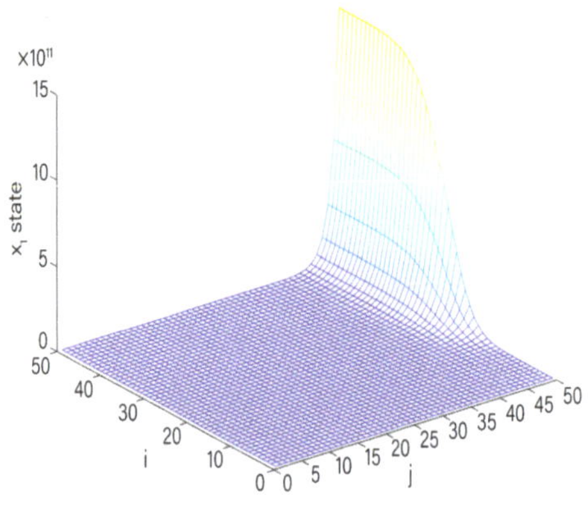

Figure 2. Open-loop state space response of $x_1(i,j)$.

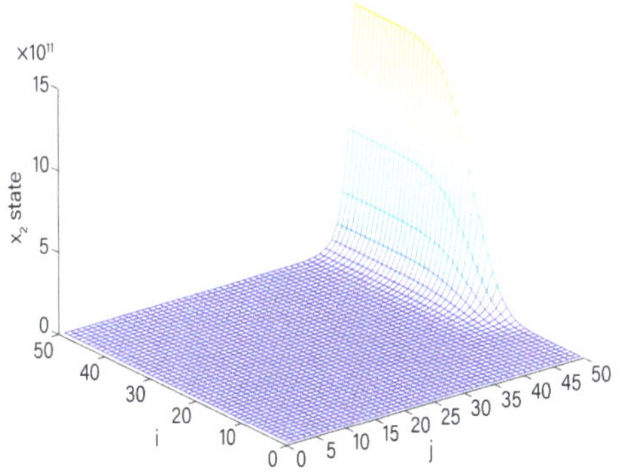

Figure 3. Open-loop state space response of $x_2(i,j)$.

Remark 5. *For better presenting the responses of the closed-loop system, we change the direction of i and j in the Figures 4–7.*

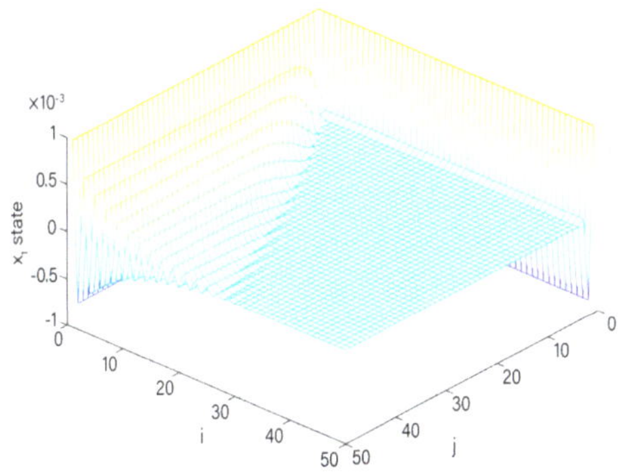

Figure 4. Closed-loop state space response of $x_1(i,j)$ when K = [0.01 0.08].

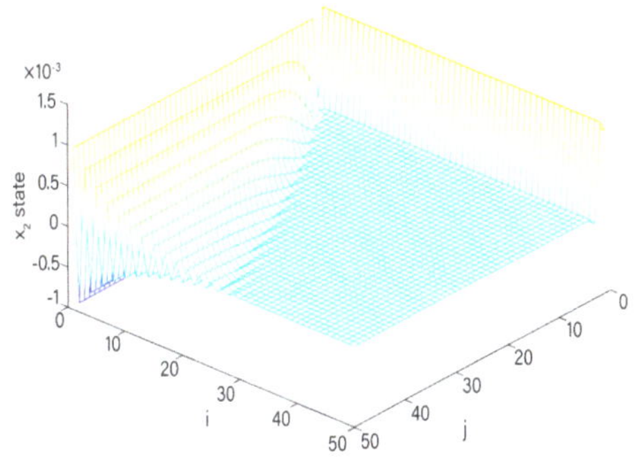

Figure 5. Closed-loop state space response of $x_2(i,j)$ when K = [0.01 0.08].

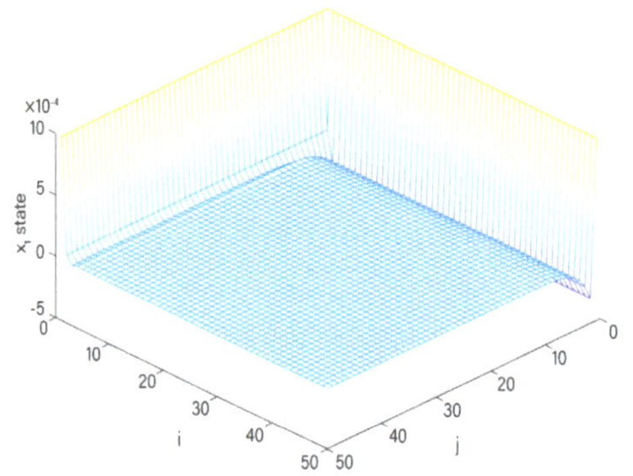

Figure 6. Closed-loop state space response of $x_1(i,j)$ when K = [0 0.05].

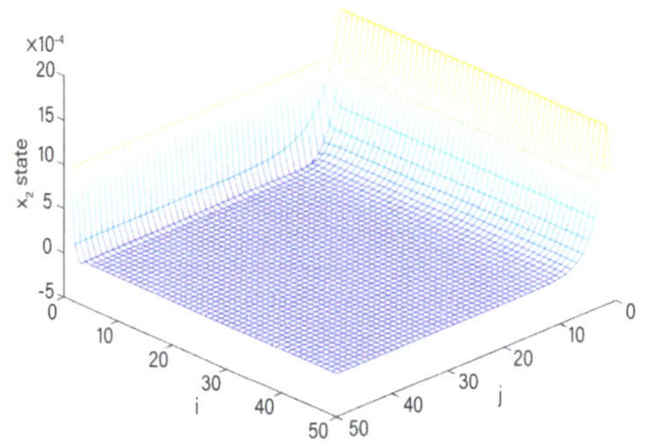

Figure 7. Closed-loop state space response of $x_2(i,j)$ when K = [0 0.05].

Remark 6. *We obtain a stable gain matrix region. All the gain matrixes of the stable gain matrix region can stabilize the system. Further, the closed-loop system (21) with state-feedback can be regarded as a 2D fractional-order system with uncertain parameters. Our method not only extends the existing methods only for positive systems, but can also solve the robust stability problem of fractional 2D systems that has not been solved by other researchers. It can obtain all the parameters that ensure the system is robust and stable, which will be discussed in the future.*

5. Conclusions

This paper has discussed the stability and stabilization problems of fractional-order 2D systems that are common in practice but rarely studied. The stability check process and the algorithm for obtaining the stable gain matrix region have been shown in the example. The method proposed in this paper can be widely used. Compared with the existing method, it is not necessary to stabilise the considered system based on the condition that the system must be positive. And the stabilization method can obtain multiple parameters of control gain to stabilize the fractional-order 2D system. The proposed methods in this paper have low computational complexity, and so are simpler and easier to use.

Author Contributions: Conceptualization, X.L.; methodology, X.L.; validation, X.L.; formal analysis, J.Y. and M.L.; investigation, X.L.; writing—original draft preparation, X.L.; writing—review and editing, X.L.; supervision, X.H. All authors have read and agreed to the published version of the manuscript.

Funding: This research was funded by the National Natural Science Foundation of China grant number 12171073 and China Postdoctoral Science Foundation grant number 2021M700703.

Institutional Review Board Statement: Not applicable.

Informed Consent Statement: Not applicable.

Data Availability Statement: Not applicable.

Conflicts of Interest: The authors declare no conflicts of interest.

References

1. Zhang, G.; Trentelman, H.L.; Wang, W.; Gao, J. Input-output finite-region stability and stabilization for discrete 2-d fornasini-marchesini models. *Syst. Control Lett.* **2017**, *99*, 9–16. [CrossRef]
2. Roesser, A. discrete state space model for linear image processing. *Autom. Control IEEE Trans.* **1975**, *20*, 1–10. [CrossRef]
3. Fornasini, E.; Marchesini, G. State-space realization theory of two-dimensional filters. *Autom. Control IEEE Trans.* **1976**, *21*, 484–492. [CrossRef]
4. Fornasini, E.; Marchesini, G. Doubly indexed dynamical systems: State space models and structural properties. *Math. Syst. Theory* **1978**, *12*, 59–72. [CrossRef]
5. Moornani, K.A. Robust stability testing function and kharitonov-like theorem for fractional order interval systems. *IET Control Theory Appl.* **2010**, *4*, 2097–2108. [CrossRef]
6. Ma, Y.; Lu, J. Robust stability and stabilization of fractional order linear systems with positive real uncertainty. *ISA Trans.* **2014**, *53*, 199–209. [CrossRef] [PubMed]
7. Lakshmikantham, V.; Leela, S. Lyapunov theory for fractional differential equations. *Commun. Appl. Anal.* **2008**, *12*, 365–376.
8. Vasundhara Devi, J.; Mc Rea, F.A.; Drici, Z. Variational lyapunov method for fractional differential equations. *Comput. Math. Appl.* **2012**, *64*, 2982–2989. [CrossRef]
9. Chen, F.L. A review of existence and stability results for discrete fractional equations. *J. Comput. Complex. Appl.* **2015**, *1*, 22–53.
10. Yang, J.; Hou, X. Robust bounds for fractional-order systems with uncertain order and structured perturbations via cylindrical algebraic decomposition method. *J. Frankl. Inst.* **2019**, *356*, 4097–4105. [CrossRef]
11. Hou, J.Y.X.; Luo, M. A cad-based algorithm for solving stable parameter region of fractional-order systems with structured perturbations. *Fract. Calc. Appl. Anal.* **2019**, *22*, 509–522.
12. Kaczorek, T. Fractional 2D linear systems. *J. Autom. Mob. Robot. Intell. Syst.* **2008**, *2*, 5–9.
13. Rogowski, K. Solution to the fractional-order 2D continuous systems described by the second Fornasini-Marchesini model. *IFAC-PapersOnLine* **2017**, *50*, 9748–9752. [CrossRef]
14. Xiang, S.H.Z. Stability of a class of fractional-order two-dimensional non-linear continuous-time syatems. *IET Control Theory Appl.* **2016**, *10*, 2559–2564.
15. Kaczorek, T. Practical stability of positive fractional 2d linear systems. *Multidimens. Syst. Signal Process.* **2009**, *57*, 289–292. [CrossRef]
16. Kaczorek, T. Practical stability and asymptotic stability of positive fractional 2d linear systems. *Asian J. Control* **2010**, *12*, 200–207. [CrossRef]
17. Kaczorek, T.; Rogowski, K. Positivity and stabilization of fractional 2d linear systems described by the roesser model. *Int. J. Appl. Math. Comput. Sci.* **2010**, *20*, 85–92. [CrossRef]
18. Dami, L.; Benhayoun, M.; Bemzaouia, A. Stabilization and positivity of 2d fractional order uncertain discrete-time systems. In Proceedings of the 14th International Multi-Conference on Systems, Signals & Devices (SSD), Marrakech, Morocco, 28–31 March 2017; pp. 545–548.
19. Yang, L.; Zeng, Z. A complete discrimination system for polynomials. *Sci. China* **1996**, *39*, 628–646.
20. Huang, T. Stability of two-dimensional recursive filters. *IEEE Trans. Audio Electroacoust.* **2003**, *20*, 158–163. [CrossRef]
21. Ebihara, Y.; Ito, Y.; Hagiwara, T. Exact stability analysis of 2-D systems using LMIs. *IEEE Trans. Autom. Control* **2006**, *51*, 1509–1513. [CrossRef]
22. Gantmacher, F.R. *The Theory of Matrices*; Chelsea Publishing Company: New York, NY, USA, 1960; Volume 2

mathematics

MDPI

Article

Chenciner Bifurcation Presenting a Further Degree of Degeneration

Sorin Lugojan [1,†], Loredana Ciurdariu [1,*,†] and Eugenia Grecu [2,†]

1 Department of Mathematics, Politehnica University of Timisoara, 300006 Timisoara, Romania; s3lugo@gmail.com
2 Department of Management, Politehnica University of Timisoara, 300006 Timisoara, Romania; eugenia.grecu@upt.ro
* Correspondence: loredana.ciurdariu@upt.ro
† These authors contributed equally to this work.

Abstract: Chenciner bifurcation appears for some two-dimensional systems with discrete time having two independent variables. Investigated here is a special case of degeneration where the implicit function theorem cannot be used around the origin, so a new approach is necessary. In this scenario, there are many more bifurcation diagrams than in the two non-degenerated cases. Several numerical simulations are presented.

Keywords: degeneracy; bifurcation; Chenciner; discrete systems

MSC: 37L10; 37G10

check for updates

Citation: Lugojan, S.; Ciurdariu, L.; Grecu, E. Chenciner Bifurcation Presenting a Further Degree of Degeneration. *Mathematics* **2022**, *10*, 1603. https://doi.org/10.3390/math10091603

Academic Editors: Jun Huang, Yueyuan Zhang, Rami Ahmad El-Nabulsi and Eva Kaslik

Received: 19 March 2022
Accepted: 5 May 2022
Published: 8 May 2022

1. Introduction

The discrete dynamical systems have an increasing role in informatics [1], computer and machine learning, and other interdisciplinary fields [2–4]. A new mathematical model was recently proposed in [5] for the dynamics of three types of phytoplankton of the Sea of Azov under the condition of salinity increase. Other examples of applied dynamical discrete systems, besides continuous ones, are given in [6–9]. Presented among them is a discrete-time epidemic model applied to the study of the COVID-19 virus [8]. The theory of discrete dynamical systems may be applied in many branches of engineering such as suspension bridges, ball bearings, and nanotechnology. The study of impact oscillators is an important source of nonlinearity in mechanical system theory [10–13]. When the impact has zero velocity, the so-called grazing impacts appear. The near-grazing systems can be described by discrete dynamical systems, and an application for harmonic oscillators is presented in [12]. The dynamics of the other two types of discrete dynamical systems, a discrete predator-prey model with group defense and nonlinear harvesting in prey and a modified Nicholson-Bailey model, were investigated, and the conditions for classical Neimark-Sacker bifurcation were given in [14,15].

Economy is another important domain of application [16]. Traditionally, economic agents are considered to have rational expectations [17], which assume that prices follow the fundamental economic value. Experiments have shown that economic agents [18] do not make rational predictions but follow empirical rules. Thus, sometimes these rules can lead them to the fundamental landmark, but other times they can be coordinated on destabilizing strategies to follow the trends. The consequences are market "bubbles" and even collapses. A "bubble" represents a strong over evaluation [19] and the duration of an asset compared to its fundamental economic value. Big "bubbles" and sudden market crashes are difficult to harmonize with the standard model of agents representing rational expectations. Some authors, for example [20], have devised a simple behavioral heuristic switching model that explains the path-dependent coordination of the individual forecast, as well as the aggregate behavior of the market. The paper analyzes the coexistence of a locally stable fundamental equilibrium state and a stable quasi-periodic orbit, created by the Chenciner bifurcation. In relation to the initial states, the economic agents will orient their individual expectations either on a stable fundamental equilibrium trajectory or on persistent price fluctuations in the vicinity of the fundamental equilibrium state.

The generalized Neimark-Sacker bifurcations or Chenciner bifurcations of discrete dynamical systems have been discovered in 1985 in [21–23], in the framework of the study of elliptic bifurcations of fixed points. Later, in 1990 in [24] this bifurcation was characterized better than before. The non-degenerate Chenciner bifurcation is one of the eleven types retrieved in the generic two-parameter discrete-time dynamical systems, according to classification from [25]. There is no other bifurcation of codimension 2 in generic discrete-time systems. The non-degeneracy condition, so called "cubic non-degeneracy", is not fulfilled in this case of the generalized Neimark-Sacker bifurcations.

In recent years, the study of degenerated discrete Chenciner bifurcation began, as seen in [26]. The singularities are always difficult to study in comparison to the regular cases. The purpose of this article is to examine the Chenciner bifurcation which doesn't check the condition (CH.1) [25] (p. 405). That is the degenerated Chenciner bifurcation. The two types of bifurcation diagrams existing in the non-degenerated variant, as seen in [25], are replaced by 32 types of bifurcation diagrams in a particular degenerated discrete Chenciner dynamical system; see [26].

The article is composed of four sections. The first section is the Introduction, where the non-degenerate Chenciner bifurcations are presented using the truncated normal form of the system (A4) and polar coordinates, and some new applications in various domains are mentioned. Section two of this paper describes the results given in [26,27] concerning the existence of bifurcation curves and their dynamics in the parametric plane (α_1, α_2) in the cases where $a_{10}b_{01}a_{01}b_{10} \neq 0$ and the linear parts of $\beta_1(\alpha)$ and $\beta_2(\alpha)$ nullify, respectively, and when $a_{10} = 0, b_{01} = 0, a_{01} = 0$ and $b_{10} = 0$. The third section is the main part of the paper, where the degeneracy case of the Chenciner bifurcation written in the truncated normal form was studied when $a_{20} = a_{11} = a_{02} = 0$ and, for b_{10} and b_{01}, two situations have been studied: $b_{10} \neq 0,\ b_{01} \neq 0$ or $b_{10} = b_{01} = 0$. In addition, some numerical simulations are presented using Matlab for checking the theoretical results. The discussions and conclusions are presented in the fourth section of the paper.

2. Methods

The study of the non-degenerated discrete Chenciner bifurcation begins by a defect of a coordinate change $(\alpha_1, \alpha_2) \to (\beta_1, \beta_2)$. The degeneration taken into account is a non-regularity of the coordinate change in the origin, which loses its quality of coordinate change. The method introduced [26] is to consider the same expression for β_1, β_2 but as functions of α_1, α_2 and not as new coordinates.

The steps of the method used in previous papers for finding the truncated normal form of generalized Neimark-Sacker bifurcation for analyzing the behavior of such general two-dimensional discrete dynamical systems in order to obtain the bifurcation diagrams are given in Appendix A. The Chenciner bifurcations imply that the center manifold for the Poincare map is two-dimensional. In [26], a new degeneration for generalized Neimark-Sacher bifurcations was introduced; therefore, the classical Chenciner bifurcations are called non-degenerate Chenciner bifurcations. This study has been continued in [27,28] and also in the present paper. In the degenerated case, there are two different approaches: the first is to work with the initial parameters α_1, α_2 in the polar form, (A6) of our system, and the second, in [28], is considered another regular transformation of parameters, when the product $a_{10}a_{01}b_{10}b_{01} \neq 0$.

The following two results, Theorems A1 and A2, which have been established in [26], play a key role in the next section and will be restated in Appendix B. Theorem A1 establishes the stability of the fix point O function of the sign of $\beta_1(\alpha)$, and then, in Theorem A2, the existence of invariant circles is discussed as a function of the sign of $\Delta(\alpha)$. From here, the generic phase portraits corresponding to different regions of the bifurcation diagrams were obtained in Figure 1 from [26] and in Appendix B, Figure A1. Table 1 from [26] gives the regions in the parametric plane defined by $\Delta(\alpha)$, $\beta_1(\alpha)$, $\beta_2(\alpha)$, and L_0. These phase portraits remain the same, but the bifurcation diagrams are different from the non-degenerate Chenciner bifurcation case in [25]. These kinds of studies represent important topics in the qualitative theory of discrete-time dynamical systems.

Now, we will write the smooth functions $\beta_{1,2}(\alpha)$ as $\beta_1(\alpha) = a_{10}\alpha_1 + a_{01}\alpha_2 + \sum_{i+j\geq 2} a_{ij}\alpha_1^i\alpha_2^j$ and $\beta_2(\alpha) = b_{10}\alpha_1 + b_{01}\alpha_2 + \sum_{i+j\geq 2} b_{ij}\alpha_1^i\alpha_2^j$ for our further goals. We recall that the transformation (A7) is not regular at $(0,0)$. That means the Chenciner bifurcation becomes degenerate, iff

$$a_{10}b_{01} - a_{01}b_{10} = 0. \tag{1}$$

The case when the linear part of $\beta_1(\alpha)$ nullifies and $\beta_2(\alpha)$ has at least one linear term was mentioned in [27] together with Theorem 2 of [27], which is an important result concerning the existence, and also the relative positions in the parametric plane, (α_1, α_2) of the bifurcation curves, function of the sign of $\beta_1(\alpha)$.

Recently, in [27], the dynamics of the system in the form (A10) and (A11) was described and studied in the case when all these coefficients $a_{10} = 0, b_{01} = 0, a_{01} = 0$ and $b_{10} = 0$, and the bifurcation diagrams obtained are different from previous situations form [26,28].

In this paper, the degeneracy condition (1) will be satisfied and the terms of degree one and two are zero in the case of $\beta_1(\alpha)$. Therefore, the functions $\beta_{1,2}(\alpha)$ become

$$\beta_1(\alpha) = a\alpha_2^3 + b\alpha_1\alpha_2^2 + c\alpha_1^2\alpha_2 + d\alpha_1^3 + \sum_{i+j=4}^{p_1} a_{ij}\alpha_1^i\alpha_2^j + O\left(|\alpha|^{p_1+1}\right) \tag{2}$$

and

$$\beta_2(\alpha) = k\alpha_1 + h\alpha_2 + \sum_{i+j=2}^{q_1} b_{ij}\alpha_1^i\alpha_2^j + O\left(|\alpha|^{q_1+1}\right) \tag{3}$$

for some $p_1 \geq 4$. $a = a_{03}$, $b = a_{12}$, $c = a_{21}$, $d = a_{30}$, respectively, and $q_1 \geq 2$, $h = b_{10}$, $k = b_{01}$.

The set $B_{1,2}$ and C will be denoted by

$$B_{1,2} = \left\{(\alpha_1, \alpha_2) \in \mathbb{R}^2, \beta_{1,2}(\alpha) = 0, |\alpha| < \varepsilon\right\} \tag{4}$$

and

$$C = \left\{(\alpha_1, \alpha_2) \in \mathbb{R}^2, \Delta(\alpha) = 0, |\alpha| < \varepsilon\right\} \tag{5}$$

for some $\varepsilon > 0$ that is sufficiently small, and then the new $\Delta(\alpha)$ is

$$\Delta(\alpha) = \beta_2^2(\alpha) - 4\beta_1(\alpha)L_2(\alpha). \tag{6}$$

3. Results

In this section, the degree of the truncated version of the first bifurcation curve, β_1, is $Deg\beta_1 = 3$, and for the second bifurcation curve, β_2, two cases will be studied: when the $Deg\beta_2 = 1$ and when the $Deg\beta_2 = 2$ in the truncated version.

3.1. Degree of the Second Bifurcation Curve Is One in the Truncated Version

Firstly, we focus on the case when Deg $\beta_2 = 1$ in the truncated version. In expression of $\beta_1(\alpha)$, we denote the coefficients a_{03}, a_{12}, a_{21}, and a_{30} by a, b, c, and d, respectively, and in expression of $\beta_2(\alpha)$, we denote the coefficients b_{01} and b_{10} by h and k, respectively.

$$\beta_1(\alpha_1, \alpha_2) = a\alpha_2^3 + b\alpha_2^2\alpha_1 + c\alpha_2\alpha_1^2 + d\alpha_1^3 + O(|\alpha|^4),$$

where $a, b, c, d \in \mathbb{R}_*$.

$$\beta_2(\alpha_1, \alpha_2) = h\alpha_2 + k\alpha_1 + O(|\alpha|^2),$$

where $h, k \in \mathbb{R}_*$.

Then

$$\Delta(\alpha) = [\beta_2(\alpha)]^2 - 4L_2(\alpha)\beta_1(\alpha), \tag{7}$$

where $\alpha = (\alpha_1, \alpha_2)$.

In the truncated version, we have:

$$\begin{aligned} \beta_1(\alpha) &= a\alpha_2^3 + b\alpha_2^2\alpha_1 + c\alpha_2\alpha_1^2 + d\alpha_1^3 \\ \beta_2(\alpha) &= h\alpha_2 + k\alpha_1 \\ \Delta(\alpha) &= [\beta_2(\alpha)]^2 \end{aligned} \tag{8}$$

Discussed below is the sign of first bifurcation curve in the truncated version.

In order to establish the sign of $\beta_1(\alpha)$, the following is used:

Remark 1. *The sign of the polynomial*

$$\beta_1(T) = aT^3 + bT^2 + cT + d \in \mathbf{R}_*[T],$$

is the same as the sign of $\beta_1(\alpha_1, \alpha_2)$, *for every* $\alpha_1, \alpha_2 \in \mathbf{R}$, *such that* $\alpha_2 = T\alpha_1$.

In order to establish the sign of $\beta_1(T)$, we denote, as usual for the third degree equation:

$$p = \frac{c}{a} - \frac{b^2}{3a^2}, \qquad q = \frac{2b^3}{27a^3} - \frac{bc}{3a^2} + \frac{d}{a},$$

and the polynomial becomes:

$$\beta_1(T) = a(T^3 + pT + q).$$

The roots of $\beta_1(T)$ are the solutions of the equation

$$T^3 + pT + q = 0.$$

For the classification of the $\beta_1(T)-$ roots, we use the notation

$$r = \left(\frac{q}{2}\right)^2 + \left(\frac{p}{3}\right)^3$$

which is called "the cubic discriminant".

1. For $p > 0,\ q \neq 0$, there is one real root e_1, and two complex conjugated ones;
2. For $p < 0,\ q = 0$, there is a triple root e_1;
3. For $p < 0,\ r > 0$, there is one real root e_1, and two complex conjugated ones;
4. For $p < 0,\ r = 0$, there are three real roots, one simple e_1, and two common;
5. For $p < 0,\ r < 0$, there are three real different roots $e_1 < e_2 < e_3$.

Lemma 1. *The following statements are true:*

1. *If* $p < 0$ *and* $r < 0$, *then*

$$sign[\beta_1(T)] = sign[a(T - e_1)(T - e_2)(T - e_3)],$$

see Table 1.

2. *If* $p > 0$ *or* ($p < 0$ *and* $r \geq 0$), *then*

$$sign[\beta_1(T)] = sign[a(T - e_1)],$$

see Table 2.

Table 1. The sign of $\beta_1(T)$ when there are three roots $e_1,\ e_2,\ e_3$.

T	$(-\infty, e_1)$	e_1	(e_1, e_2)	e_2	(e_2, e_3)	e_3	(e_2, ∞)
sign$\beta_1(T)$	sign$(-a)$	0	sign(a)	0	sign(-a)	0	sign(a)

Table 2. The sign of $\beta_1(T)$ when there is one root e_1.

T	$(-\infty, e_1)$	e_1	(e_1, ∞)
sign$\beta_1(T)$	sign(-a)	0	sign(a)

The case when p and r are strictly negative are rendered below.

From Appendix A, $\theta_0 = \theta(0)$ and $L_0 = L_2(0) \neq 0$. The case $p < 0,\ r < 0$ involves four cases to analyze, impossing that $hk > 0$.

1. $L_0 > 0,\ k > 0$
2. $L_0 > 0,\ k < 0$
3. $L_0 < 0,\ k > 0$
4. $L_0 < 0,\ k < 0$.

The bifurcation diagrams are respectively given in Figures 1–4.

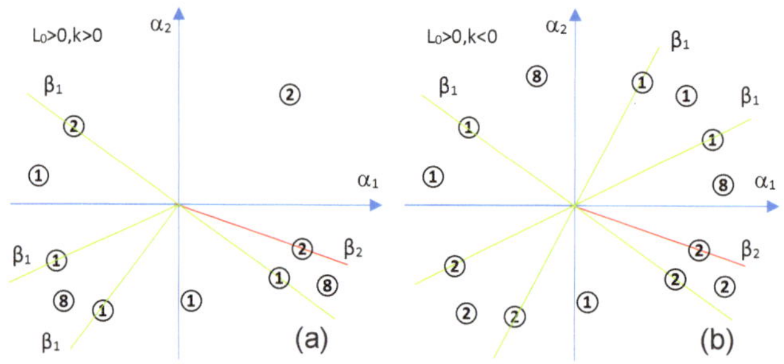

Figure 1. Bifurcation diagrams when $p < 0,\ r < 0$, and $hk > 0$: **(a)** $L_0 > 0,\ k > 0$; **(b)** $L_0 > 0,\ k < 0$.

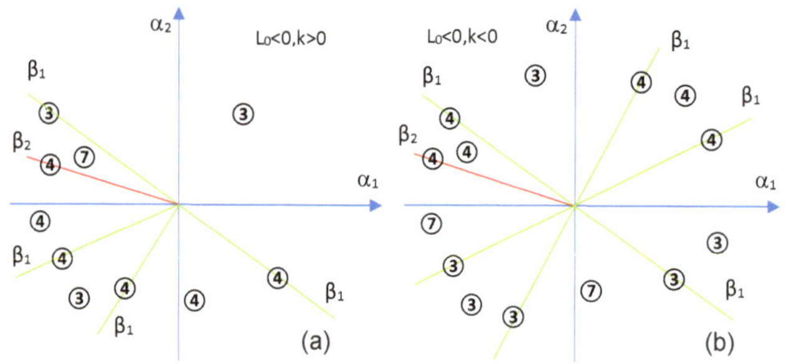

Figure 2. Bifurcation diagrams when $p < 0,\ r < 0$, and $hk > 0$: **(a)** $L_0 < 0,\ k > 0$; **(b)** $L_0 < 0,\ k < 0$.

Remark 2. *When $\beta_2(\alpha) = 0$, then the sign of $\Delta(\alpha)$ is given by the relation (7), instead of (8).*

The case when p is strictly positive or (p is strictly negative and r is positive) will be studied below.

In the case $p > 0$ or ($p < 0$ and $r \geq 0$), from Lemma 1 (2), it results that $sign[\beta_1(T)] = sign[a(T - e_1)]$, see Table 2, where e_1 is the unique real root of $\beta_1(\alpha) = 0$.

From $\beta_2(\alpha) = h\alpha_2 + k\alpha_1$, it results that $m_2 = -\frac{k}{h}$. In our case, $\Delta(\alpha) = [\beta_2(\alpha)]^2$. We impose that $hk > 0$.

Therefore we will have the following two bifurcation diagrams presented in Figure 3.

Remark 3. *In the case $p > 0$ or ($p < 0$ and $r \geq 0$), we will obtain only two distinct figures; that means the following Figure 3a,b:*

1. *if $a > 0,\ k > 0,\ L_0 > 0$ or $a > 0,\ k < 0,\ L_0 > 0$ or $a < 0,\ k > 0,\ L_0 > 0$ or $a < 0,\ k < 0,\ L_0 > 0$, we get Figure 3a;*
2. *if $a > 0,\ k > 0,\ L_0 < 0$ or $a > 0,\ k < 0,\ L_0 < 0$ or $a < 0,\ k > 0,\ L_0 < 0$ or $a < 0,\ k < 0,\ L_0 < 0$, we get Figure 3b.*

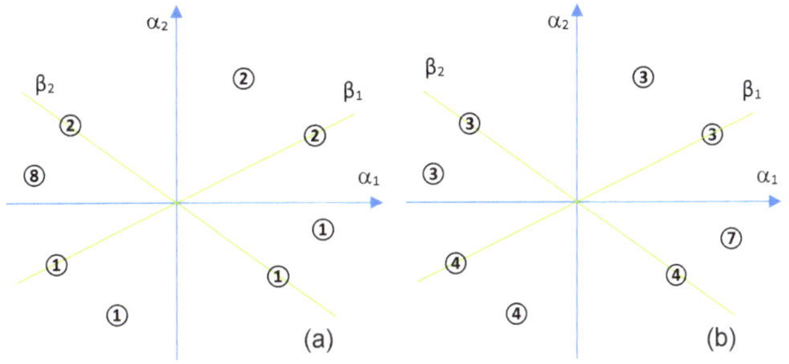

Figure 3. Bifurcation diagrams when $p > 0$ or ($p < 0$ and $r > 0$) and $hk > 0$: (**a**) $a > 0,\ k > 0,\ L_0 > 0$ or $a > 0,\ k < 0,\ L_0 > 0$ or $a < 0,\ k > 0,\ L_0 > 0$ or $a < 0,\ k < 0,\ L_0 > 0$; (**b**) $a > 0,\ k > 0,\ L_0 < 0$ or $a > 0,\ k < 0,\ L_0 < 0$ or $a < 0,\ k > 0,\ L_0 < 0$ or $a < 0,\ k < 0,\ L_0 < 0$.

3.2. Degree of the Second Bifurcation Curve Is Two

If Deg $\beta_2 = 2$, then its first three coefficients will be denoted as below.

$$\beta_1(\alpha_1, \alpha_2) = a\alpha_2^3 + b\alpha_2^2\alpha_1 + c\alpha_2\alpha_1^2 + d\alpha_1^3 + O(|\alpha|^4),$$

$$\beta_2(\alpha_1, \alpha_2) = h\alpha_2^2 + k\alpha_1\alpha_2 + l\alpha_1^2 + O(|\alpha|^3),$$

where $h, k, l \in \mathbf{R}_*$.

$$\Delta(\alpha_1, \alpha_2) = (h\alpha_2^2 + k\alpha_1\alpha_2 + l\alpha_1^2)^2 - 4L_2(\alpha)[a\alpha_2^3 + b\alpha_2^2\alpha_1 + c\alpha_2\alpha_1^2 + d\alpha_1^3 + O(|\alpha|^4)]$$
$$= -4L_0(a\alpha_2^3 + b\alpha_2^2\alpha_1 + c\alpha_2\alpha_1^2 + d\alpha_1^3) + O(|\alpha|^4).$$

Truncated, that is:

$$\beta_1(\alpha) = a\alpha_2^3 + b\alpha_2^2\alpha_1 + c\alpha_2\alpha_1^2 + d\alpha_1^3$$
$$\beta_2(\alpha) = h\alpha_2^2 + k\alpha_1\alpha_2 + l\alpha_1^2,$$

having $\Delta_2 = k^2 - 4hl$, $\Delta(\alpha) = -4L_0\beta_1(\alpha)$.

The sign of β_1 was previously analyzed.

The case when p and r are strictly negative and Δ_2 is strictly positive are considered below.

In the case $p < 0,\ r < 0,\ \Delta_2 > 0$, the polynomial $\beta_1(T)$. This has the real roots $e_1 < e_2 < e_3$ (and the polynomial $\beta_2(T)$ has the real roots $m_1 < m_2$).

There are three cases that must be considered:

I $e_1 < m_1 < m_2 < e_2 < e_3$;
II $e_1 < m_1 < e_2 < m_2 < e_3$;
III $e_1 < m_1 < e_2 < e_3 < m_2$.

In each of those cases, there are four sub-cases depending on the signs of h and L_0.

The bifurcation diagrams are given below, in Figures 4–7.

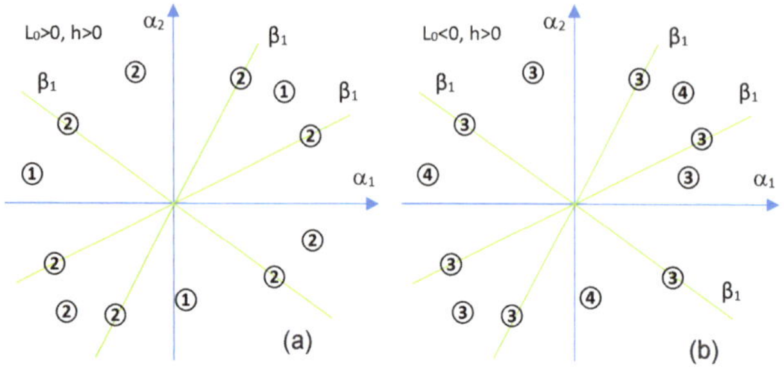

Figure 4. Bifurcation diagrams in the Case I when $p < 0,\ r < 0$, and $\Delta_2 > 0$: (**a**) $L_0 > 0,\ h > 0$; (**b**) $L_0 < 0,\ h > 0$.

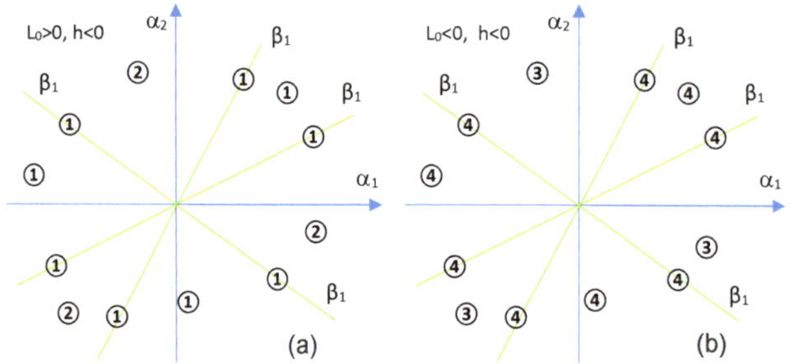

Figure 5. Bifurcation diagrams in the Case I when $p < 0,\ r < 0$, and $\Delta_2 > 0$: (**a**) $L_0 > 0,\ h < 0$; (**b**) $L_0 < 0,\ h < 0$.

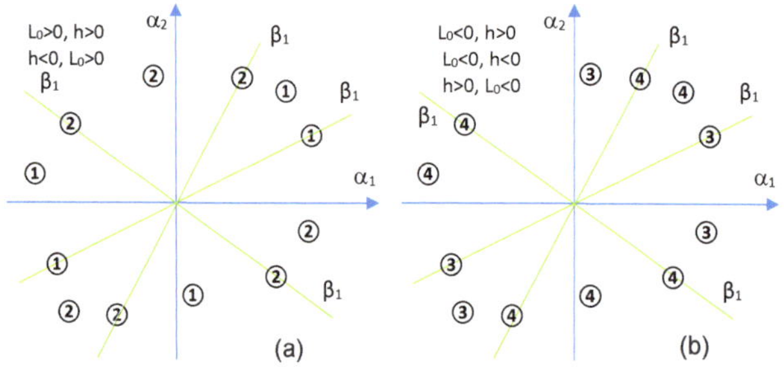

Figure 6. Bifurcation diagrams in the Case II and III when $p < 0,\ r < 0$, and $\Delta_2 > 0$: (**a**) $L_0 > 0,\ h > 0$ or $h < 0,\ L_0 > 0$; (**b**) $L_0 < 0,\ h > 0$ or $L_0 < 0,\ h < 0$ or $h > 0,\ L_0 < 0$.

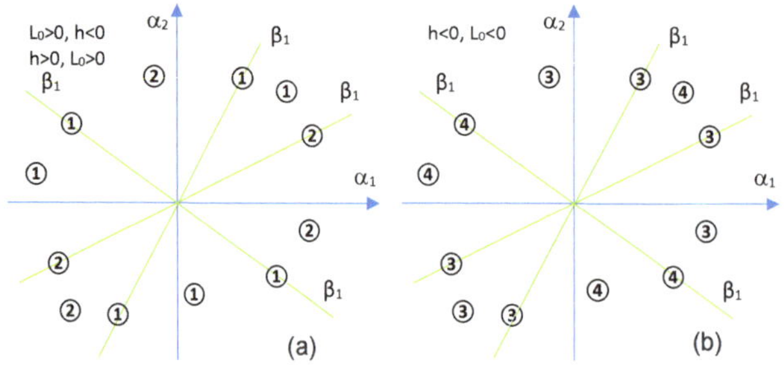

Figure 7. Bifurcation diagrams in the Case II and III when $p < 0,\ r < 0$, and $\Delta_2 > 0$: (**a**) $L_0 > 0,\ h < 0$ or $h > 0,\ L_0 > 0$; (**b**) $h < 0,\ L_0 < 0$.

The case when p, r, and Δ_2 are strictly negative is presented in the following.

In the case $p < 0,\ r < 0,\ \Delta_2 < 0$, we see that $\beta_1(T)$ has the real roots $e_1 < e_2 < e_3$ and $\beta_2(T)$ has no real roots ($\Delta_2 < 0$); therefore, $sign\ \beta_2(\alpha) = sign(h)$.

We know that $sign\ \delta(\alpha) = -sign(L_0)sign\ \beta_1(\alpha)$.

According to Lemma 1, (1), when $p < 0$ and $r < 0$, the $sign\ \beta_1(T) = sign[a(T - e_1)(T - e_2)(T - e_3)]$; see Table 1.

From the information presented above, we obtain the following:

Remark 4. *When $p < 0,\ r < 0,\ \Delta_2 < 0$, the bifurcation diagrams are given in the following:*

(1) *If $a > 0,\ h > 0,\ L_0 > 0$ or $a < 0,\ h > 0,\ L_0 > 0$, then we get the Figure 8a.*
(2) *If $a > 0,\ h > 0,\ L_0 < 0$ or $a < 0,\ h > 0,\ L_0 < 0$, then we get the Figure 8b.*
(3) *If $a > 0,\ h < 0,\ L_0 > 0$ or $a < 0,\ h < 0,\ L_0 > 0$, then we get the Figure 9a.*
(4) *If $a > 0,\ h < 0,\ L_0 < 0$ or $a < 0,\ h < 0,\ L_0 < 0$, then we get the Figure 9b.*

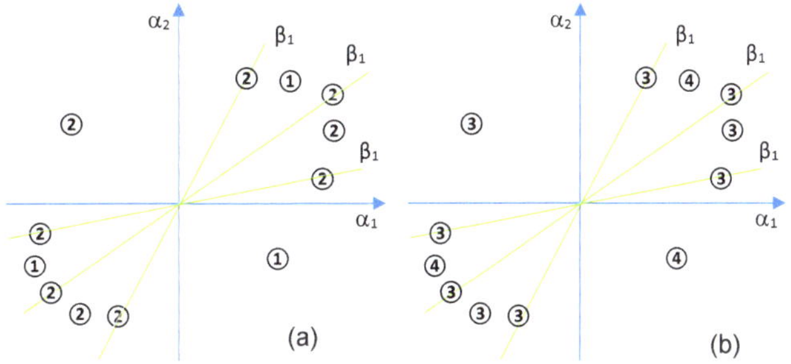

Figure 8. Bifurcation diagrams when $p < 0,\ r < 0$, and $\Delta_2 < 0$: (**a**) $a > 0,\ h > 0,\ L_0 > 0$ or $a < 0,\ h > 0,\ L_0 > 0$; (**b**) $a > 0,\ h > 0,\ L_0 < 0$ or $a < 0,\ h > 0,\ L_0 < 0$.

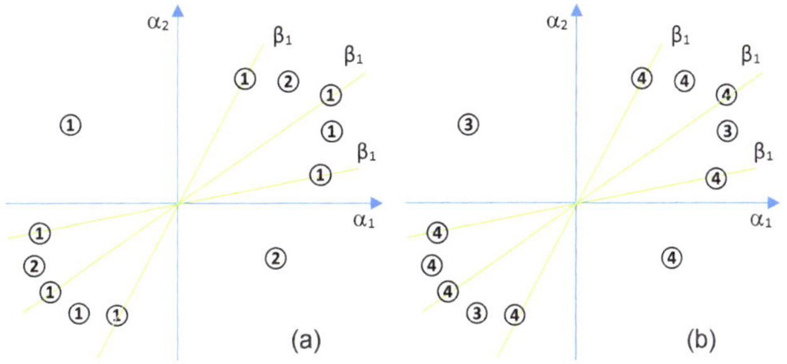

Figure 9. Bifurcation diagrams when $p < 0,\ r < 0$, and $\Delta_2 < 0$: (**a**) $a > 0,\ h < 0,\ L_0 > 0$ or $a < 0,\ h < 0,\ L_0 > 0$; (**b**) $a > 0,\ h < 0,\ L_0 < 0$ or $a < 0,\ h < 0,\ L_0 < 0$.

The case when p is strictly positive or (p is strictly negative and r is positive) will be investigated next.

In the case when $p > 0$ or ($p < 0$ and $r \geq 0$), from Lemma 1, (2) we have,

$$sign\ \beta_1(T) = sign[a(T - e_1)],$$

see Table 2.

(a) There is one real root e_1 and two complex conjugates roots of $\beta_1(T)$ when $r > 0$;

(b) When $p < 0$ and $r = 0$, there are three real roots, one simple e_1 and two common;

(c) Then $p > 0,\ q \neq 0$, there is one real root e_1 and two complex conjugates;

(d) If $p > 0,\ q = 0$, there is a triple root e_1.

From (a)–(d), we see that, in all these cases, $\beta_1(T) = 0$ has a single real root e_1 and then $sign\ \beta_1(T) = sign[a(T - e_1)]$.

$\Delta(\alpha) = -4L_0\beta_1(\alpha)$ and then $sign\Delta(\alpha) = -sign(L_0)sign[\beta_1(\alpha)]$.
For the sign of $\beta_2(\alpha)$, we have two cases:

1. $\Delta_2 < 0$ implies $sign\ \beta_2(\alpha) = sign(h)$;
2. $\Delta_2 > 0$, then there is $m_1,\ m_2$, two distinct real roots of $\beta_2(\alpha) = 0$ and

$$sign\ \beta_2(\alpha) = \begin{cases} sign(h), & \text{if} \quad m \in (-\infty, m_1) \cup (m_2, \infty) \\ -sign(h), & \text{if} \quad m \in (m_1, m_2). \end{cases}$$

Remark 5. *When $p > 0$ or ($p < 0$ and $r \geq 0$) and $\Delta_2 < 0$, then only two cases will appear:*

1. *If $a > 0,\ L_0 > 0,\ h > 0$ or $a < 0,\ L_0 > 0,\ h > 0$, see Figure 10a;*
2. *If $a > 0,\ L_0 < 0,\ h > 0$ or $a < 0,\ L_0 < 0,\ h > 0$, see Figure 10b;*
3. *If $a > 0,\ L_0 > 0,\ h < 0$ or $a < 0,\ L_0 > 0,\ h < 0$, see Figure 11a;*
4. *If $a > 0,\ L_0 < 0,\ h < 0$ or $a < 0,\ L_0 < 0,\ h < 0$, see Figure 11b.*

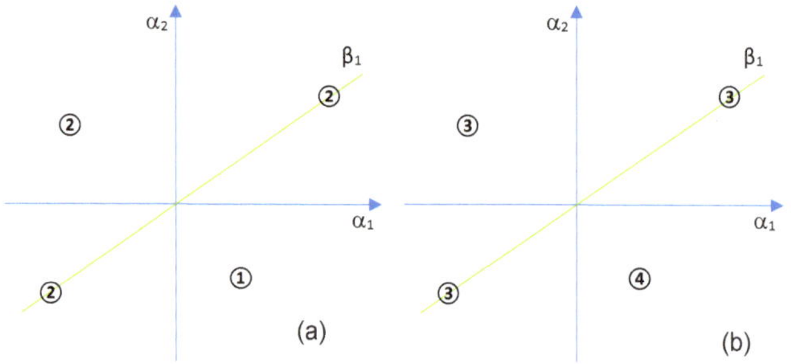

Figure 10. Bifurcation diagrams when $p > 0$ or ($p < 0$ and $r > 0$): (**a**) $a > 0,\ L_0 > 0,\ h > 0$ or $a < 0,\ L_0 > 0,\ h > 0$; (**b**) $a > 0,\ L_0 < 0,\ h > 0$ or $a < 0,\ L_0 < 0,\ h > 0$.

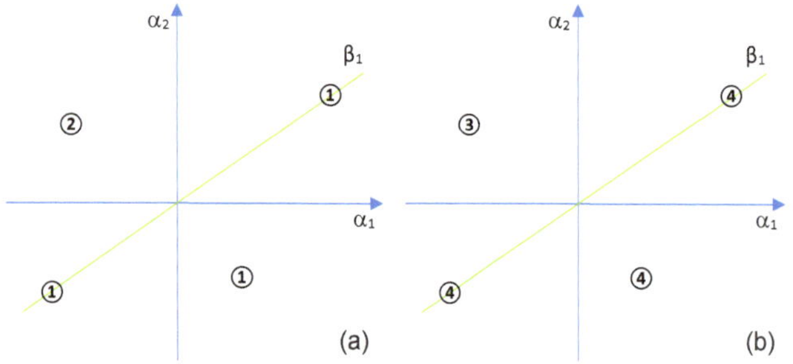

Figure 11. Bifurcation diagrams when $p > 0$ or ($p < 0$ and $r > 0$): (**a**) $a > 0,\ L_0 > 0,\ h < 0$ or $a < 0,\ L_0 > 0,\ h < 0$; (**b**) $a > 0,\ L_0 < 0,\ h < 0$ or $a < 0,\ L_0 < 0,\ h < 0$.

When $\Delta_2 > 0$, we have $e_1,\ m_1,\ m_2$, so we write the following situations: $e_1 < m_1 < m_2$, $m_1 < e_1 < m_2$, and $m_1 < m_2 < e_1$. We notice that, in the case $m_1 < m_2 < e_1$, the bifurcations diagrams will be obtained by a rotation from the bifurcation diagrams obtained in the case $e_1 < m_1 < m_2$ because e_1 is not in the interval (m_1, m_2). In addition, we will draw below only β_1 because the two lines of β_2 do not produce the changing of the region of bifurcation in this case.

Remark 6. *When $\Delta_2 > 0$ and $p > 0$ or ($p < 0$ and $r \geq 0$), then the bifurcation diagrams will be obtained as in previous remark, as follows:*

1. *If $e_1 < m_1 < m_2$ and $a > 0,\ L_0 > 0,\ h > 0$ or $a < 0,\ L_0 > 0,\ h > 0$ or if $m_1 < e_1 < m_2$ and $a > 0,\ L_0 > 0,\ h < 0$ or $a < 0,\ L_0 > 0,\ h < 0$, then will obtain Figure 10a.*

2. *If* $e_1 < m_1 < m_2$ *and* $a > 0,\ L_0 < 0,\ h > 0$ *or* $a < 0,\ L_0 < 0,\ h > 0$ *or if* $m_1 < e_1 < m_2$ *and* $a > 0,\ L_0 < 0,\ h < 0$ *or* $a < 0,\ L_0 < 0,\ h < 0$*, then will obtain Figure 10b.*
3. *If* $e_1 < m_1 < m_2$ *and* $a > 0,\ L_0 > 0,\ h < 0$ *or* $a < 0,\ L_0 > 0,\ h < 0$ *or if* $m_1 < e_1 < m_2$ *and* $a > 0,\ L_0 > 0,\ h < 0$ *or* $a < 0,\ L_0 > 0,\ h > 0$*, then will obtain Figure 11a.*
4. *If* $e_1 < m_1 < m_2$ *and* $a > 0,\ L_0 < 0,\ h < 0$ *or* $a < 0,\ L_0 < 0,\ h > 0$ *or if* $m_1 < e_1 < m_2$ *and* $a > 0,\ L_0 < 0,\ h > 0$ *or* $a < 0,\ L_0 < 0,\ h < 0$*, then will obtain Figure 11b.*

3.3. Numerical Simulations

Some numerical examples are given below in order to illustrate the theoretical approach. Matlab simulations are presented for the regions in Figure 11b, but first we have to check the conditions of Remark 5, i.e., $p > 0,\ \Delta_2 < 0$, and $a > 0,\ L_0 < 0,\ h < 0$ for the example given below. Considering $\beta_1(\alpha) = 2\alpha_1^3 + \alpha_2 + \alpha_1^2\alpha_2,\ \beta_2(\alpha) = -(\alpha_1^2 + \alpha_1\alpha_2 + \alpha_2^2)$, with $|\alpha|$ being sufficiently small and $\theta_0 = 0.1,\ L_0 = -1$, we notice that $a = 1,\ b = 0, c = 1,\ d = 2,\ h = -1,\ k = -2,\ l = -1$, and $p = 1 > 0, \Delta_2 < 0, a > 0,\ h < 0$. We find different orbits (x_n, y_n), where $x_n = \rho_n \sin\varphi_n,\ x_n = \rho_n \cos\varphi_n$, when $n = 1, \ldots, N$, N being a fixed number. Then the two-dimensional map, in polar coordinates, becomes,

$$\rho_{n+1} = \rho_n + \rho_n\beta_1(\alpha) + \rho_n^3\beta_2(\alpha) - \rho_n^5,\ \varphi_{n+1} = \varphi_n + \theta_0. \qquad (9)$$

It is obvious that the Chenciner bifurcation is degenerated here.

Figures 12a,b and 13a give the generic portrait phase 3, and Figure 13b gives the generic portrait phase 4.

First consider $\alpha_1 = 0.1,\ \alpha_2 = 0.1,\ N = 2000$, and $(\rho_1, \varphi_1) = (0.3, 0)$ (for green curve), $(\rho_1, \varphi_1) = (0.01, 0)$ (for blue curve), and $(\rho_1, \varphi_1) = (0.03, 0)$ (for red curve), respectively; the discrete orbits can be seen in Figures 12a,b and 13a. The orbits for blue, red, and green curves tend to an invariant stable closed curve. Moreover, in Figure 14a, the red, blue, and green sequence of points represent the ρ_n sequence corresponding to the previous three orbits, respectively, when $N = 2000$ in $(nO\rho_n)$ axis. We can notice that the results from Figures 12 and 13a are checked because ρ_n tends to the same constant number when n tends to infinity, and then the orbits will be on the same circle. In Figure 14b, the red, blue, and green sequence of points represent the ρ_n sequence corresponding to previous three orbits, respectively, when $N = 2000$ in $(nO\rho_n)$ axis. This time, these sequences tend to zero, so the three orbits tend to origin and the result from Figure 13b is checked. Here, $\alpha_1 = 0.5,\ \alpha_2 = -0.513,\ N = 2000$ are taken, and the start points are the same as in Figure 13b. It can be observed that the orbit tends to the origin, therefore region 4 will appear; see Figure 13b.

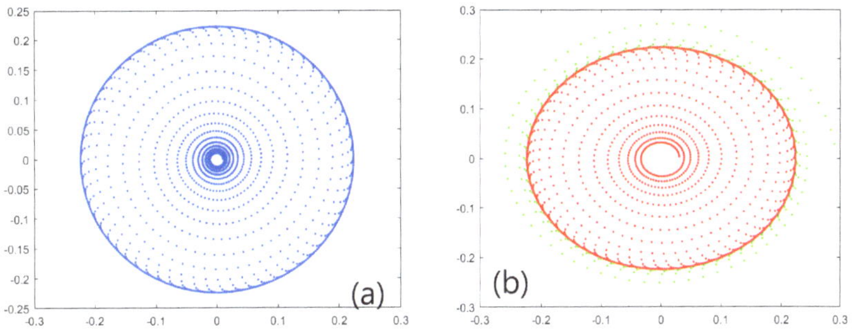

Figure 12. Numerical simulation for the map (9) when $\beta_1(\alpha) = 2\alpha_1^3 + \alpha_2^3 + \alpha_1^2\alpha_2,\ \beta_2(\alpha) = -\alpha_1^2 - \alpha_1\alpha_2 - \alpha_2^2$, with $\alpha_1 = 0.1,\ \alpha_2 = 0.1$: (**a**) blue orbit starts from $(\rho_1, \varphi_1) = (0.01, 0)$; (**b**) red orbit starts from $(\rho_1, \varphi_1) = (0.03, 0)$.

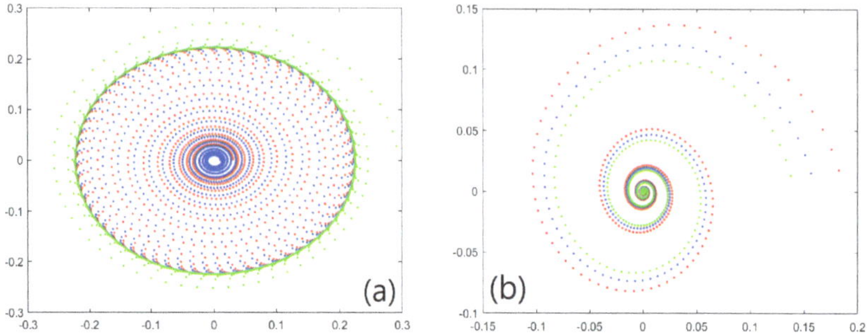

Figure 13. Numerical simulation for the map (9) when $\beta_1(\alpha) = 2\alpha_1^3 + \alpha_2^3 + \alpha_1^2\alpha_2$, $\beta_2(\alpha) = -\alpha_1^2 - \alpha_1\alpha_2 - \alpha_2^2$: (**a**) the three orbits are represented here with $(\rho_1, \varphi_1) = (0.01, 0)$, $(\rho_1, \varphi_1) = (0.03, 0)$ and $(\rho_1, \varphi_1) = (0.3, 0)$, respectively, and $\alpha_1 = 0.1$, $\alpha_2 = 0.1$; (**b**) the three orbits are represented here with $(\rho_1, \varphi_1) = (0.183, 0)$, $(\rho_1, \varphi_1) = (0.16, 0)$ and $(\rho_1, \varphi_1) = (0.14, 0)$, respectively, and $\alpha_1 = 0.5$, $\alpha_2 = -0.513$.

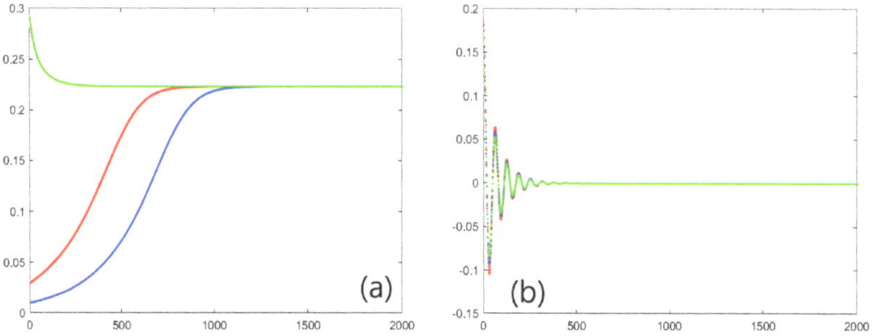

Figure 14. The discrete sequence ρ_n given by the map (9) in the plane $(nO\rho_n)$ when $\beta_1(\alpha) = 2\alpha_1^3 + \alpha_2^3 + \alpha_1^2\alpha_2$, $\beta_2(\alpha) = -\alpha_1^2 - \alpha_1\alpha_2 - \alpha_2^2$: (**a**) when $(\rho_1, \varphi_1) = (0.01, 0)$, $(\rho_1, \varphi_1) = (0.03, 0)$, $(\rho_1, \varphi_1) = (0.284, 0)$ and $a_1 = 0.1$, $a_2 = 0.1$; (**b**) when starting points are as in Figure 13b.

Now choosing $\beta_1(\alpha) = \alpha_1^3 + 2\alpha_2^3 - \alpha_1\alpha_2^2 - \alpha_1^2\alpha_2$, $\beta_2(\alpha) = -\alpha_1 - 3\alpha_2$, $\alpha_1 = 0.1$, $\alpha_2 = -0.1$, and $(\rho_1, \varphi_1) = (0.06)$, $N = 700$, the orbit (green color) tends to origin and will depart from the inner invariant curve (magenta color). However, when $(\rho_1, \varphi_1) = (0.187, 0)$, the orbit (blue color) will tend from interior to the outer invariant curve (red color). When $(\rho_1, \varphi_1) = (0.3, 0)$, the orbit (in red) will tend from exterior to the outer invariant curve. Thus, here, in Figure 15a, appears the phase portrait for the region 7, see Appendix A, and this is confirmed also from theoretical conditions from Figure 2b. In Figure 15b, the sequence ρ_n in $(nO\rho_n)$ axis is shown for green orbit from Figure 15a, where $N = 6000$, observing that this sequence tends to zero when n tends to infinity. In Figure 16a, the sequence x_n is given in the axis (nOx_n), for N = 15,000, and also tends to zero.

In Figure16b is considered the case when $(\alpha_1, \alpha_2) = (0.9, -0.9)$ are on $\beta_1(\alpha) = 0$. Here $\beta_1(\alpha) = 2\alpha_1^3 + \alpha_2^3 + \alpha_1^2\alpha_2$, $\beta_2(\alpha) = -\alpha_1^2 - \alpha_1\alpha_2 - \alpha_2^2$, $\theta_0 = 0.1$. Now $\beta_1(\alpha_1, \alpha_2) = 0$, but $\Delta_2 < 0$ and $(\rho_1, \varphi_1) = (0.187, 0)$ for red orbit, $(\rho_1, \varphi_1) = (0.16, 0)$ for blue orbit, and $(\rho_1, \varphi_1) = (0.14, 0)$ for green orbit, respectively, which tend to the origin. Therefore, the region 4 corresponds to the phase portrait, see Figure 11b, this being the third and last case analyzed for Figure 11b.

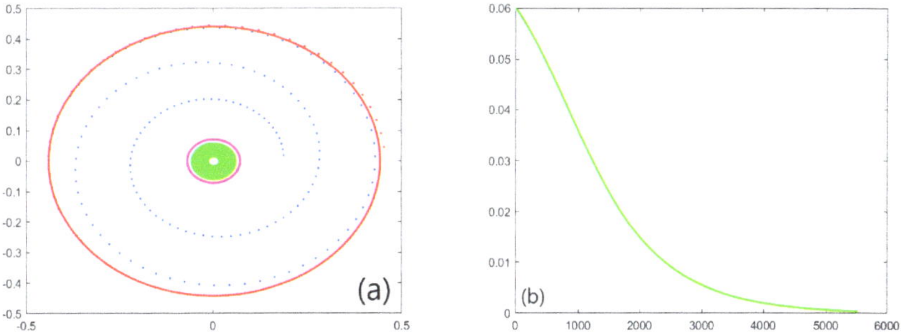

Figure 15. Numerical simulations for the map (9) when $\beta_1(\alpha) = \alpha_1^3 + 2\alpha_2^3 - \alpha_1\alpha_2^2 - \alpha_1^2\alpha_2$, $\beta_2(\alpha) = -\alpha_1 - 3\alpha_2$ and $(\alpha_1, \alpha_2) = (0.1, -0.1)$: (**a**) four orbits corresponding to $(\rho_1, \varphi_1) = (0.06, 0)$ (the orbit in red), $(\rho_1, \varphi_1) = (0.187, 0)$ (the orbit in blue), $(\rho_1, \varphi_1) = (0.0716, 0)$ (the orbit in magenta), $(\rho_1, \varphi_1) = (0.06)$ (the orbit in green); (**b**) the sequence ρ_n in the plane $(nO\rho_n)$ corresponding to the green orbit, when $N = 6000$ from (**a**).

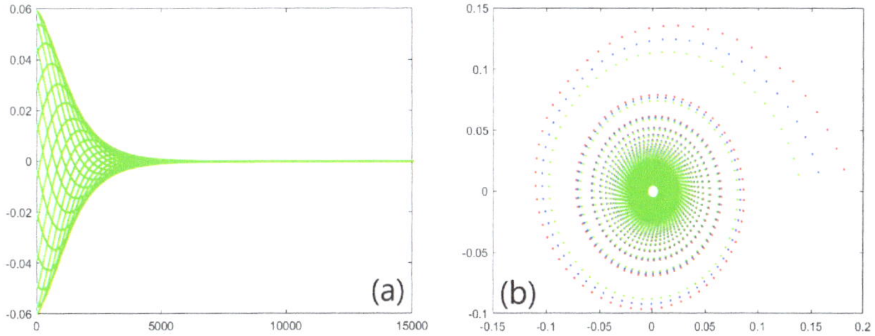

Figure 16. Numerical simulations for the map (9): (**a**) sequence x_n in the plane $(nO\rho)$ from Figure 15b; (**b**) numerical simulations for the map (9) when $\beta_1(\alpha) = 2\alpha_1^3 + \alpha_2^3 + \alpha_1^2\alpha_2$, $\beta_2(\alpha) = -\alpha_1^2 - \alpha_1\alpha_2 - \alpha_2^2$, $(\alpha_1, \alpha_2) = (0.9, -0.9)$ and $(\rho_1, \varphi_1) = (0.183, 0)$ (red orbit), $(\rho_1, \varphi_1) = (0.16, 0)$ (blue orbit), $(\rho_1, \varphi_1) = (0.14, 0)$, (green orbit), respectively.

Moreover, in Figure 17a,b appear the phase portraits 2 and 1 from Figure 11a, when $p > 0,\ a > 0,\ L_0 > 0,\ h < 0$ for the map,

$$\rho_{n+1} = \rho_n + \rho_n\beta_1(\alpha) + \rho_n^3\beta_2(\alpha) + \rho_n^5, \varphi_{n+1} = \varphi_n + \theta_0, \tag{10}$$

i.e., $L_0 = 1$. Here we take $\theta_0 = 0.1$, $\beta_1(\alpha) = 2\alpha_1^3 + \alpha_2^3 + \alpha_1^2\alpha_2$, $\beta_2(\alpha) = -\alpha_1^2 - \alpha_2^2 - \alpha_1\alpha_2$, and $\alpha_1 = 0.1,\ \alpha_2 = 0.1$ for Figure 17a. The starting points of the three orbits correspond to $(\rho_1, \varphi_1) = (0.2, 0)$ for the red color, $(\rho_1, \varphi_1) = (0.16, 0)$ for the blue color, and $(\rho_1, \varphi_1) = (0.11, 0)$ for the green color, respectively, and $N = 100$ step for the red orbit and $N = 150$ step for the blue and green orbits. The orbits depart from the origin and escape to infinity. This situation corresponds to phase portrait 2.

When $\alpha_1 = 0.1,\ \alpha_2 = -0.112$, and the same starting points are taken for the red and green orbits, but $\theta_0 = 0.003$, $N = 1500$ for the blue and green orbits, and, for the blue orbit, $(\rho_1, \varphi_1) = (0.1711, 0)$, $N = 200$, and $\theta_0 = 0.1$, then the red orbit departs from the invariant circle, which is the blue orbit, and the green orbit departs from the invariant circle and tends to origin. That corresponds to the phase portrait 1, and this happens in region 1 from Figure 11a.

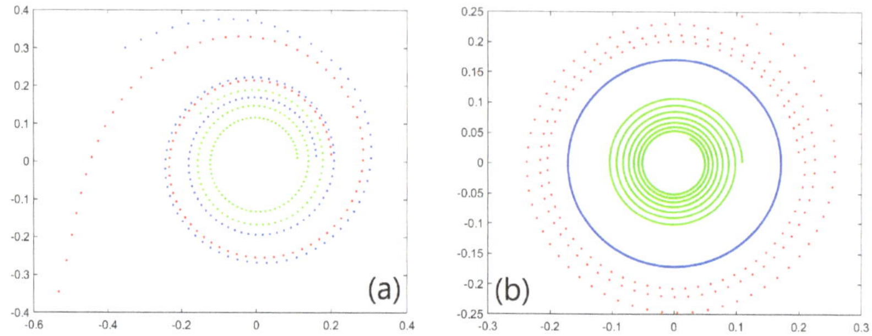

Figure 17. Numerical simulations for the map (10) when $\beta_1(\alpha) = 2\alpha_1^3 + \alpha_2^3 + \alpha_1^2\alpha_2$, $\beta_2(\alpha) = -\alpha_1^2 - \alpha_1\alpha_2 - \alpha_2^2$: (**a**) when $(\alpha_1, \alpha_2) = (0.1, 0.1)$, three orbits having $(\rho_1, \varphi_1) = (0.2, 0)$ (red color), $(\rho_1, \varphi_1) = (0.16, 0)$ (blue color), and $(\rho_1, \varphi_1) = (0.11, 0)$ (green color) are given, corresponding this case to region 2 from Figure 11a; (**b**) when $(\alpha_1, \alpha_2) = (0.1, -0.112)$ and the three starting points of the orbits correspond to $(\rho_1, \varphi_1) = (0.2, 0)$ (red orbit), $(\rho_1, \varphi_1) = (0.1711, 0)$ (blue orbit), $(\rho_1, \varphi_1) = (0.11, 0)$, (green orbit), respectively, we obtain the phase portrait corresponding to region 1 from Figure 11a.

4. Discussions and Conclusions

This paper contributes to the enrichment of the literature related to the Chenciner bifurcation. This study may be useful in biology, medicine, and economics, where discrete Chenciner bifurcation occurs.

The degeneracy case of the Chenciner bifurcation written in the truncated normal form, which was analyzed here, takes place when $a_{20} = a_{11} = a_{02} = 0$, and for b_{10} and b_{01}, we have two situations: $b_{10} \neq 0$, $b_{01} \neq 0$ or $b_{10} = b_{01} = 0$. This is a further degeneration of β_1. It appears here a symmetry and an asymmetry of some regions from bifurcation diagrams in this case studied.

The proposed approach is different from that of [28], being similar to that of [26,27], solving the problem in a more general framework than in [28]. This paper continues the study realized in [26,27], which is shortly described in Appendixes A and B, by considering the following new assumption $a_{10} = a_{01} = a_{20} = a_{11} = a_{02} = 0$. A different method is necessary than that used in [26], based on the sign of Δ and Δ_2 when degree of $\beta_1(\alpha)$ is three and degree of $\beta_2(\alpha)$ is one or two.

This article highlights 18 different bifurcation diagrams, which is more than the two obtained in the case of non-degeneration [25]. Those 18 different bifurcation diagrams come from the first case, Case 3.1, when $\text{Deg}\beta_1(\alpha) = 3$ and $\text{Deg}\beta_2(\alpha) = 1$, here having six bifurcation diagrams, and from the second case, Case 3.2, when $\text{Deg}\beta_1(\alpha) = 3$ and $\text{Deg}\beta_2(\alpha) = 2$, where 12 different bifurcation diagrams appear. The study we conducted in this article confirms the hypothesis. Therefore, in a case of degeneration that does not involve resonance, there is an increase in the number of bifurcation diagrams. This study answers a part of the open problem from [26], and a new open problem would be to study the behavior of the system when $Deg\beta_1(\alpha) = 3$ and $Deg\beta_2(\alpha) = 3$ in the truncated form.

There are more cases of possible degeneration of Chenciner bifurcation, and each of them requires a special characteristic method of solving, especially developed for each case. Matlab simulations verify the theoretical conclusions.

Author Contributions: Conceptualization, S.L. and L.C.; formal analysis, S.L., L.C., and E.G.; investigation, S.L., L.C., and E.G.; methodology, S.L. and L.C.; resources, L.C. and E.G.; data curation, L.C. and E.G.; writing—original draft preparation, L.C.; writing—review and editing, L.C.; visualization, S.L., L.C., and E.G.; supervision, S.L., L.C., and E.G.; project administration, S.L., L.C., and E.G.; funding acquisition, L.C. and E.G. All authors have read and agreed to the published version of the manuscript.

Funding: This research received no external funding.

Institutional Review Board Statement: Not applicable.

Informed Consent Statement: Not applicable.

Data Availability Statement: Data sharing not applicable.

Acknowledgments: This research was partially supported by the Horizon 2020-2017-RISE-777911 project.

Conflicts of Interest: The authors declare no conflict of interest.

Appendix A. Chenciner Bifurcations

Below is written the normal form of Neimark-Sacker bifurcation with cubic degeneracy, i.e., Chenciner bifurcation for the system (A1). A discrete dynamical system:"

$$x_{n+1} = f(x_n, \alpha) \tag{A1}$$

with $\alpha = (\alpha_1, \alpha_2) \in \mathbb{R}^2$, $x_n \in \mathbb{R}^2$, $n \in \mathbb{N}$, $f \in C^r$, and $r \geq 2$ can be written as

$$x \longmapsto f(x, \alpha) \tag{A2}$$

"Ref. [26]. By using the same methods as in [25–27], (A2) becomes

$$z \longmapsto \mu(\alpha)z + g(z, \bar{z}, \alpha), \tag{A3}$$

and"

$$\begin{aligned} w &\longmapsto \left(r(\alpha)e^{i\theta(\alpha)} + a_1(\alpha)w\bar{w} + a_2(\alpha)w^2\bar{w}^2\right)w + O\left(|w|^6\right) \\ &= \left(r(\alpha) + b_1(\alpha)w\bar{w} + b_2(\alpha)w^2\bar{w}^2\right)we^{i\theta(\alpha)} + O\left(|w|^6\right) \end{aligned} \tag{A4}$$

respectively, taking into account that g can be written as

$$g(z, \bar{z}, \alpha) = \sum_{k+l\geq 2} \frac{1}{k!l!} g_{kl}(\alpha) z^k \bar{z}^l,$$

where μ, g, $g_{kl}(\alpha)$ are smooth functions, $b_k(\alpha) = a_k(\alpha)e^{-i\theta(\alpha)}$, $k = 1, 2.$, $\mu(\alpha) = r(\alpha)e^{i\theta(\alpha)}$, $r(0) = 1$, and $\theta(0) = \theta_0$" [26]. The following notations were used:

$$\beta_1(\alpha) = r(\alpha) - 1 \text{ and } \beta_2(\alpha) = Re(b_1(\alpha)) \tag{A5}$$

in [26,27] and (A4) was"

$$\begin{cases} \rho_{n+1} = & \rho_n\left(1 + \beta_1(\alpha) + \beta_2(\alpha)\rho_n^2 + L_2(\alpha)\rho_n^4\right) + \rho_n O(\rho_n^6) \\ \varphi_{n+1} = & \varphi_n + \theta(\alpha) + \rho_n^2\left(\frac{Im(b_1(\alpha))}{\beta_1(\alpha)+1} + O(\rho_n, \alpha)\right) \end{cases}, \tag{A6}$$

$L_2(\alpha) = \frac{Im^2(b_1(\alpha)) + 2(1+\beta_1(\alpha))Re(b_2(\alpha))}{2(\beta_1(\alpha)+1)}$" [26–28]).

When $r(0) = 1$, $Re(b_1(0)) = 0$, but $L_2(0) \neq 0$ in (A6), the generalized Neimark–Sacker bifurcation appears and the transformation of parameters

$$(\alpha_1, \alpha_2) \longmapsto (\beta_1(\alpha), \beta_2(\alpha)) \tag{A7}$$

is regular at $(0, 0)$. These types of bifurcations have been studied in [25], and there they are called Chenciner bifurcations. It is easy to see from above that, for $\beta_1(0) = 0$, we have $L_2(0) = \frac{1}{2}\left(Im^2(b_1(0)) + 2Re(b_2(0))\right)$. The idea is "to change these coordinates and to work only using the initial parameters (α_1, α_2) in the form (A6)" [26].

It is known from [26], relation (13), page 4 that

$$\beta_1(\alpha) = \sum_{i+j=1}^{p} a_{ij}\alpha_1^i\alpha_2^j + O(|\alpha|^{p+1}), \quad \beta_2(\alpha) = \sum_{i+j=1}^{q} b_{ij}\alpha_1^i\alpha_2^j + O(|\alpha|^{q+1}) \tag{A8}$$

for $p, q \geq 1$, $a_{10} = \frac{\partial \beta_1}{\partial \alpha_1}|_{\alpha=0}$, $a_{01} = \frac{\partial \beta_1}{\partial \alpha_2}|_{\alpha=0}$, $b_{01} = \frac{\partial \beta_2}{\partial \alpha_2}|_{\alpha=0}$, $b_{10} = \frac{\partial \beta_2}{\partial \alpha_1}|_{\alpha=0}$, and so on.

The transformation (A7) is not regular at $(0,0)$, i.e., the Chenciner bifurcation degenerates, if and only if

$$a_{10}b_{01} - a_{01}b_{10} = 0. \tag{A9}$$

Knowing the "truncated form of the ρ-map of (A6),

$$\rho_{n+1} = \rho_n\left(1 + \beta_1(\alpha) + \beta_2(\alpha)\rho_n^2 + L_2(\alpha)\rho_n^4\right), \tag{A10}$$

the φ-map of the system (A6) describes a rotation by an angle depending on α and ρ and can be approximated by,

$$\varphi_{n+1} = \varphi_n + \theta(\alpha), \tag{A11}$$

being assumed that $0 < \theta(0) < \pi$" [26]. The truncated normal form of (A4) is (A10) and (A11).

Appendix B. Literature Review

It is known that "the one dimensional dynamic system for the ρ-map (A10) has a fixed point in origin for all values of α, which corresponds to the fixed point $O(0, 0)$ in the system (A10) and (A11), and that a positive nonzero fixed point of the one-dimensional ρ-map (A10), corresponds to a closed invariant curve in the truncated two-dimensional map (A10)–(A11)" [26].

On the other hand, $sign(L_2(\alpha)) = sign(L_0)$ for $|\alpha| = \sqrt{\alpha_1^2 + \alpha_2^2}$ sufficiently small because $L_2(\alpha) = L_0(1 + O(|\alpha|))$ and $L_0 \neq 0$. It is considered $O(|\alpha|^n)$ for $n \geq 1$ to be the series, $O(|\alpha|^n) = \sum_{i+j\geq n} c_{ij}\alpha_1^i\alpha_2^j$.

Theorem A1. *The fixed point O is (linearly) stable if $\beta_1(\alpha) < 0$ and unstable if $\beta_1(\alpha) > 0$, for all values α with $|\alpha|$ sufficiently small. On the bifurcation curve $\beta_1(\alpha) = 0$, O is (non-linearly) stable if $\beta_2(\alpha) < 0$ and unstable if $\beta_2(\alpha) > 0$, when $|\alpha|$ is sufficiently small. At $\alpha = 0$, O is (non-linearly) stable if $L_0 < 0$ and unstable if $L_0 > 0$ [26].*

The positive nonzero fixed points of (A10) are solutions of the following equation:

$$L_2(\alpha)y^2 + \beta_2(\alpha)y + \beta_1(\alpha) = 0 \tag{A12}$$

where $y = \rho_n^2$. The roots of (A12) will be denoted by $y_1 = \frac{1}{2L_2}\left(\sqrt{\Delta} - \beta_2\right)$ and $y_2 = -\frac{1}{2L_2}\left(\sqrt{\Delta} + \beta_2\right)$ when these roots are real, and $\Delta(\alpha) = \beta_2^2(\alpha) - 4\beta_1(\alpha)L_2(\alpha)$ [26].

Theorem A2. *"(1) When $\Delta(\alpha) < 0$ for all $|\alpha|$ sufficiently small, the system* (A10) *and* (A11) *has no invariant circles.*

(2) When $\Delta(\alpha) > 0$ for all $|\alpha|$ sufficiently small, the system (A10) *and* (A11) *has:*

(a) one invariant unstable circle $\rho_n = \sqrt{y_1}$ if $L_0 > 0$ and $\beta_1(\alpha) < 0$;
(b) one invariant stable circle $\rho_n = \sqrt{y_2}$ if $L_0 < 0$ and $\beta_1(\alpha) > 0$;
(c) two invariant circles, $\rho_n = \sqrt{y_1}$ unstable and $\rho_n = \sqrt{y_2}$ stable, if $L_0 > 0$, $\beta_1(\alpha) > 0$, $\beta_2(\alpha) < 0$ or $L_0 < 0$, $\beta_1(\alpha) < 0$, $\beta_2(\alpha) > 0$; in addition, $y_1 < y_2$ if $L_0 < 0$ and $y_2 < y_1$ if $L_0 > 0$;
(d) no invariant circles if $L_0 > 0$, $\beta_1(\alpha) > 0$, $\beta_2(\alpha) > 0$ or $L_0 < 0$, $\beta_1(\alpha) < 0$, $\beta_2(\alpha) < 0$.

(3) On the bifurcation curve $\Delta(\alpha) = 0$*, the system* (A10) *and* (A11) *has one invariant unstable circle* $\rho_n = \sqrt{y_1}$ *for all* $L_0 \neq 0$*. Moreover, if* $L_0 < 0$*, the invariant circle is stable from the exterior and unstable from the interior, while if* $L_0 > 0$ *it is vice versa.*

(4) When $\beta_1(\alpha) = 0$*, the system* (A10) *and* (A11) *has one invariant circle* $\rho_n = \sqrt{-\frac{\beta_2(\alpha)}{L_0}}$ *whenever* $L_0\beta_2(\alpha) < 0$*. It is stable if* $L_0 < 0$ *and* $\beta_2(\alpha) > 0$*, respectively, unstable if* $L_0 > 0$ *and* $\beta_2(\alpha) < 0$*" [26–28].*

Corresponding to the studies we have carried out previously [26,27], the following phase portraits can be highlighted below. In this case, the phase portraits for the curves of bifurcation when $\Delta(\alpha) = 0$ are shown.

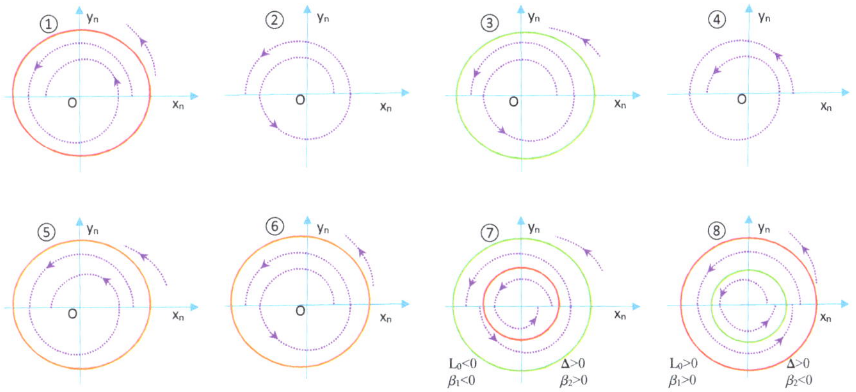

Figure A1. Generic portraits phase when $\theta_0 > 0$.

References

1. Sanchez-Ruiz, L.M.; Moll-Lopez, S.; Morano-Fernandez, J.A.; Rosello, M.D. Dynamical Continuous Discrete Assessment of Competencies Achievement: An Approach to Continuous Assessment. *Mathematics* **2021**, *9*, 2082. [CrossRef]
2. Elhassan, C.; Zulkifli, S.A.; Iliya, S.Z.; Bevrani, H.; Kabir, M.; Jackson, R.; Khan, M.H.; Ahmed, M. Deadbeat Current Control in Grid-Connected Inverters: A Comprehensive Discussion. *IEEE Access* **2022**, *10*, 3990–4014. [CrossRef]
3. Chang, X.H.; Jin, X. Observer-based fuzzy feedback control for nonlinear systems subject to transmission signal quantization. *Appl. Math. Comput.* **2022**, *414*, 126657. [CrossRef]
4. Park, S.J.; Cho, K.H. Discrete Event Dynamic Modeling and Analysis of the Democratic Progress in a Society Controlled by Networked Agents. *IEEE Trans. Autom. Control* **2022**, *1*, 359–365. [CrossRef]
5. Sukhinov, A.; Belova, Y.; Chistyakov, A.; Beskopylny, A.; Meskhi, B. Mathematical Modeling of the Phytoplankton Populations Geografic Dynamics for Possible Scenarios of Changes in the Azov Sea Hydrological Regime. *Mathematics* **2021**, *9*, 3025. [CrossRef]
6. Niu, L.; Ruiz-Herrera, A. Simple dynamics in non-monotone Kolmogorov systems. *Proc. R. Soc. Edinb. Sect. A-Math.* **2021**, 1–16. [CrossRef]
7. Wang, X.; Lu, J.; Wang, Z.; Li, Y. Dynamics of discrete epidemic models on heterogeneous networks. *Physica A* **2020**, *539*, 122991. [CrossRef]
8. Abdel-Gawad, H.I.; Abdel-Gawad, A.H. Discrete and continuum models of COVID-19 virus, formal solutions, stability and comparison with real data. *Math. Comput. Simul.* **2021**, *190*, 222–230. [CrossRef]
9. Miranda, L.K.A.; Kuwana, C.M.; Huggler, Y.H.; da Fonseca, A.K.P.; Yoshida, M.; de Oliveira, J.A.; Leonel, E.D. A short review of phase transition in a chaotic system. *Eur. Phys. J.-Spec. Top.* **2021**, 1–11. [CrossRef]
10. Nordmark, A.B. Non-periodic motion caused by grazing incidence in an impact oscillator. *J. Sound Vib.* **1991**, *145*, 279–297. [CrossRef]
11. Water, W.; Molenaar, J. Dynamics of vibrating atomic force microscopy. *Nanotechnology* **2000**, *11*, 192–199. [CrossRef]
12. Molenaar, J.; Weger, J.G.; Water, W. Mappings of grazing-impact oscillators. *Nonlinearity* **2001**, *14*, 301–321. [CrossRef]
13. Mercinger, M.; Fercec, B.; Oliveire, R.; Pagon, D. Cyclicity of some analytic maps. *Appl. Math. Comput.* **2017**, *295*, 114–125.
14. Yao, W.B.; Li, X.Y. Complicate bifurcation behaviors of a discrete predator-prey model with group defense and nonlinear harvesting in prey. *Appl. Anal.* **2022**, 1–16. [CrossRef]
15. Akrami, M.H.; Atabaigi, A. Dynamics and Neimark-Sacker Bifurcation of a Modified Nicholson-Bailey Model. *J. Math. Ext.* **2022**, *16*, 1–18.
16. Deng, S.F. Bifurcations of a Bouncing Ball Dynamical System. *Int. Bifurc. Chaos* **2019**, *29*, 1950191. [CrossRef]

17. Lines, M.; Westerhoff, F. Effects of inflation expectations on macroeconomics dynamics:extrapolative versus regressive expectations. *Stud. Nonlinear Dyn. Econom.* **2012**, *16*, 7. [CrossRef]
18. Neugart, M.; Tuinstra, J. Endogenous fluctuations in the demand for education. *J. Evol. Econ.* **2003**, *13*, 29–51. [CrossRef]
19. Palan, S. A Review of Bubbles and Crashes in Experimental Asset Markets. *J. Econ. Surv.* **2013**, *27*, 570–588. [CrossRef]
20. Agliari, A.; Hommes, C.H.; Pecora, N. Path dependent coordination of expectations in asset pricing experiments: A behavioral explanation. *J. Econ. Behav. Organ.* **2016**, *121*, 15–28. [CrossRef]
21. Chenciner, A. Bifurcations de points fixes elliptiques. III. Orbites periodiques de "petites periodes" et elimination resonnante des couples de courbes invariantes. *Inst. Hautes Etudes Sci. Publ. Math.* **1988**, *66*, 5–91. [CrossRef]
22. Chenciner, A. Bifurcations de points fixes elliptiques. I. Courbes invariantes. *IHES-Publ. Math.* **1985**, *61*, 67–127. [CrossRef]
23. Chenciner, A. Bifurcations de points fixes elliptiques. II. Orbites periodiques et ensembles de Cantor invariants. *Invent. Math.* **1985**, *80*, 81–106. [CrossRef]
24. Arrowsmith, D.; Place C. *An Introduction to Dynamical Systems*; Cambridge University Press: Cambridge, UK, 1990.
25. Kuznetsov, Y.A. *Elements of Applied Bifurcation Theory*, 3rd ed.; Springer: New York, NY, USA, 2004.
26. Tigan, G.; Lugojan, S.; Ciurdariu, L. Analysis of Degenerate Chenciner Bifurcation. *Int. J. Bifurc. Chaos* **2020**, *30*, 2050245. [CrossRef]
27. Lugojan, S.; Ciurdariu, L.; Grecu, E. New Elements of analysis of a degenerate Chenciner bifurcation. *Symmetry* **2022**, *14*, 77. [CrossRef]
28. Tigan, G.; Brandibur, O.; Kokovics, E.A.; Vesa, L.F. Degenerate Chenciner Bifurcation Revisited. *Int. J. Bifurc. Chaos* **2021**, *31*, 2150160. [CrossRef]

mathematics

Article

Event-Triggered Asynchronous Filter of Nonlinear Switched Positive Systems with Output Quantization

Shitao Zhang [1], Peng Lin [1,2,*] and Junfeng Zhang [1,2,*]

[1] School of Automation, Hangzhou Dianzi University, Hangzhou 310018, China; shitaozhang@hdu.edu.cn
[2] Digital Economy Research Institute, Hangzhou Dianzi University, Hangzhou 310018, China
* Correspondence: penglin@hdu.edu.cn (P.L.); jfz@hdu.edu.cn (J.Z.)

Abstract: This paper deals with a static/dynamic event-triggered asynchronous filter of nonlinear switched positive systems with output quantization. The nonlinear function is located in a sector. Both static and dynamic event-triggering conditions are established based on the 1-norm form. By virtue of the event-triggering mechanism, the error system is transformed into an interval uncertain system. An event-triggered asynchronous filter is designed by employing a matrix decomposition approach. The positivity and L_1-gain stability of the error system are guaranteed by means of linear copositive Lyapunov functions and a linear programming approach. Finally, two examples are given to verify the effectiveness of the design.

Keywords: switched nonlinear positive systems; event-triggered filter; asynchronous switching; linear programming

check for updates

Citation: Zhang, S.; Lin, P.; Zhang, J. Event-Triggered Asynchronous Filter of Nonlinear Switched Positive Systems with Output Quantization. *Mathematics* **2022**, *10*, 599. https://doi.org/10.3390/math10040599

Academic Editor: Yueyuan Zhang

Received: 4 January 2022
Accepted: 3 February 2022
Published: 15 February 2022

1. Introduction

As an important class of hybrid systems, switched positive systems composing of a series of positive subsystems and a switching rule to coordinate the operation of subsystems have attracted extensive attention [1,2]. Compared with the general (non-positive) switched systems [3–5], switched positive systems are more suitable to accurately model a kind of practical system consisting of nonnegative quantities, such as communication networks [6], chemical engineering [7], and water systems [8]. In [9,10], the stability and stabilization of switched positive systems were investigated based on linear copositive Lyapunov functions.

The study [11] dealt with the issue of L_1-gain characterization for switched positive systems by virtue of copositive Lyapunov function and linear programming approach. The L_1-gain analysis and control synthesis of switched positive systems was investigated in [12] using multiple linear copositive Lyapunov functions incorporated with the average dwell time approach. More results on a switched positive systems can be found in [13–15]. The above literature mainly investigated linear switched positive systems. In fact, nonlinear processes exist in most practical systems. A linear model cannot describe the nonlinear processes accurately.

Modeling such systems via nonlinear switched systems will have less error than linear switched systems. In [16], the distributed filter was proposed for nonlinear switched positive systems with stochastic nonlinearities and missing measurements based on switched Lyapunov function and linear programming. A sector nonlinearity was first introduced to ensure the positivity of nonlinear switched positive systems in [17,18], where the considered nonlinear functions are located in a sector.

A robust fault detection filter was designed for nonlinear switched systems with time-varying delay based on the average dwell time approach and the Lyapunov functional technique [19]. In [20], the issue of H_∞ filter of nonlinear switched systems with stable and unstable subsystems was solved by means of the mode-dependent average dwell time technique. Further results about nonlinear switched systems can refer to [21–25].

The filter design for switched systems in the literature mentioned above is mainly based on synchronous switching. It should be pointed out that it takes time to identify which subsystem is activated and which matched filter is activated. The transmission time delay of the switching signal or the impact of the external factors may result in asynchronous switching between the filters and the switched systems [26,27]. Therefore, an asynchronous filter is more practical than a synchronous one.

An asynchronous ℓ_1 positive filter of switched positive systems with modal-dwell-time was proposed in [28]. Using the average dwell time and linear matrix inequality, the study [29] was concerned with the H_∞ filtering problem of linear switched systems with asynchronous switching. An L_1-gain filter of switched positive systems was investigated by introducing a clock-dependent Lyapunov function [30]. In addition, the issue of quantization was considered in [27], which can deal with the failure phenomenon of elements. The quantization can also guarantee the safety of information transmission [31–33].

The study [31] was concerned with feedback stabilization problems for linear time-invariant control systems with saturating quantized measurements. In [32], the authors investigated a design method of a time-varying quantizer to stabilize switched systems with quantized output and switching delays based on a dwell-time assumption and level sets of a common Lyapunov function. Using the sojourn probability-based switching law and parameterized Lyapunov functional, the literature [33] addressed the issue of quantized H_∞ filtering for switched linear parameter-varying systems with both sojourn probabilities and unreliable communication channels. How can we establish an asynchronous filter framework of nonlinear switched positive systems and solve the signal quantization based on a linear approach? These questions motivate the current investigation.

Up to now, many related results on event-triggering mechanism have been reported in [34–38]. Event-triggered communication mechanism provides a more effective and practical method for solving the control issues than time-triggered sample to reduce unnecessary signal transmission. The study [39] investigated the event-triggered L_1-gain filter of switched positive systems subject to state saturation using linear programming and linear copositive Lyapunov function.

In [40], an event-triggered filter of positive systems was designed by adopting a matrix decomposition approach and linear copositive Lyapunov functions. An event-triggered filter of switched positive systems subject to state saturation was investigated by resorting to linear programming and average dwell time technology [41]. More recently, a dynamic event-triggered mechanism, which was developed from the static one, has been presented in [42,43]. In [44], the issue of recursive distributed filtering was investigated for nonlinear time-varying systems under a dynamic event-triggered mechanism.

With the help of the mathematical induction and Lyapunov theorem, the study [45] presented a dynamic event-triggered control scheme for linear time-invariant systems. The study [46] dealt with the stability of linear stochastic systems based on the dynamic event-triggered mechanism with an impulsive switched system approach. To the best of authors' knowledge, there have been no research achievements regarding the asynchronous filter design of nonlinear switched positive systems under a static/dynamic event-triggering mechanism. Therefore, applying static and dynamic event-triggering communication mechanisms to the asynchronous filter design of switched nonlinear positive systems is one motivation of this work.

In this paper, we focus on the event-triggered L_1-gain asynchronous filter of nonlinear switched positive systems with output quantization. Static and dynamic event-triggering schemes based on 1-norm inequality are constructed for the considered systems, respectively. The filter gain matrices are designed by using the matrix decomposition technique to guarantee the positivity and L_1-gain stability of the underlying systems. The outline of the paper is as follows: Section 2 provides the problem formulation; Section 3 presents the main results; Two examples are given in Section 4; and Section 5 concludes this paper.

Notation 1. *Let $\mathbb{R}^n$ (or $\mathbb{R}^n_+$) and $\mathbb{R}^{n\times m}$ be sets of n-dimensional vectors (or, nonnegative) and $n \times m$ matrices, respectively. The symbols $\mathbb{N}$ and $\mathbb{N}_+$ denote the sets of nonnegative and positive integers, respectively. For a matrix $A = [a_{ij}]$, $A \succeq 0$ ($\succ 0$) indicates that $a_{ij} \geq 0$ ($a_{ij} > 0$), $\forall i, j = 1, \cdots, n$, where a_{ij} is the element in the ith row and jth column of A. $A^\top$ stands for the transpose of matrix A.*

For $v \in \mathbb{R}^n$, $v^{(\iota)}$ is the ιth element of the vector. $v \succeq 0$ ($\succ 0$) means $v^{(\iota)} \succeq$ ($\succ 0$), $\forall \iota = 1, \cdots, n$. The 1-norm of $x = (x_1, x_2, \ldots, x_n)$ is defined by $\|x\|_1 = \sum_{\iota=1}^n |x_\iota|$, and the ℓ_1 norm of the vector is $\sum_{k=0}^\infty \|x(k)\|_1$. Define $1_n = (\underbrace{1, \ldots, 1}_{n})^\top \in \mathbb{R}^n$ and $1_n^{(\iota)} = (\underbrace{0, \ldots, 0, 1}_{\iota}, \underbrace{0, \ldots, 0}_{n-\iota})^\top$. A matrix I denotes the identity matrix with appropriate dimensions, and $1_{n\times n} \in \mathbb{R}^{n\times n}$ is a matrix with all the elements being 1. The logic operator $a \vee b$ means that a is valid or b is valid.

2. Preliminaries

Consider the discrete-time nonlinear switched system:

$$\begin{aligned} x(k+1) &= A_{\sigma(k)} f(x(k)) + B_{\sigma(k)} g(\omega(k)), \\ y(k) &= C_{\sigma(k)} h(x(k)) + D_{\sigma(k)} l(\omega(k)), \\ z(k) &= E_{\sigma(k)} p(x(k)) + F_{\sigma(k)} q(\omega(k)), \end{aligned} \tag{1}$$

where $x(k) = (x_1(k), \ldots, x_n(k))^\top \in \mathbb{R}^n$, $y(k) \in \mathbb{R}^m$, $\omega(k) \in \mathbb{R}^m_+$, and $z(k) \in \mathbb{R}^s$ are the system state, measurable output, disturbance, and output to be estimated, respectively. The nonlinear functions satisfy that

$$\begin{aligned} f(x) &= (f_1(x_1), \ldots, f_n(x_n))^\top, h(x) = (h_1(x_1), \ldots, h_n(x_n))^\top, \\ p(x) &= (p_1(x_1), \ldots, p_n(x_n))^\top, g(\omega) = (g_1(\omega_1), \ldots, g_m(\omega_m))^\top, \\ l(\omega) &= (l_1(\omega_1), \ldots, l_m(\omega_m))^\top, q(\omega) = (q_1(\omega_1), \ldots, q_m(\omega_m))^\top. \end{aligned}$$

The function $\sigma(k)$ denotes the switching law taking values at a finite set $S = \{1, 2, \ldots, N\}$, $N \in \mathbb{N}_+$, where N represents the number of subsystems. Assume that the ith subsystem is invoked when $\sigma(k) = i$.

Assumption 1. *The system matrices satisfy that $A_i \succeq 0, B_i \succeq 0, C_i \succeq 0, D_i \succeq 0, E_i \succeq 0$, and $F_i \succeq 0$ for each $i \in S$.*

Assumption 2. *The nonlinear functions $f(x), g(\omega), h(x), l(\omega), p(x)$, and $q(\omega)$ are located in some sector fields with*

$$\varpi_1 x_i^2 \leq f_i(x_i)x_i \leq \varpi_2 x_i^2, \varpi_3 x_i^2 \leq h_i(x_i)x_i \leq \varpi_4 x_i^2, \varpi_5 x_i^2 \leq p_i(x_i)x_i \leq \varpi_6 x_i^2, \tag{2}$$

$$\varepsilon_1 \omega_\iota^2 \leq g_\iota(\omega_\iota)\omega_\iota \leq \varepsilon_2 \omega_\iota^2, \varepsilon_3 \omega_\iota^2 \leq l_\iota(\omega_\iota)\omega_\iota \leq \varepsilon_4 \omega_\iota^2, \varepsilon_5 \omega_\iota^2 \leq q_\iota(\omega_\iota)\omega_\iota \leq \varepsilon_6 \omega_\iota^2, \tag{3}$$

where $i = 1, 2, \cdots, n$, $\iota = 1, 2, \cdots, m$, $0 < \varpi_1 \leq \varpi_2, 0 < \varpi_3 \leq \varpi_4, 0 < \varpi_5 \leq \varpi_6, 0 < \varepsilon_1 \leq \varepsilon_2, 0 < \varepsilon_3 \leq \varepsilon_4, 0 < \varepsilon_5 \leq \varepsilon_6$, and $f_i(0) = 0$.

Some preliminaries about positive systems are introduced.

Definition 1 ([1,2])**.** *A system is said to be positive if all its states and outputs are nonnegative for any nonnegative initial conditions and nonnegative inputs.*

Remark 1. *There indeed exist some systems whose states and outputs are nonnegative for some non-positive initial conditions and inputs. The nonnegativity of these systems only holds for some of initial conditions and inputs rather than any nonnegative initial conditions and inputs. In* *this paper, the definition of positive system means that the states and outputs are nonnegative for any nonnegative initial conditions and inputs. The definition follows the notions in [1,2]. Such a definition is to guarantee the essential nonnegativity of a system for any nonnegative initial conditions and inputs. Thus, the nonnegative initial conditions are required.*

Lemma 1 ([1,2])**.** *A system $x(k+1) = Ax(k)$ is positive if and only if $A \succeq 0$.*

Lemma 2 ([1,2])**.** *Given a matrix $A \succeq 0$, the following conditions are equivalent:*
(i) The matrix A is a Schur matrix.
(ii) There exists some vector $v \succ 0$ such that $(A - I)v \prec 0$.

Definition 2 ([3])**.** *For a switching signal $\sigma(k)$ and $0 \leq k_1 \leq k_2$, denote the number of the switching of $\sigma(k)$ by $N_\sigma(k_2, k_1)$. If there exist $N_0 \geq 0$ and $\tau_a > 0$ such that*

$$N_\sigma(k_2, k_1) \leq N_0 + (k_2 - k_1)/\tau_a,$$

then τ_a is an average dwell time of the switching signal $\sigma(k)$.

Definition 3 ([41])**.** *The system (1) is said to be ℓ_1-gain stable if the following statements hold:*
(i) For $\omega(k) = 0$, the system (1) is asymptotically stable.
(ii) Under zero-initial conditions, the following inequality holds for $\omega(k) \neq 0$,

$$\textstyle\sum_{k=0}^{\infty} e^{-\hbar k}\|e(k)\|_1 \leq \gamma \sum_{k=0}^{\infty} \|\omega(k)\|_1,$$

where $\gamma > 0$ is the ℓ_1-gain value and $\hbar > 0$.

3. Main Results

This section first explores the positivity of system (1). Then, a nonlinear asynchronous filter is designed under static event-triggering mechanism for system (1) with output quantization. Finally, a dynamic event-triggering filter for system (1) is proposed.

3.1. Positivity

Lemma 3. *Under Assumption 2, system (1) is positive if and only if Assumption 1 holds.*

Proof. *Necessity*. Let $x(0) = 0$. Then, $x(1) = B_i g(\omega(0))$ for some $i \in S$. By Assumption 2, $g(\omega(0)) \succeq 0$ for any $\omega(0) \succeq 0$. Since $x(1) \succeq 0$ for any $g(\omega(0)) \succeq 0$, then $B_i \succeq 0$.

Now, we prove that $A_i \succeq 0$ via reductio ad absurdum. Let $g(\omega(k)) = 0$. Suppose there exists an element $a_i^{(\imath\jmath)} < 0$, then we find

$$x_\imath(k+1) = \textstyle\sum_{j=1, j\neq \jmath}^{n} (a_i^{(\imath j)} f_j(x_j(k)) + a_i^{(\imath\jmath)} f_\jmath(x_\jmath(k))). \tag{4}$$

It is possible that $x_\imath(k+1) < 0$ if $a_i^{(\imath\jmath)}$ takes a small value enough, which yields a contradiction with the positivity of system (1). Thus, $A_i \succeq 0$.

Sufficiency. Denote by $\digamma$ the set of indices that satisfies $x_\imath(k) = 0$ for $\imath \in \digamma$. Then, for some $i \in S$

$$x_\imath(k+1) = \textstyle\sum_{j \notin \Omega} a_i^{(\imath j)} f_j(x_j(k)) + \sum_{\iota=1}^{m} b_i^{(\imath\iota)} g_\iota(\omega_\iota(k)), \imath \in \digamma, \tag{5}$$

where $a_i^{(\imath\jmath)}$ is the $\imath$th row $\jmath$th column element of A_i and $b_i^{(\imath\iota)}$ is the $\imath$th row ιth column element of B_i. Note the condition (2), it follows that $f_j(x_j(k)) \geq 0$ for $k \in [0, +\infty)$. By Assumption 1, $a_i^{(\imath\jmath)} \geq 0$ and $b_i^{(\imath\iota)} \geq 0$. From (3), we have $g(\omega(k)) \succeq 0, l(\omega(k)) \succeq 0$, and $q(\omega(k)) \succeq 0$ for $\omega(k) \in \mathbb{R}_+^m$. So, we have $x_\imath(k+1) \geq 0$ for $g(\omega(k)) \succeq 0$. By (2a), we have $h(x(k)) \succeq 0$ and $p(x(k)) \succeq 0$ for $x(k) \succeq 0$. Together these with $C_i \succeq 0, D_i \succeq 0, E_i \succeq 0$, and $F_i \succeq 0$ give $y(k) \succeq 0$ and $z(k) \succeq 0$. □

The proof of Lemma 3 follows the proof of the positivity in [1,2]. The sector conditions in Assumption 2 are key to the positivity of system (1).

3.2. Static Event-Triggering Case

This subsection aims to design an event-triggered nonlinear asynchronous filter for a switched nonlinear positive system (1). A more general case is investigated, where the nonlinear function is unknown. This implies that the nonlinear function $f(x(k))$ cannot be used for the filter design. Thus, a nonlinear function $\hat{f}(x)$ is introduced to estimate the unknown nonlinear function $f(x)$. Then, the corresponding filter design is more difficult compared with the former filter design.

First, we introduce an event-triggering mechanism to detect and manage the transmission of output variables. Define $e_y(k) = \tilde{y}(k) - y(k)$, where $\tilde{y}(k) = y(k_\wp), y(k_\wp)$ is the output of event generator at the event-triggering instant $k_\wp, \wp \in \mathbb{N}$. The measurement output will be released only when the following event-triggering condition is satisfied:

$$\|e_y(k)\|_1 > \beta \|y(k)\|_1, \tag{6}$$

where $k \in [k_\wp, k_{\wp+1})$ and $\beta \in [0, 1)$ is the event-triggering coefficient.

To further reduce the design cost and increase the practicability of the filter, we introduce a quantization technique to measure the output signal. Figure 1 is the event-triggered nonlinear quantization filter framework of switched nonlinear positive systems. The model of the quantized output signal is given as:

$$\bar{y}(k) = \bar{\mathcal{U}}(\tilde{y}(k)) = \left(\bar{\mathcal{U}}_1(\tilde{y}_1(k)), \bar{\mathcal{U}}_2(\tilde{y}_2(k)), \cdots, \bar{\mathcal{U}}_m(\tilde{y}_m(k))\right)^\top,$$

where $\bar{y}(k) \in \mathbb{R}^m$ denotes the quantized signal of the event generator's output signal $\tilde{y}(k)$ and $\bar{\mathcal{U}}(\tilde{y}(k))$ is the logarithmic quantizer. Moreover, the subquantizer $\bar{\mathcal{U}}_c(\tilde{y}_c(k))$ $(1 \leq c \leq m)$ is characterized by the set of quantization levels:

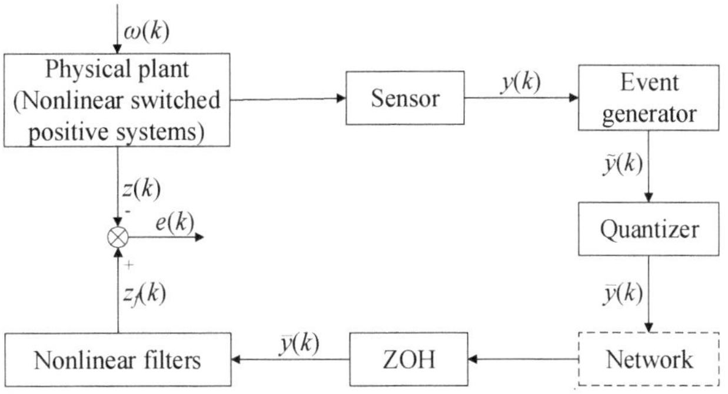

Figure 1. The event-triggered nonlinear filter with output quantization.

$$u_c = \{\phi_c | \phi_c = \kappa_c \phi_{c0}\},$$

where $0 < \kappa_c < 1, \phi_{c0} > 0, \phi_c$ denotes the quantization level corresponding to a segment of cth component of the output signal $\tilde{y}(k)$. Then, the subquantizer $\bar{\mathcal{U}}_c(\tilde{y}_c(k))$ is defined as follows:

$$\bar{\mathcal{U}}_c(\tilde{y}_c(k)) = \begin{cases} \phi_c, & \dfrac{1}{1+\epsilon_c}\phi_c < \tilde{y}_c(k) \leq \dfrac{1}{1-\epsilon_c}\phi_c, \\ 0, & \tilde{y}_c(k) = 0, \\ \tilde{y}_c(k), & 0 < \tilde{y}_c(k) < \dfrac{1}{1+\epsilon_c}\phi_c, \tilde{y}_c(k) > \dfrac{1}{1-\epsilon_c}\phi_c, \end{cases} \tag{7}$$

where $\epsilon_c = \frac{1-\kappa_c}{1+\kappa_c}$. For any quantization error, the following sector-bound expression can be obtained:

$$\bar{U}(\tilde{y}(k)) - \tilde{y}(k) = \Delta(k)\tilde{y}(k),$$

where $\Delta(k) = \text{diag}\{\Delta_1(k), \Delta_2(k), \cdots, \Delta_m(k)\}$ and $|\Delta_c(k)| \leq \epsilon_c$. Then, the output quantization signal based on the event-triggering strategy received by the filter can be described as: $\bar{y}(k) = (I + \Delta(k))\tilde{y}(k)$.

A nonlinear asynchronous filter with output quantization is constructed as follows:

$$\begin{aligned} x_f(k+1) &= A_{f\sigma_f(k)}\hat{f}(x_f(k)) + B_{f\sigma_f(k)}\bar{y}(k), \\ z_f(k) &= E_{f\sigma_f(k)}p(x_f(k)) + F_{f\sigma_f(k)}\bar{y}(k), \end{aligned} \tag{8}$$

where $\hat{f}(x_f(k)) = (\hat{f}_1(x_1), \ldots, \hat{f}_n(x_n))^\top$ satisfies $\vartheta_1 x_i^2 \leq \hat{f}_i(x_i)x_i \leq \vartheta_2 x_i^2$ with $\vartheta_2 > \vartheta_1 > 0$; $x_f(k)$ is the state of the filter; $z_f(k)$ is the estimation of $z(k)$; $\sigma_f(k)$ is switching law of the filter taking values in $S = \{1, 2, \ldots, N\}$; The matrices $A_{f\sigma_f(k)}$, $B_{f\sigma_f(k)}$, $E_{f\sigma_f(k)}$, and $F_{f\sigma_f(k)}$ are to be determined.

Remark 2. *Consider the interval $[k_r, k_{r+1})$, $r = 0, 1, \cdots$, where the asynchronous phenomenon occurs in $[k_r, k_r + \Delta_r)$ and the synchronous switching arises in $[k_r + \Delta_r, k_{r+1})$. This indicates that the ith subsystem and the jth filter are active in $k \in [k_r, k_r + \Delta_r)$, and then the ith filter is active in $[k_r + \Delta_r, k_{r+1})$.*

Remark 3. *The filter (8) is a switched system, and the matrices of the filter depend on the system modes. In this paper, the filter is assumed to be switched asynchronously with the subsystems, which means that the switching instant of the filter lags behind the system (1) by Δ_r, where $\Delta_0 = 0$, $\Delta_r < k_{r+1} - k_r$, $r = 1, 2, \cdots$, and k_r is the switching time instant. Therefore, $\sigma_f(k_r) = \sigma(k_r) + \Delta_r$.*

Let $\tilde{x}(k) = (x^\top(k)\ x_f^\top(k) - x^\top(k))^\top$ and $e(k) = z_f(k) - z(k)$. Based on system (1) and filter (8), the following error system is obtained: For $k \in [k_r, k_r + \Delta_r)$,

$$\begin{aligned} \tilde{x}(k+1) &= \begin{pmatrix} A_i f(x(k)) + B_i g(\omega(k)) \\ A_{fj}\hat{f}(x_f(k)) + B_{fj}(I + \Delta(k))(C_i h(x(k)) + D_i l(\omega(k)) + e_y(k)) - A_i f(x(k)) - B_i g(\omega(k)) \end{pmatrix}, \\ e(k) &= E_{fj}p(x_f(k)) + F_{fj}(I + \Delta(k))(C_i h(x(k)) + D_i l(\omega(k)) + e_y(k)) - E_i p(x(k)) - F_i q(\omega(k)), \end{aligned} \tag{9}$$

and for $k \in [k_r + \Delta_r, k_{r+1})$,

$$\begin{aligned} \tilde{x}(k+1) &= \begin{pmatrix} A_i f(x(k)) + B_i g(\omega(k)) \\ A_{fi}\hat{f}(x_f(k)) + B_{fi}(I + \Delta(k))(C_i h(x(k)) + D_i l(\omega(k)) + e_y(k)) - A_i f(x(k)) - B_i g(\omega(k)) \end{pmatrix}, \\ e(k) &= E_{fi}p(x_f(k)) + F_{fi}(I + \Delta(k))(C_i h(x(k)) + D_i l(\omega(k)) + e_y(k)) - E_i p(x(k)) - F_i q(\omega(k)). \end{aligned} \tag{10}$$

Let $\Lambda = \text{diag}\{\epsilon_1, \epsilon_2, \cdots, \epsilon_m\}$. Thus, we have $0 \preceq L \preceq I + \Delta(k) \preceq J$, where $L = I - \Lambda$ and $J = I + \Lambda$. Based on Assumption 2, we have that, for $k \in [k_r, k_r + \Delta_r)$,

$$\begin{aligned} \tilde{x}(k+1) &\succeq \widetilde{A}_{1ij}\tilde{x}(k) + \widetilde{B}_{1ij}\omega(k) + \widetilde{D}_{1j}e_y(k), \\ e(k) &\succeq \widetilde{E}_{1ij}\tilde{x}(k) + \widetilde{F}_{1ij}\omega(k) + F_{fj}Le_y(k), \end{aligned} \tag{11}$$

and

$$\begin{aligned} \tilde{x}(k+1) &\preceq \widetilde{A}_{2ij}\tilde{x}(k) + \widetilde{B}_{2ij}\omega(k) + \widetilde{D}_{2j}e_y(k), \\ e(k) &\preceq \widetilde{E}_{2ij}\tilde{x}(k) + \widetilde{F}_{2ij}\omega(k) + F_{fj}Je_y(k), \end{aligned} \tag{12}$$

where

$$\widetilde{A}_{1ij} = \begin{pmatrix} \varpi_1 A_i & 0 \\ \vartheta_1 A_{fj} + \varpi_3 B_{fj} L C_i - \varpi_2 A_i & \vartheta_1 A_{fj} \end{pmatrix},$$
$$\widetilde{A}_{2ij} = \begin{pmatrix} \varpi_2 A_i & 0 \\ \vartheta_2 A_{fj} + \varpi_4 B_{fj} J C_i - \varpi_1 A_i & \vartheta_2 A_{fj} \end{pmatrix},$$
$$\widetilde{E}_{1ij} = \begin{pmatrix} \varpi_5 E_{fj} + \varpi_3 F_{fj} L C_i - \varpi_6 E_i & \varpi_5 E_{fj} \end{pmatrix},$$
$$\widetilde{E}_{2ij} = \begin{pmatrix} \varpi_6 E_{fj} + \varpi_4 F_{fj} J C_i - \varpi_5 E_i & \varpi_6 E_{fj} \end{pmatrix},$$
$$\widetilde{B}_{1ij} = \begin{pmatrix} \varepsilon_1 B_i \\ \varepsilon_3 B_{fj} L D_i - \varepsilon_2 B_i \end{pmatrix}, \widetilde{B}_{2ij} = \begin{pmatrix} \varepsilon_2 B_i \\ \varepsilon_4 B_{fj} J D_i - \varepsilon_1 B_i \end{pmatrix},$$
$$\widetilde{F}_{1ij} = \begin{pmatrix} \varepsilon_3 F_{fj} L D_i - \varepsilon_6 F_i \end{pmatrix}, \widetilde{F}_{2ij} = \begin{pmatrix} \varepsilon_4 F_{fj} J D_i - \varepsilon_5 F_i \end{pmatrix},$$
$$\widetilde{D}_{1j} = \begin{pmatrix} 0 \\ B_{fj} L \end{pmatrix}, \widetilde{D}_{2j} = \begin{pmatrix} 0 \\ B_{fj} J \end{pmatrix},$$

and for $k \in [k_r + \Delta_r, k_{r+1})$,

$$\begin{aligned} \widetilde{x}(k+1) &\succeq \widetilde{A}_{1i}\widetilde{x}(k) + \widetilde{B}_{1i}\omega(k) + \widetilde{D}_{1i}e_y(k), \\ e(k) &\succeq \widetilde{E}_{1i}\widetilde{x}(k) + \widetilde{F}_{1i}\omega(k) + F_{fi}Le_y(k), \end{aligned} \tag{13}$$

and

$$\begin{aligned} \widetilde{x}(k+1) &\preceq \widetilde{A}_{2i}\widetilde{x}(k) + \widetilde{B}_{2i}\omega(k) + \widetilde{D}_{2i}e_y(k), \\ e(k) &\preceq \widetilde{E}_{2i}\widetilde{x}(k) + \widetilde{F}_{2i}\omega(k) + F_{fi}Je_y(k), \end{aligned} \tag{14}$$

where

$$\widetilde{A}_{1i} = \begin{pmatrix} \varpi_1 A_i & 0 \\ \vartheta_1 A_{fi} + \varpi_3 B_{fi} L C_i - \varpi_2 A_i & \vartheta_1 A_{fi} \end{pmatrix},$$
$$\widetilde{A}_{2i} = \begin{pmatrix} \varpi_2 A_i & 0 \\ \vartheta_2 A_{fi} + \varpi_4 B_{fi} J C_i - \varpi_1 A_i & \vartheta_2 A_{fi} \end{pmatrix},$$
$$\widetilde{E}_{1i} = \begin{pmatrix} \varpi_5 E_{fi} + \varpi_3 F_{fi} L C_i - \varpi_6 E_i & \varpi_5 E_{fi} \end{pmatrix},$$
$$\widetilde{E}_{2i} = \begin{pmatrix} \varpi_6 E_{fi} + \varpi_4 F_{fi} J C_i - \varpi_5 E_i & \varpi_6 E_{fi} \end{pmatrix},$$
$$\widetilde{B}_{1i} = \begin{pmatrix} \varepsilon_1 B_i \\ \varepsilon_3 B_{fi} L D_i - \varepsilon_2 B_i \end{pmatrix}, \widetilde{B}_{2i} = \begin{pmatrix} \varepsilon_2 B_i \\ \varepsilon_4 B_{fi} J D_i - \varepsilon_1 B_i \end{pmatrix},$$
$$\widetilde{F}_{1i} = \begin{pmatrix} \varepsilon_3 F_{fi} L D_i - \varepsilon_6 F_i \end{pmatrix}, \widetilde{F}_{2i} = \begin{pmatrix} \varepsilon_4 F_{fi} J D_i - \varepsilon_5 F_i \end{pmatrix},$$
$$\widetilde{D}_{1i} = \begin{pmatrix} 0 \\ B_{fi} L \end{pmatrix}, \widetilde{D}_{2i} = \begin{pmatrix} 0 \\ B_{fi} J \end{pmatrix}.$$

Theorem 1. *If there exist constants* $0 < \varpi_1 \leq \varpi_2, 0 < \varpi_3 \leq \varpi_4, 0 < \varpi_5 \leq \varpi_6, 0 < \varepsilon_1 \leq \varepsilon_2$, $0 < \varepsilon_3 \leq \varepsilon_4, 0 < \varepsilon_5 \leq \varepsilon_6, 0 < \vartheta_1 \leq \vartheta_2$, $\gamma > 0$, $\lambda > 1$, $0 \leq \beta < 1$, $0 < \mu_1 < 1$, $\mu_2 > 1$, $\mathbb{R}^n$ *vectors* $\zeta_i \succ 0, \zeta_{(i,j)} \succ 0$, $\varphi_i \succ 0, \varphi_{(i,j)} \succ 0$, $\xi_i \succeq 0$, $\xi_{i\iota} \succeq 0$, $\xi_j \succeq 0$, $\xi_{j\iota} \succeq 0$, $\rho_{ij} \succeq 0$, *and* $\mathbb{R}^m$ *vectors* $\delta_i \succeq 0$, $\delta_{i\iota} \succeq 0$, $\delta_j \succeq 0$, $\delta_{j\iota} \succeq 0$, $\theta_{ij} \succeq 0$ *such that*

$$\vartheta_1 \textstyle\sum_{\iota=1}^{n} 1_n^{(\iota)} \xi_{i\iota}^\top + \varpi_3 \sum_{\iota=1}^{n} 1_n^{(\iota)} \delta_{i\iota}^\top LMC_i - \varpi_2 1_n^\top \varphi_i A_i \succeq 0, \tag{15}$$

$$\varepsilon_3 \textstyle\sum_{\iota=1}^{n} 1_n^{(\iota)} \delta_{i\iota}^\top LMD_i - \varepsilon_2 1_n^\top \varphi_i B_i \succeq 0, \tag{16}$$

$$\varpi_5 \textstyle\sum_{\jmath=1}^{s} 1_s^{(\jmath)} \rho_{i\jmath}^\top + \varpi_3 \sum_{\jmath=1}^{s} 1_s^{(\jmath)} \theta_{i\jmath}^\top LMC_i - \varpi_6 E_i \succeq 0, \tag{17}$$

$$\varepsilon_3 \textstyle\sum_{\jmath=1}^{s} 1_s^{(\jmath)} \theta_{i\jmath}^\top LMD_i - \varepsilon_6 F_i \succeq 0, \tag{18}$$

$$\vartheta_1 \textstyle\sum_{\iota=1}^{n} 1_n^{(\iota)} \xi_{j\iota}^\top + \varpi_3 \sum_{\iota=1}^{n} 1_n^{(\iota)} \delta_{j\iota}^\top LMC_i - \varpi_2 1_n^\top \varphi_j A_i \succeq 0, \tag{19}$$

$$\varepsilon_3 \textstyle\sum_{\iota=1}^{n} 1_n^{(\iota)} \delta_{j\iota}^\top LMD_i - \varepsilon_2 1_n^\top \varphi_j B_i \succeq 0, \tag{20}$$

$$\varpi_5 \textstyle\sum_{\jmath=1}^{s} 1_s^{(\jmath)} \rho_{j\jmath}^\top + \varpi_3 \sum_{\jmath=1}^{s} 1_s^{(\jmath)} \theta_{j\jmath}^\top LMC_i - \varpi_6 E_i \succeq 0, \tag{21}$$

$$\varepsilon_3 \textstyle\sum_{\jmath=1}^{s} 1_s^{(\jmath)} \theta_{j\jmath}^\top LMD_i - \varepsilon_6 F_i \succeq 0, \tag{22}$$

$$\begin{aligned}&\varpi_2 A_i^\top \zeta_i + \vartheta_2 \xi_i + \varpi_4 C_i^\top HJ\delta_i - \varpi_1 A_i^\top \varphi_i - \mu_1 \zeta_i + (\varpi_6 \textstyle\sum_{\jmath=1}^{s} 1_s^{(\jmath)} \rho_{i\jmath}^\top \\ &\quad + \varpi_4 \textstyle\sum_{\jmath=1}^{s} 1_s^{(\jmath)} \theta_{i\jmath}^\top JHC_i - \varpi_5 E_i)^\top 1_s \preceq 0,\end{aligned} \tag{23}$$

$$\varpi_6 (\textstyle\sum_{\jmath=1}^{s} 1_s^{(\jmath)} \rho_{i\jmath}^\top)^\top 1_s + \vartheta_2 \xi_i - \mu_1 \varphi_i \preceq 0, \tag{24}$$

$$\varepsilon_2 B_i^\top \zeta_i + \varepsilon_4 D_i^\top HJ\delta_i - \varepsilon_1 B_i^\top \varphi_i + \varepsilon_4 D_i^\top HJ(\textstyle\sum_{\jmath=1}^{s} 1_s^{(\jmath)} \theta_{i\jmath}^\top)^\top 1_s - \varepsilon_5 F_i^\top 1_s - \gamma 1_m \preceq 0, \tag{25}$$

$$\begin{aligned}&\varpi_2 A_i^\top \zeta_{(i,j)} + \vartheta_2 \xi_j + \varpi_4 C_i^\top HJ\delta_j - \varpi_1 A_i^\top \varphi_{(i,j)} - \mu_2 \zeta_{(i,j)} + (\varpi_6 \textstyle\sum_{\jmath=1}^{s} 1_s^{(\jmath)} \rho_{j\jmath}^\top \\ &\quad + \varpi_4 \textstyle\sum_{\jmath=1}^{s} 1_s^{(\jmath)} \theta_{j\jmath}^\top JHC_i - \varpi_5 E_i)^\top 1_s \preceq 0,\end{aligned} \tag{26}$$

$$\varpi_6 (\textstyle\sum_{\jmath=1}^{s} 1_s^{(\jmath)} \rho_{j\jmath}^\top)^\top 1_s + \vartheta_2 \xi_j - \mu_2 \varphi_{(i,j)} \preceq 0, \tag{27}$$

$$\begin{aligned}&\varepsilon_2 B_i^\top \zeta_{(i,j)} + \varepsilon_4 D_i^\top HJ\delta_j - \varepsilon_1 B_i^\top \varphi_{(i,j)} + \varepsilon_4 D_i^\top HJ(\textstyle\sum_{\jmath=1}^{s} 1_s^{(\jmath)} \theta_{j\jmath}^\top)^\top 1_s \\ &\quad - \varepsilon_5 F_i^\top 1_s - \gamma 1_m \preceq 0,\end{aligned} \tag{28}$$

$$\begin{aligned}&\zeta_i \preceq \lambda \zeta_{(i,j)}, \zeta_i \preceq \lambda \zeta_{(j,i)}, \zeta_{(i,j)} \preceq \lambda \zeta_i, \zeta_{(j,i)} \preceq \lambda \zeta_i, \\ &\varphi_i \preceq \lambda \varphi_{(i,j)}, \varphi_i \preceq \lambda \varphi_{(j,i)}, \varphi_{(i,j)} \preceq \lambda \varphi_i, \varphi_{(j,i)} \preceq \lambda \varphi_i,\end{aligned} \tag{29}$$

$$\xi_{i\imath} \preceq \xi_i, \delta_{i\imath} \preceq \delta_i, \xi_{j\imath} \preceq \xi_j, \delta_{j\imath} \preceq \delta_j, \tag{30}$$

hold $\forall i, j \in S, i \neq j, \imath = 1, 2, \cdots, n$ *and* $\jmath = 1, 2, \cdots, s$, *then the error systems (9) and (10) are positive and stable with filter gain matrices*

$$A_{fi} = \frac{\sum_{\imath=1}^{n} 1_n^{(\imath)} \xi_{i\imath}^\top}{1_n^\top \varphi_i}, B_{fi} = \frac{\sum_{\imath=1}^{n} 1_n^{(\imath)} \delta_{i\imath}^\top}{1_n^\top \varphi_i}, \tag{31}$$

$$E_{fi} = \sum_{\jmath=1}^{s} 1_s^{(\jmath)} \rho_{i\jmath}^\top, F_{fi} = \sum_{\jmath=1}^{s} 1_s^{(\jmath)} \theta_{i\jmath}^\top, \tag{32}$$

and the switching law satisfying

$$\frac{\Gamma^-(k_0,k)}{\Gamma^+(k_0,k)} \geq \frac{\ln \mu_2 - \ln \mu_1}{\ln \mu_1^* - \ln \mu_1}, \mu_1^* \in (\mu_1, 1), \tag{33}$$

$$\tau_a \geq \tau_a^* = -\frac{2 \ln \lambda + (\ln \mu_2 - \ln \mu_1)\Delta_{max}}{\ln \mu_1^*}, \tag{34}$$

where $M = I - \beta 1_{m\times m}, H = I + \beta 1_{m \times m}$, *and* Δ_{max} *denotes the maximum of time lag* Δ_r.

Proof. First, the positivity of the error systems (9) and (10) are considered. For $x(k_0) \succeq 0$, the output satisfies $y(k_0) \succeq 0$. Using event-triggering condition (6) gives

$$\|e_y(k_0)\|_1 \leq \beta 1_m^\top y(k_0), \tag{35}$$

which gives that

$$-\beta 1_{m\times m} y(k_0) \preceq e_y(k_0) \preceq \beta 1_{m\times m} y(k_0). \tag{36}$$

For $k \in [k_r, k_r + \Delta_r)$, it follows from (11) and (36) that

$$\begin{aligned}&\tilde{x}(k_0+1) \succeq \underline{\tilde{A}}_{1ij} \tilde{x}(k_0) + \underline{\tilde{B}}_{1ij} \omega(k_0), \\ &e(k_0) \succeq \underline{\tilde{E}}_{1ij} \tilde{x}(k_0) + \underline{\tilde{F}}_{1ij} \omega(k_0),\end{aligned} \tag{37}$$

where

$$\begin{aligned}&\underline{\tilde{A}}_{1ij} = \begin{pmatrix} \varpi_1 A_i & 0 \\ \vartheta_1 A_{fj} + \varpi_3 B_{fj} LMC_i - \varpi_2 A_i & \vartheta_1 A_{fj} \end{pmatrix}, \underline{\tilde{B}}_{1ij} = \begin{pmatrix} \varepsilon_1 B_i \\ \varepsilon_3 B_{fj} LMD_i - \varepsilon_2 B_i \end{pmatrix}, \\ &\underline{\tilde{E}}_{1ij} = \begin{pmatrix} \varpi_5 E_{fj} + \varpi_3 F_{fj} LMC_i - \varpi_6 E_i & \varpi_5 E_{fj} \end{pmatrix}, \underline{\tilde{F}}_{1ij} = \begin{pmatrix} \varepsilon_3 F_{fj} LMD_i - \varepsilon_6 F_i \end{pmatrix}.\end{aligned}$$

For $k \in [k_r + \Delta_r, k_{r+1})$,

$$\begin{aligned} \widetilde{x}(k_0+1) &\succeq \underline{\widetilde{A}}_{1i}\widetilde{x}(k_0) + \underline{\widetilde{B}}_{1i}\omega(k_0), \\ e(k_0) &\succeq \underline{\widetilde{E}}_{1i}\widetilde{x}(k_0) + \underline{\widetilde{F}}_{1i}\omega(k_0), \end{aligned} \tag{38}$$

where

$$\begin{aligned} \underline{\widetilde{A}}_{1i} &= \begin{pmatrix} \varpi_1 A_i & 0 \\ \vartheta_1 A_{fi} + \varpi_3 B_{fi} LMC_i - \varpi_2 A_i & \vartheta_1 A_{fi} \end{pmatrix}, \underline{\widetilde{B}}_{1i} = \begin{pmatrix} \varepsilon_1 B_i \\ \varepsilon_3 B_{fi} LMD_i - \varepsilon_2 B_i \end{pmatrix}, \\ \underline{\widetilde{E}}_{1i} &= \begin{pmatrix} \varpi_5 E_{fi} + \varpi_3 F_{fi} LMC_i - \varpi_6 E_i & \varpi_5 E_{fi} \end{pmatrix}, \underline{\widetilde{F}}_{1i} = \begin{pmatrix} \varepsilon_3 F_{fi} LMD_i - \varepsilon_6 F_i \end{pmatrix}. \end{aligned}$$

Using (15), (16), (19) and (20) gives

$$\frac{\vartheta_1 \sum_{\imath=1}^{n} 1_n^{(\imath)} \xi_{i\imath}^{\top}}{1_n^{\top} \varphi_i} + \frac{\varpi_3 \sum_{\imath=1}^{n} 1_n^{(\imath)} \delta_{i\imath}^{\top}}{1_n^{\top} \varphi_i} LMC_i - \varpi_2 A_i \succeq 0, \tag{39}$$

$$\frac{\varepsilon_3 \sum_{\imath=1}^{n} 1_n^{(\imath)} \delta_{i\imath}^{\top}}{1_n^{\top} \varphi_i} LMD_i - \varepsilon_2 B_i \succeq 0, \tag{40}$$

$$\frac{\vartheta_1 \sum_{\imath=1}^{n} 1_n^{(\imath)} \xi_{j\imath}^{\top}}{1_n^{\top} \varphi_j} + \frac{\varpi_3 \sum_{\imath=1}^{n} 1_n^{(\imath)} \delta_{j\imath}^{\top}}{1_n^{\top} \varphi_j} LMC_i - \varpi_2 A_i \succeq 0, \tag{41}$$

$$\frac{\varepsilon_3 \sum_{\imath=1}^{n} 1_n^{(\imath)} \delta_{j\imath}^{\top}}{1_n^{\top} \varphi_j} LMD_i - \varepsilon_2 B_i \succeq 0. \tag{42}$$

Together with (17), (18), (21), (22), (31), and (32), we have

$$\vartheta_1 A_{fi} + \varpi_3 B_{fi} LMC_i - \varpi_2 A_i \succeq 0, \tag{43}$$

$$\varepsilon_3 B_{fi} LMD_i - \varepsilon_2 B_i \succeq 0, \tag{44}$$

$$\varpi_5 E_{fi} + \varpi_3 F_{fi} LMC_i - \varpi_6 E_i \succeq 0, \tag{45}$$

$$\varepsilon_3 F_{fi} LMD_i - \varepsilon_6 F_i \succeq 0, \tag{46}$$

$$\vartheta_1 A_{fj} + \varpi_3 B_{fj} LMC_i - \varpi_2 A_i \succeq 0, \tag{47}$$

$$\varepsilon_3 B_{fj} LMD_i - \varepsilon_2 B_i \succeq 0, \tag{48}$$

$$\varpi_5 E_{fj} + \varpi_3 F_{fj} LMC_i - \varpi_6 E_i \succeq 0, \tag{49}$$

$$\varepsilon_3 F_{fj} LMD_i - \varepsilon_6 F_i \succeq 0. \tag{50}$$

Due to $\xi_{i\imath} \succeq 0, \delta_{i\imath} \succeq 0, \rho_{ij} \succeq 0$, and $\theta_{ij} \succeq 0$, this yields $A_{fi} \succeq 0, B_{fi} \succeq 0, E_{fi} \succeq 0$, and $F_{fi} \succeq 0$. Thus, we have $\underline{\widetilde{A}}_{1i} \succeq 0, \underline{\widetilde{B}}_{1i} \succeq 0, \underline{\widetilde{E}}_{1i} \succeq 0$, and $\underline{\widetilde{F}}_{1i} \succeq 0$. Similarly, we can obtain $\underline{\widetilde{A}}_{1ij} \succeq 0, \underline{\widetilde{B}}_{1ij} \succeq 0, \underline{\widetilde{E}}_{1ij} \succeq 0$, and $\underline{\widetilde{F}}_{1ij} \succeq 0$. By (37), (38), and Lemma 1, we have $\widetilde{x}(k_0+1) \succeq 0$ and $e(k_0) \succeq 0$. Using recursive derivation gives $\widetilde{x}(k) \succeq 0$ and $e(k) \succeq 0$, that is to say, the error systems (9) and (10) are positive.

Next, we will analyze the ℓ_1-gain stability of the considered error systems. Construct a piecewise multiple copositive Lyapunov function candidate:

$$V_i(k) = \begin{cases} \widetilde{x}^{\top}(k) v_i, \ \forall k \in [k_{r-1} + \Delta_{r-1}, k_r), \\ \widetilde{x}^{\top}(k) v_{(i,j)}, \ \forall k \in [k_r, k_r + \Delta_r), \end{cases} \tag{51}$$

where $v_i = (\zeta_i^{\top} \ \varphi_i^{\top})^{\top}$ and $v_{(i,j)} = (\zeta_{(i,j)}^{\top} \ \varphi_{(i,j)}^{\top})^{\top}$. From (12) and (36), for $k \in [k_r, k_r + \Delta_r)$, it follows that

$$\begin{aligned} \widetilde{x}(k+1) &\preceq \overline{\widetilde{A}}_{1ij}\widetilde{x}(k) + \overline{\widetilde{B}}_{1ij}\omega(k), \\ e(k) &\preceq \overline{\widetilde{E}}_{1ij}\widetilde{x}(k) + \overline{\widetilde{F}}_{1ij}\omega(k), \end{aligned} \tag{52}$$

where

$$\overline{\overline{A}}_{1ij} = \begin{pmatrix} \varpi_2 A_i & 0 \\ \vartheta_2 A_{fj} + \varpi_4 B_{fj} JHC_i - \varpi_1 A_i & \vartheta_2 A_{fj} \end{pmatrix}, \overline{\overline{B}}_{1ij} = \begin{pmatrix} \varepsilon_2 B_i \\ \varepsilon_4 B_{fj} JHD_i - \varepsilon_1 B_i \end{pmatrix},$$
$$\overline{\overline{E}}_{1ij} = \begin{pmatrix} \varpi_6 E_{fj} + \varpi_4 F_{fj} JHC_i - \varpi_5 E_i & \varpi_6 E_{fj} \end{pmatrix}, \overline{\overline{F}}_{1ij} = \begin{pmatrix} \varepsilon_4 F_{fj} JHD_i - \varepsilon_5 F_i \end{pmatrix}.$$

For $k \in [k_r + \Delta_r, k_{r+1})$, we have

$$\begin{aligned} \tilde{x}(k+1) &\preceq \overline{\overline{A}}_{1i}\tilde{x}(k) + \overline{\overline{B}}_{1i}\omega(k), \\ e(k) &\preceq \overline{\overline{E}}_{1i}\tilde{x}(k) + \overline{\overline{F}}_{1i}\omega(k), \end{aligned} \tag{53}$$

where

$$\overline{\overline{A}}_{1i} = \begin{pmatrix} \varpi_2 A_i & 0 \\ \vartheta_2 A_{fi} + \varpi_4 B_{fi} JHC_i - \varpi_1 A_i & \vartheta_2 A_{fi} \end{pmatrix}, \overline{\overline{B}}_{1i} = \begin{pmatrix} \varepsilon_2 B_i \\ \varepsilon_4 B_{fi} JHD_i - \varepsilon_1 B_i \end{pmatrix},$$
$$\overline{\overline{E}}_{1i} = \begin{pmatrix} \varpi_6 E_{fi} + \varpi_4 F_{fi} JHC_i - \varpi_5 E_i & \varpi_6 E_{fi} \end{pmatrix}, \overline{\overline{F}}_{1i} = \begin{pmatrix} \varepsilon_4 F_{fi} JHD_i - \varepsilon_5 F_i \end{pmatrix}.$$

From the upper bound systems (52) and (53), the forward difference of (51) along the trajectories satisfies

$$\Delta V_i(k) \leq \begin{cases} \tilde{x}^\top(k)\Upsilon_1 + \omega^\top(k)\Theta_1, \ \forall k \in [k_{r-1} + \Delta_{r-1}, k_r), \\ \tilde{x}^\top(k)\Upsilon_2 + \omega^\top(k)\Theta_2, \ \forall k \in [k_r, k_r + \Delta_r), \end{cases} \tag{54}$$

where

$$\Upsilon_1 = \overline{\overline{A}}_{1i}^\top v_i - v_i = \begin{pmatrix} \varpi_2 A_i^\top \zeta_i + (\vartheta_2 A_{fi}^\top + \varpi_4 C_i^\top HJB_{fi}^\top - \varpi_1 A_i^\top)\varphi_i - \zeta_i \\ \vartheta_2 A_{fi}^\top \varphi_i - \varphi_i \end{pmatrix},$$
$$\Upsilon_2 = \overline{\overline{A}}_{1ij}^\top v_{(i,j)} - v_{(i,j)} = \begin{pmatrix} \varpi_2 A_i^\top \zeta_{(i,j)} + (\vartheta_2 A_{fj}^\top + \varpi_4 C_i^\top HJB_{fj}^\top - \varpi_1 A_i^\top)\varphi_{(i,j)} - \zeta_{(i,j)} \\ \vartheta_2 A_{fj}^\top \varphi_{(i,j)} - \varphi_{(i,j)} \end{pmatrix},$$
$$\Theta_1 = \overline{\overline{B}}_{1i}^\top v_i = \left(\varepsilon_2 B_i^\top \zeta_i + (\varepsilon_4 D_i^\top HJB_{fi}^\top - \varepsilon_1 B_i^\top)\varphi_i\right),$$
$$\Theta_2 = \overline{\overline{B}}_{1ij}^\top v_{(i,j)} = \left(\varepsilon_2 B_i^\top \zeta_{(i,j)} + (\varepsilon_4 D_i^\top HJB_{fj}^\top - \varepsilon_1 B_i^\top)\varphi_{(i,j)}\right).$$

Using (30) and (31), it derives that

$$A_{fi} \preceq \frac{\sum_{\iota=1}^n 1_n^{(\iota)} \xi_i^\top}{1_n^\top \varphi_i} = \frac{1_n \xi_i^\top}{1_n^\top \varphi_i}, \ B_{fi} \preceq \frac{\sum_{\iota=1}^n 1_n^{(\iota)} \delta_i^\top}{1_n^\top \varphi_i} = \frac{1_n \delta_i^\top}{1_n^\top \varphi_i}.$$

Together with the fact $\varphi_i \succ 0$ gives

$$A_{fi}^\top \varphi_i \preceq \frac{\xi_i 1_n^\top}{1_n^\top \varphi_i}\varphi_i = \xi_i, \ B_{fi}^\top \varphi_i \preceq \frac{\delta_i 1_n^\top}{1_n^\top \varphi_i}\varphi_i = \delta_i. \tag{55}$$

Similarly,

$$A_{fj}^\top \varphi_{(i,j)} \preceq \frac{\xi_j 1_n^\top}{1_n^\top \varphi_{(i,j)}}\varphi_{(i,j)} = \xi_j, \ B_{fj}^\top \varphi_{(i,j)} \preceq \frac{\delta_j 1_n^\top}{1_n^\top \varphi_{(i,j)}}\varphi_{(i,j)} = \delta_j. \tag{56}$$

Define $\Xi(k) = \gamma\|\omega(k)\|_1 - \|e(k)\|_1$. From (23)–(28), (54), (55), and (56), we can obtain

$$V_i(k) \leq \begin{cases} \mu_1 V_i(k-1) + \Xi(k-1), \ \forall k \in [k_{r-1} + \Delta_{r-1}, k_r), \\ \mu_2 V_{(i,j)}(k-1) + \Xi(k-1), \ \forall k \in [k_r, k_r + \Delta_r). \end{cases} \tag{57}$$

Thus,

$$V_i(k) \leq \begin{cases} \mu_1^{k-k_{r-1}-\Delta_{r-1}} V_i(k_{r-1}+\Delta_{r-1}) + \sum_{\varsigma=k_{r-1}+\Delta_{r-1}}^{k-1} \mu_1^{k-1-\varsigma}\Xi(\varsigma),\ \forall k \in [k_{r-1}+\Delta_{r-1}, k_r), \\ \mu_2^{k-k_r} V_{(i,j)}(k_r) + \sum_{\varsigma=k_r}^{k-1} \mu_2^{k-1-\varsigma}\Xi(\varsigma),\ \forall k \in [k_r, k_r+\Delta_r). \end{cases} \tag{58}$$

Noting the condition (29), then

$$V_i(k) \leq \begin{cases} \lambda\mu_1^{k-k_{r-1}-\Delta_{r-1}} V_i(k_{r-1}+\Delta_{r-1}) + \sum_{\varsigma=k_{r-1}+\Delta_{r-1}}^{k-1} \mu_1^{k-1-\varsigma}\Xi(\varsigma),\ \forall k \in [k_{r-1}, k_{r-1}+\Delta_{r-1}), \\ \lambda\mu_2^{k-k_r} V_{(i,j)}(k_r) + \sum_{\varsigma=k_r}^{k-1} \mu_2^{k-1-\varsigma}\Xi(\varsigma),\ \forall k \in [k_{r-1}+\Delta_{r-1}, k_r). \end{cases} \tag{59}$$

For $T \in [k_{N_\sigma(T,k_0)} + \Delta_{N_\sigma(T,k_0)}, k_{N_\sigma(T,k_0)+1})$, repeating (58) and (59) follows that:

$$\begin{aligned} V_{\sigma(k_{\aleph+1})}(T) &\leq \mu_1^{T-k_\aleph-\Delta_\aleph} V_{\sigma(k_{\aleph+1})}(k_\aleph+\Delta_\aleph) + \sum_{\varsigma=k_\aleph+\Delta_\aleph}^{T-1} \mu_1^{T-1-\varsigma}\Xi(\varsigma) \\ &\leq \lambda\mu_1^{T-k_\aleph-\Delta_\aleph}\mu_2^{\Delta_\aleph} V_{\sigma(k_\aleph+\Delta_\aleph)}(k_\aleph) + \sum_{\varsigma=k_\aleph+\Delta_\aleph}^{T-1} \mu_1^{T-1-\varsigma}\Xi(\varsigma) \\ &+ \lambda\mu_1^{T-k_\aleph-\Delta_\aleph} \sum_{\varsigma=k_\aleph}^{k_\aleph+\Delta_\aleph-1} \mu_2^{k_\aleph+\Delta_\aleph-1-\varsigma} \\ &\leq \lambda^2\mu_1^{T-k_{\aleph-1}-\Delta_\aleph-\Delta_{\aleph-1}}\mu_2^{\Delta_\aleph} V_{\sigma(k_{\aleph-1})}(k_{\aleph-1}+\Delta_{\aleph-1}) + \sum_{\varsigma=k_\aleph+\Delta_\aleph}^{T-1} \mu_1^{T-1-\varsigma}\Xi(\varsigma) \\ &+ \lambda\mu_1^{T-k_\aleph-\Delta_\aleph} \sum_{\varsigma=k_\aleph}^{k_\aleph+\Delta_\aleph-1} \mu_2^{k_\aleph+\Delta_\aleph-1-\varsigma} \\ &+ \lambda^2\mu_2^{\Delta_\aleph} \sum_{\varsigma=k_{\aleph-1}+\Delta_{\aleph-1}}^{k_\aleph-1} \mu_1^{T-1-\Delta_\aleph-\varsigma}\Xi(\varsigma) \\ &= \lambda^2 e^{(T-k_{\aleph-1}-\Delta_\aleph-\Delta_{\aleph-1})\ln\mu_1} e^{\Delta_\aleph \ln\mu_2} V_{\sigma(k_{\aleph-1})}(k_{\aleph-1}+\Delta_{\aleph-1}) + \sum_{\varsigma=k_\aleph+\Delta_\aleph}^{T-1} e^{(T-1-\varsigma)\ln\mu_1}\Xi(\varsigma) \\ &+ \lambda e^{(T-k_\aleph-\Delta_\aleph)\ln\mu_1} \sum_{\varsigma=k_\aleph}^{k_\aleph+\Delta_\aleph-1} e^{(k_\aleph+\Delta_\aleph-1-\varsigma)\ln\mu_2} \\ &+ \lambda^2 e^{\Delta_\aleph \ln\mu_2} \sum_{\varsigma=k_{\aleph-1}+\Delta_{\aleph-1}}^{k_\aleph-1} e^{(T-1-\Delta_\aleph-\varsigma)\ln\mu_1}\Xi(\varsigma) \\ &\leq \cdots \\ &\leq e^{2\aleph\ln\lambda} e^{(T-k_0-\aleph\Delta_{\max})\ln\mu_1} e^{\aleph\Delta_{\max}\ln\mu_2} V_{\sigma(k_0)}(k_0) \\ &+ \sum_{\varsigma=k_0}^{T-1} e^{2N_\sigma(T,\varsigma)\ln\lambda} e^{N_\sigma(T-1,\varsigma)\Delta_{\max}\ln\mu_2} e^{(T-1-N_\sigma(T-1,\varsigma)\Delta_{\max}-\varsigma)\ln\mu_1}\Xi(\varsigma), \end{aligned} \tag{60}$$

where $\aleph = N_\sigma(T,k_0)$ and $\Delta_{\max} = \max\{\Delta_1, \Delta_2, \ldots, \Delta_\aleph\}$. Under zero initial conditions, we have

$$0 \leq \sum_{\varsigma=k_0}^{T-1} e^{2N_\sigma(T,\varsigma)\ln\lambda} e^{N_\sigma(T-1,\varsigma)\Delta_{\max}\ln\mu_2} e^{(T-1-N_\sigma(T-1,\varsigma)\Delta_{\max}-\varsigma)\ln\mu_1}\Xi(\varsigma), \tag{61}$$

that is,

$$\begin{aligned} &\sum_{\varsigma=k_0}^{T-1} e^{2N_\sigma(T,\varsigma)\ln\lambda} e^{N_\sigma(T-1,\varsigma)\Delta_{\max}\ln\mu_2} e^{(T-1-N_\sigma(T-1,\varsigma)\Delta_{\max}-\varsigma)\ln\mu_1} \|e(\varsigma)\|_1 \\ &\leq \gamma \sum_{\varsigma=k_0}^{T-1} e^{2N_\sigma(T,\varsigma)\ln\lambda} e^{N_\sigma(T-1,\varsigma)\Delta_{\max}\ln\mu_2} e^{(T-1-N_\sigma(T-1,\varsigma)\Delta_{\max}-\varsigma)\ln\mu_1} \|\omega(\varsigma)\|_1. \end{aligned} \tag{62}$$

Multiplying both sides of the inequality (62) with $e^{(\ln\mu_1-\ln\mu_2)N_\sigma(T,k_0)\Delta_{\max}-2N_\sigma(T,k_0)\ln\lambda}$ gives

$$\begin{aligned} &\sum_{\varsigma=k_0}^{T-1} e^{(T-1-\varsigma)\ln\mu_1+(\ln\mu_1-\ln\mu_2)N_\sigma(\varsigma,0)\Delta_{\max}-2N_\sigma(\varsigma,0)\ln\lambda} \|e(\varsigma)\|_1 \\ &\leq \gamma \sum_{\varsigma=k_0}^{T-1} e^{(T-1-\varsigma)\ln\mu_1+(\ln\mu_1-\ln\mu_2)N_\sigma(\varsigma,0)\Delta_{\max}-2N_\sigma(\varsigma,0)\ln\lambda} \|\omega(\varsigma)\|_1. \end{aligned} \tag{63}$$

From Definition 2 and (34), it is clear that

$$N_\sigma(\varsigma, k_0) \leq N_0 + \frac{(\varsigma-k_0)\ln\mu_1^*}{2\ln\lambda+(\ln\mu_2-\ln\mu_1)\Delta_{\max}}. \tag{64}$$

Then, (63) can be transformed into

$$\begin{aligned} &\sum_{\varsigma=k_0}^{T-1} e^{(T-1-\varsigma)\ln\mu_1-[2\ln\lambda+(\ln\mu_2-\ln\mu_1)\Delta_{\max}](N_0+\frac{(\varsigma-k_0)\ln\mu_1^*}{2\ln\lambda+(\ln\mu_2-\ln\mu_1)\Delta_{\max}})} \|e(\varsigma)\|_1 \\ &\leq \gamma \sum_{\varsigma=k_0}^{T-1} e^{(T-1-\varsigma)\ln\mu_1-[2\ln\lambda+(\ln\mu_2-\ln\mu_1)\Delta_{\max}](N_0+\frac{(\varsigma-k_0)\ln\mu_1^*}{2\ln\lambda+(\ln\mu_2-\ln\mu_1)\Delta_{\max}})} \|\omega(\varsigma)\|_1. \end{aligned} \tag{65}$$

The above inequality can be further written as

$$\begin{aligned}&\sum_{\varsigma=k_0}^{T-1} e^{(T-1)\ln\mu_1-k_0\ln\mu_1^*}e^{(\ln\mu_1^*-\ln\mu_1)\varsigma}\|e(\varsigma)\|_1\\ &\leq \gamma\sum_{\varsigma=k_0}^{T-1} e^{(T-1)\ln\mu_1-k_0\ln\mu_1^*}e^{(\ln\mu_1^*-\ln\mu_1)\varsigma}\|\omega(\varsigma)\|_1.\end{aligned} \tag{66}$$

From (66), we can obtain

$$\sum_{\varsigma=k_0}^{T-1} e^{-(\ln\mu_1^*-\ln\mu_1)\varsigma}\|e(\varsigma)\|_1 \leq \gamma\sum_{\varsigma=k_0}^{T-1}\|\omega(\varsigma)\|_1. \tag{67}$$

Summing from 0 to ∞ for both sides of (67), it yields that

$$\sum_{k=0}^{\infty} e^{-\hbar k}\|e(k)\|_1 \leq \gamma\sum_{k=0}^{\infty}\|\omega(k)\|_1, \tag{68}$$

where $\hbar = \ln\mu_1^* - \ln\mu_1$. By Definition 3, the error systems (9) and (10) satisfy the ℓ_1-gain performance index (68). □

Remark 4. *Generally, the signal can be quantized during the actual transmission process due to various reasons, such as energy consumption issues, intermittent sensor fault, limited digital communication resource, and so on. It is necessary to incorporate the quantization technique with event-triggering mechanism to generate the quantized output. Theorem 1 first introduces the quantization technique to the filter design of positive systems, where a sector restriction is adopted to analyze and mitigate the quantization effect [47–49]. Currently, few efforts have been devoted to positive systems, though the quantization approach is effective and practical for dealing with many practical problems. It is interesting to develop the quantization approach in Theorem 1 for other issues of positive systems.*

3.3. Dynamic Event-Triggering Case

In this section, we propose a dynamic event-triggering mechanism as an alternative of the static event-triggering mechanism for system (1). Define the sampling error of the event generator as $e_y(k) = \tilde{y}(k) - y(k)$, where $\tilde{y}(k) = y(k_\wp)$, $y(k_\wp)$ is the output signal of the event generator at the event-triggering instant $k_\wp$, $\wp \in \mathbb{N}$. Following this, the output will be released by the following dynamic event-triggering condition:

$$\|e_y(k)\|_1 > \beta\|y(k)\|_1 + \frac{1}{\psi}\eta(k) \vee \eta(k) > \|y(k)\|_1, \tag{69}$$

where β and ψ are given positive constants, and $\eta(k)$ is an internal dynamic variable satisfying

$$\eta(k+1) = \varrho\eta(k) + \beta\|y(k)\|_1 - \|e_y(k)\|_1, \tag{70}$$

with $\eta(k_0) = \eta_0$ as the initial value and $\varrho \in (0,1)$ as a given constant.

Theorem 2. *If there exist constants $0 < \varpi_1 \leq \varpi_2, 0 < \varpi_3 \leq \varpi_4, 0 < \varpi_5 \leq \varpi_6, 0 < \varepsilon_1 \leq \varepsilon_2$, $0 < \varepsilon_3 \leq \varepsilon_4, 0 < \varepsilon_5 \leq \varepsilon_6, 0 < \vartheta_1 \leq \vartheta_2$, $\psi > 0$, $\gamma > 0$, $\lambda > 1, 0 \leq \beta < 1, 0 < \mu_1 < 1$, $\mu_2 > 1$, $\mathbb{R}^n$ vectors $\zeta_i \succ 0, \zeta_{(i,j)} \succ 0$, $\varphi_i \succ 0, \varphi_{(i,j)} \succ 0$, $\xi_i \succeq 0$, $\xi_{\imath i} \succeq 0$, $\xi_j \succeq 0$, $\xi_{j\imath} \succeq 0$, $\rho_{ij} \succeq 0$, and $\mathbb{R}^m$ vectors $\delta_i \succeq 0$, $\delta_{\imath i} \succeq 0$, $\delta_j \succeq 0$, $\delta_{j\imath} \succeq 0$, $\theta_{ij} \succeq 0$ such that*

$$\vartheta_1\sum_{\imath=1}^{n} 1_n^{(\imath)}\xi_{\imath i}^\top + \varpi_3\sum_{\imath=1}^{n} 1_n^{(\imath)}\delta_{\imath i}^\top L\Psi C_i - \varpi_2 1_n^\top\varphi_i A_i \succeq 0, \tag{71}$$

$$\varepsilon_3\sum_{\imath=1}^{n} 1_n^{(\imath)}\delta_{\imath i}^\top L\Psi D_i - \varepsilon_2 1_n^\top\varphi_i B_i \succeq 0, \tag{72}$$

$$\varpi_5\sum_{\jmath=1}^{s} 1_s^{(\jmath)}\rho_{ij}^\top + \varpi_3\sum_{\jmath=1}^{s} 1_s^{(\jmath)}\theta_{ij}^\top L\Psi C_i - \varpi_6 E_i \succeq 0, \tag{73}$$

$$\varepsilon_3\sum_{\jmath=1}^{s} 1_s^{(\jmath)}\theta_{ij}^\top L\Psi D_i - \varepsilon_6 F_i \succeq 0, \tag{74}$$

$$\vartheta_1\sum_{\imath=1}^{n} 1_n^{(\imath)}\xi_{j\imath}^\top + \varpi_3\sum_{\imath=1}^{n} 1_n^{(\imath)}\delta_{j\imath}^\top L\Psi C_i - \varpi_2 1_n^\top\varphi_j A_i \succeq 0, \tag{75}$$

$$\varepsilon_3 \sum_{\imath=1}^{n} 1_n^{(\imath)} \delta_{j\imath}^\top L\Psi D_i - \varepsilon_2 1_n^\top \varphi_j B_i \succeq 0, \tag{76}$$

$$\varpi_5 \sum_{\jmath=1}^{s} 1_s^{(\jmath)} \rho_{j\jmath}^\top + \varpi_3 \sum_{\jmath=1}^{s} 1_s^{(\jmath)} \theta_{j\jmath}^\top L\Psi C_i - \varpi_6 E_i \succeq 0, \tag{77}$$

$$\varepsilon_3 \sum_{\jmath=1}^{s} 1_s^{(\jmath)} \theta_{j\jmath}^\top L\Psi D_i - \varepsilon_6 F_i \succeq 0, \tag{78}$$

$$\begin{aligned} &\varpi_2 A_i^\top \zeta_i + \vartheta_2 \xi_i + \varpi_4 C_i^\top \Phi J \delta_i - \varpi_1 A_i^\top \varphi_i - \mu_1 \zeta_i + (\varpi_6 \textstyle\sum_{\jmath=1}^{s} 1_s^{(\jmath)} \rho_{i\jmath}^\top \\ &+\varpi_4 \textstyle\sum_{\jmath=1}^{s} 1_s^{(\jmath)} \theta_{i\jmath}^\top J\Phi C_i - \varpi_5 E_i)^\top 1_s + \beta\varpi_4 C_i^\top 1_m \preceq 0, \end{aligned} \tag{79}$$

$$\varpi_6 (\textstyle\sum_{\jmath=1}^{s} 1_s^{(\jmath)} \rho_{i\jmath}^\top)^\top 1_s + \vartheta_2 \xi_i - \mu_1 \varphi_i \preceq 0, \tag{80}$$

$$\begin{aligned} &\varepsilon_2 B_i^\top \zeta_i + \beta\varepsilon_4 D_i^\top 1_m + \varepsilon_4 D_i^\top \Phi J \delta_i - \varepsilon_1 B_i^\top \varphi_i + \varepsilon_4 D_i^\top \Phi J (\textstyle\sum_{\jmath=1}^{s} 1_s^{(\jmath)} \theta_{i\jmath}^\top)^\top 1_s \\ &-\varepsilon_5 F_i^\top 1_s - \gamma 1_m \preceq 0, \end{aligned} \tag{81}$$

$$\begin{aligned} &\varpi_2 A_i^\top \zeta_{(i,j)} + \vartheta_2 \xi_j + \varpi_4 C_i^\top \Phi J \delta_j - \varpi_1 A_i^\top \varphi_{(i,j)} - \mu_2 \zeta_{(i,j)} + (\varpi_6 \textstyle\sum_{\jmath=1}^{s} 1_s^{(\jmath)} \rho_{j\jmath}^\top \\ &+\varpi_4 \textstyle\sum_{\jmath=1}^{s} 1_s^{(\jmath)} \theta_{j\jmath}^\top J\Phi C_i - \varpi_5 E_i)^\top 1_s + \beta\varpi_4 C_i^\top 1_m \preceq 0, \end{aligned} \tag{82}$$

$$\varpi_6 (\textstyle\sum_{\jmath=1}^{s} 1_s^{(\jmath)} \rho_{j\jmath}^\top)^\top 1_s + \vartheta_2 \xi_j - \mu_2 \varphi_{(i,j)} \preceq 0, \tag{83}$$

$$\begin{aligned} &\varepsilon_2 B_i^\top \zeta_{(i,j)} + \beta\varepsilon_4 D_i^\top 1_m + \varepsilon_4 D_i^\top \Phi J \delta_j - \varepsilon_1 B_i^\top \varphi_{(i,j)} + \varepsilon_4 D_i^\top \Phi J (\textstyle\sum_{\jmath=1}^{s} 1_s^{(\jmath)} \theta_{j\jmath}^\top)^\top 1_s \\ &-\varepsilon_5 F_i^\top 1_s - \gamma 1_m \preceq 0, \end{aligned} \tag{84}$$

$$\begin{aligned} &\zeta_i \preceq \lambda \zeta_{(i,j)}, \zeta_i \preceq \lambda \zeta_{(j,i)}, \zeta_{(i,j)} \preceq \lambda \zeta_i, \zeta_{(j,i)} \preceq \lambda \zeta_i, \\ &\varphi_i \preceq \lambda \varphi_{(i,j)}, \varphi_i \preceq \lambda \varphi_{(j,i)}, \varphi_{(i,j)} \preceq \lambda \varphi_i, \varphi_{(j,i)} \preceq \lambda \varphi_i, \end{aligned} \tag{85}$$

$$\xi_{i\imath} \preceq \xi_i, \delta_{i\imath} \preceq \delta_i, \xi_{j\imath} \preceq \xi_j, \delta_{j\imath} \preceq \delta_j, \tag{86}$$

hold $\forall i, j \in S, i \neq j, \imath = 1, 2, \cdots, n$ *and* $\jmath = 1, 2, \cdots, s$, *then the error systems (9) and (10) are positive and stable with filter gain matrices (31) and (32) and the switching law satisfying*

$$\frac{\Gamma^-(k_0, k)}{\Gamma^+(k_0, k)} \geq \frac{\ln \mu_2 - \ln \mu_1}{\ln \mu_1^* - \ln \mu_1}, \; \mu_1^* \in (\mu_1, 1), \tag{87}$$

$$\tau_a \geq \tau_a^* = -\frac{2\ln\lambda + (\ln \mu_2 - \ln \mu_1)\Delta_{max}}{\ln \mu_1^*}, \tag{88}$$

where $\Psi = I - (\beta + \frac{1}{\psi}) 1_{m\times m}, \Phi = I + (\beta + \frac{1}{\psi}) 1_{m\times m}$, *and* Δ_{max} *denotes the maximum of time lag* Δ_r.

Proof. First, the positivity of the error systems (9) and (10) are considered. For $x(k_0) \succeq 0$, the output satisfies $y(k_0) \succeq 0$. We can obtain from the dynamic event-triggering condition (69) and (70) that

$$\|e_y(k_0)\|_1 \leq \beta \|y(k_0)\|_1 + \frac{1}{\psi}\eta(k_0) \leq (\beta + \frac{1}{\psi}) \mathbf{1}_m^\top y(k_0), \tag{89}$$

which leads to

$$-(\beta + \frac{1}{\psi}) \mathbf{1}_{m\times m} y(k_0) \preceq e_y(k_0) \preceq (\beta + \frac{1}{\psi}) \mathbf{1}_{m\times m} y(k_0). \tag{90}$$

From (11), (13), and (90), it is clear that, for $k \in [k_r, k_r + \Delta_r)$,

$$\begin{aligned} \tilde{x}(k_0 + 1) &\succeq \underline{\tilde{A}}_{2ij} \tilde{x}(k_0) + \underline{\tilde{B}}_{2ij} \omega(k_0), \\ e(k_0) &\succeq \underline{\tilde{E}}_{2ij} \tilde{x}(k_0) + \underline{\tilde{F}}_{2ij} \omega(k_0), \end{aligned} \tag{91}$$

where

$$\widetilde{\underline{A}}_{2ij} = \begin{pmatrix} \varpi_1 A_i & 0 \\ \vartheta_1 A_{fj} + \varpi_3 B_{fj} L\Psi C_i - \varpi_2 A_i & \vartheta_1 A_{fj} \end{pmatrix}, \widetilde{\underline{B}}_{2ij} = \begin{pmatrix} \varepsilon_1 B_i \\ \varepsilon_3 B_{fj} L\Psi D_i - \varepsilon_2 B_i \end{pmatrix},$$
$$\widetilde{\underline{E}}_{2ij} = \begin{pmatrix} \varpi_5 E_{fj} + \varpi_3 F_{fj} L\Psi C_i - \varpi_6 E_i & \varpi_5 E_{fj} \end{pmatrix}, \widetilde{\underline{F}}_{2ij} = \begin{pmatrix} \varepsilon_3 F_{fj} L\Psi D_i - \varepsilon_6 F_i \end{pmatrix}.$$

and for $k \in [k_r + \Delta_r, k_{r+1})$,

$$\begin{aligned} \widetilde{x}(k_0+1) &\succeq \widetilde{\underline{A}}_{2i}\widetilde{x}(k_0) + \widetilde{\underline{B}}_{2i}\omega(k_0), \\ e(k_0) &\succeq \widetilde{\underline{E}}_{2i}\widetilde{x}(k_0) + \widetilde{\underline{F}}_{2i}\omega(k_0), \end{aligned} \tag{92}$$

where

$$\widetilde{\underline{A}}_{2i} = \begin{pmatrix} \varpi_1 A_i & 0 \\ \vartheta_1 A_{fi} + \varpi_3 B_{fi} L\Psi C_i - \varpi_2 A_i & \vartheta_1 A_{fi} \end{pmatrix}, \widetilde{\underline{B}}_{2i} = \begin{pmatrix} \varepsilon_1 B_i \\ \varepsilon_3 B_{fi} L\Psi D_i - \varepsilon_2 B_i \end{pmatrix},$$
$$\widetilde{\underline{E}}_{2i} = \begin{pmatrix} \varpi_5 E_{fi} + \varpi_3 F_{fi} L\Psi C_i - \varpi_6 E_i & \varpi_5 E_{fi} \end{pmatrix}, \widetilde{\underline{F}}_{2i} = \begin{pmatrix} \varepsilon_3 F_{fi} L\Psi D_i - \varepsilon_6 F_i \end{pmatrix}.$$

Using a similar method in Theorem 1 gives that the error systems (9) and (10) are positive. From (12) and (90), for $k \in [k_r, k_r + \Delta_r)$, we have

$$\begin{aligned} \widetilde{x}(k+1) &\preceq \overline{\overline{A}}_{2ij}\widetilde{x}(k) + \overline{\overline{B}}_{2ij}\omega(k), \\ e(k) &\preceq \overline{\overline{E}}_{2ij}\widetilde{x}(k) + \overline{\overline{F}}_{2ij}\omega(k), \end{aligned} \tag{93}$$

where

$$\overline{\overline{A}}_{2ij} = \begin{pmatrix} \varpi_2 A_i & 0 \\ \vartheta_2 A_{fj} + \varpi_4 B_{fj} J\Phi C_i - \varpi_1 A_i & \vartheta_2 A_{fj} \end{pmatrix}, \overline{\overline{B}}_{2ij} = \begin{pmatrix} \varepsilon_2 B_i \\ \varepsilon_4 B_{fj} J\Phi D_i - \varepsilon_1 B_i \end{pmatrix},$$
$$\overline{\overline{E}}_{2ij} = \begin{pmatrix} \varpi_6 E_{fj} + \varpi_4 F_{fj} J\Phi C_i - \varpi_5 E_i & \varpi_6 E_{fj} \end{pmatrix}, \overline{\overline{F}}_{2ij} = \begin{pmatrix} \varepsilon_4 F_{fj} J\Phi D_i - \varepsilon_5 F_i \end{pmatrix}.$$

For $k \in [k_r + \Delta_r, k_{r+1})$, we have

$$\begin{aligned} \widetilde{x}(k+1) &\preceq \overline{\overline{A}}_{2i}\widetilde{x}(k) + \overline{\overline{B}}_{2i}\omega(k), \\ e(k) &\preceq \overline{\overline{E}}_{2i}\widetilde{x}(k) + \overline{\overline{F}}_{2i}\omega(k), \end{aligned} \tag{94}$$

where

$$\overline{\overline{A}}_{2i} = \begin{pmatrix} \varpi_2 A_i & 0 \\ \vartheta_2 A_{fi} + \varpi_4 B_{fi} J\Phi C_i - \varpi_1 A_i & \vartheta_2 A_{fi} \end{pmatrix}, \overline{\overline{B}}_{2i} = \begin{pmatrix} \varepsilon_2 B_i \\ \varepsilon_4 B_{fi} J\Phi D_i - \varepsilon_1 B_i \end{pmatrix},$$
$$\overline{\overline{E}}_{2i} = \begin{pmatrix} \varpi_6 E_{fi} + \varpi_4 F_{fi} J\Phi C_i - \varpi_5 E_i & \varpi_6 E_{fi} \end{pmatrix}, \overline{\overline{F}}_{2i} = \begin{pmatrix} \varepsilon_4 F_{fi} J\Phi D_i - \varepsilon_5 F_i \end{pmatrix}.$$

Choose a linear copositive Lyapunov function:

$$V_i(k) = \begin{cases} \widetilde{x}^\top(k)v_i + \eta(k), \ \forall k \in [k_{r-1} + \Delta_{r-1}, k_r), \\ \widetilde{x}^\top(k)v_{(i,j)} + \eta(k), \ \forall k \in [k_r, k_r + \Delta_r). \end{cases}$$

Combining (93) and (94), we find

$$\Delta V_i(k) \le \begin{cases} \widetilde{x}^\top(k)\Omega_1 + \omega^\top(k)\Gamma_1 + (\varrho - 1)\eta(k), \ \forall k \in [k_{r-1} + \Delta_{r-1}, k_r), \\ \widetilde{x}^\top(k)\Omega_2 + \omega^\top(k)\Gamma_2 + (\varrho - 1)\eta(k), \ \forall k \in [k_r, k_r + \Delta_r), \end{cases}$$

where

$$\Omega_1 = \overline{\widetilde{A}}_{2i}^{\top} v_i - v_i = \begin{pmatrix} \varpi_2 A_i^{\top}\zeta_i + (\vartheta_2 A_{fi}^{\top} + \varpi_4 C_i^{\top}\Phi J B_{fi}^{\top} - \varpi_1 A_i^{\top})\varphi_i + \beta\varpi_4 C_i^{\top} 1_m - \zeta_i \\ \vartheta_2 A_{fi}^{\top}\varphi_i - \varphi_i \end{pmatrix},$$

$$\Omega_2 = \overline{\widetilde{A}}_{2ij}^{\top} v_{(i,j)} - v_{(i,j)} = \begin{pmatrix} \varpi_2 A_i^{\top}\zeta_{(i,j)} + (\vartheta_2 A_{fj}^{\top} + \varpi_4 C_i^{\top}\Phi J B_{fj}^{\top} - \varpi_1 A_i^{\top})\varphi_{(i,j)} + \beta\varpi_4 C_i^{\top} 1_m - \zeta_{(i,j)} \\ \vartheta_2 A_{fj}^{\top}\varphi_{(i,j)} - \varphi_{(i,j)} \end{pmatrix},$$

$$\Gamma_1 = \overline{\widetilde{B}}_{2i}^{\top} v_i = \varepsilon_2 B_i^{\top}\zeta_i + \beta\varepsilon_4 D_i^{\top} 1_m + (\varepsilon_4 D_i^{\top}\Phi J B_{fi}^{\top} - \varepsilon_1 B_i^{\top})\varphi_i,$$

$$\Gamma_2 = \overline{\widetilde{B}}_{2ij}^{\top} v_{(i,j)} = \varepsilon_2 B_i^{\top}\zeta_{(i,j)} + \beta\varepsilon_4 D_i^{\top} 1_m + (\varepsilon_4 D_i^{\top}\Phi J B_{fj}^{\top} - \varepsilon_1 B_i^{\top})\varphi_{(i,j)}.$$

Using a similar method to Theorem 1, it is not difficult to obtain (68) under average dwell time (88), which means that the error systems (9) and (10) are ℓ_1-gain stable with performance γ. □

Remark 5. *Compared with the existing results on static event-triggering strategies of positive systems [39–41], the dynamic strategy proposed in Theorem 2 is more flexible and releases fewer data. For the dynamic event-triggered issues of general systems [42–46], it is clear that dynamic event-triggering conditions cannot be directly used for positive systems. Therefore, we proposed a dynamic event-triggering condition (69), and the lower bound of the error systems can be obtained from an interval system. Based on this point, the dynamic event-triggering condition can be applied to other issues of positive systems, such as output feedback control, observer design, etc.*

4. Illustrative Examples

Two examples are provided to verify the effectiveness of the proposed design.

Example 1. *Consider the system (1) with two subsystems:*

$$A_1 = \begin{pmatrix} 0.1503 & 0.1452 & 0.0231 \\ 0.2046 & 0.1729 & 0.1967 \\ 0.1050 & 0.2130 & 0.1714 \end{pmatrix}, B_1 = \begin{pmatrix} 0.0519 & 0.2034 \\ 0.1107 & 0.0257 \\ 0.0173 & 0.1322 \end{pmatrix}, C_1 = \begin{pmatrix} 0.3 & 0.3 & 0.2 \\ 0.4 & 0.1 & 0.1 \end{pmatrix},$$

$$D_1 = \begin{pmatrix} 0.3 & 0.2 \\ 0.2 & 0.4 \end{pmatrix}, E_1 = \begin{pmatrix} 0.1 & 0.2 & 0.2 \end{pmatrix}, F_1 = \begin{pmatrix} 0.3 & 0.6 \end{pmatrix},$$

$$A_2 = \begin{pmatrix} 0.2214 & 0.0496 & 0.1183 \\ 0.1426 & 0.1651 & 0.1742 \\ 0.1357 & 0.2073 & 0.0137 \end{pmatrix}, B_2 = \begin{pmatrix} 0.2641 & 0.2026 \\ 0.1273 & 0.1797 \\ 0.2065 & 0.1842 \end{pmatrix}, C_2 = \begin{pmatrix} 0.2 & 0.4 & 0.2 \\ 0.2 & 0.3 & 0.1 \end{pmatrix},$$

$$D_2 = \begin{pmatrix} 0.2 & 0.3 \\ 0.1 & 0.2 \end{pmatrix}, E_2 = \begin{pmatrix} 0.1 & 0.2 & 0.2 \end{pmatrix}, F_2 = \begin{pmatrix} 0.4 & 0.5 \end{pmatrix},$$

where $f_i(x_i(k)) = 2e^{-0.2k}x_i(k), \hat{f}_i(x_i(k)) = e^{-k}x_{fi}(k), h_i(x_i(k)) = x_i(k) + \frac{x_i(k)}{x_i^2(k)+1}, p_i(x_i(k)) = x_i(k) + \frac{x_i(k)}{x_i^2(k)+5}$, *the disturbance signal is* $\omega_\iota(k) = \begin{pmatrix} \frac{0.4}{(k+1)^{2/3}} & 0.4e^{-0.1k} \end{pmatrix}^{\top}$, *and the nonlinear disturbance is given as* $g_\iota(\omega_\iota(k)) = 0.3e^{-0.03k}\omega_\iota(k), l_\iota(\omega_\iota(k)) = \omega_\iota(k), q_\iota(\omega_\iota(k)) = 0.75\omega_\iota(k)$. *Then,* $\varpi_1 = 0.20, \varpi_2 = 0.30, \varpi_3 = 0.30, \varpi_4 = 0.50, \varpi_5 = 0.20, \varpi_6 = 0.30, \varepsilon_1 = 0.10, \varepsilon_2 = 0.20, \varepsilon_3 = 1, \varepsilon_4 = 1, \varepsilon_5 = 0.30, \varepsilon_6 = 0.50$. *Choose* $\mu_1 = 0.69, \mu_1^* = 0.80, \mu_2 = 1.30, \beta = 0.15$, *and* $\lambda = 1.20$. *By Theorem 1, the filter gain matrices are:*

$$A_{f1} = \begin{pmatrix} 0.3132 & 0.1982 & 0.2287 \\ 0.2897 & 0.2154 & 0.2727 \\ 0.2698 & 0.2372 & 0.2524 \end{pmatrix}, B_{f1} = \begin{pmatrix} 0.5950 & 0.0317 \\ 0.5807 & 0.0258 \\ 0.5845 & 0.0259 \end{pmatrix}, E_{f1} = \begin{pmatrix} 0.0013 \\ 0.0014 \\ 0.0190 \end{pmatrix}^{\top}, F_{f1} = \begin{pmatrix} 1.2703 \\ 0.9366 \end{pmatrix}^{\top},$$

$$A_{f2} = \begin{pmatrix} 0.1622 & 0.0921 & 0.2278 \\ 0.1468 & 0.0894 & 0.2468 \\ 0.1493 & 0.0946 & 0.2309 \end{pmatrix}, B_{f2} = \begin{pmatrix} 1.3793 & 1.5699 \\ 1.5275 & 1.4030 \\ 1.5288 & 1.3936 \end{pmatrix}, E_{f2} = \begin{pmatrix} 0.0055 \\ 0.0046 \\ 0.0126 \end{pmatrix}^{\top}, F_{f2} = \begin{pmatrix} 1.4756 \\ 0.6147 \end{pmatrix}^{\top},$$

and the ℓ_1*-gain value is* $\gamma = 0.9644$ *and average dwell time switching satisfies* $\tau_a \geq 4.4728$. *Figures 2 and 3 denote the event-triggered output signal* $\tilde{y}(k)$ *and the quantified output signal* $\bar{y}(k)$ *for nonlinear switched positive systems. The simulations of the output signal* $\tilde{y}(k)$ *and* $\bar{y}(k)$

under different initial conditions are shown in Figure 4. Figure 5 provides the simulations of the output $z(k)$ and the estimated output $z_f(k)$ under the asynchronous switching signal. Figure 6 shows the event-triggering release interval. The simulations of $z(k)$ and $z_f(k)$ under different initial conditions are given in Figure 7.

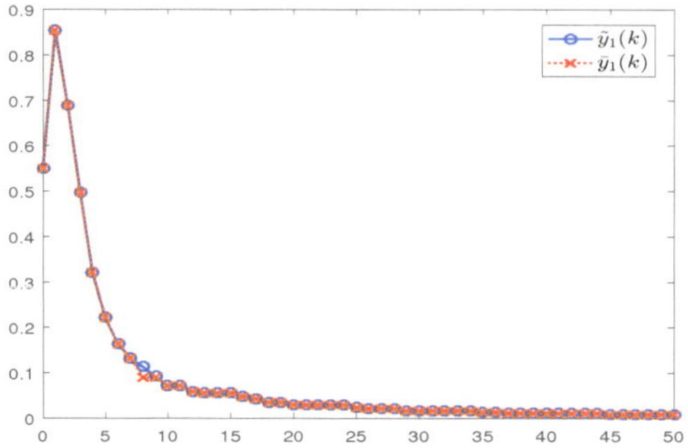

Figure 2. Event-triggered output signal $\tilde{y}_1(k)$ and quantified output signal $\bar{y}_1(k)$.

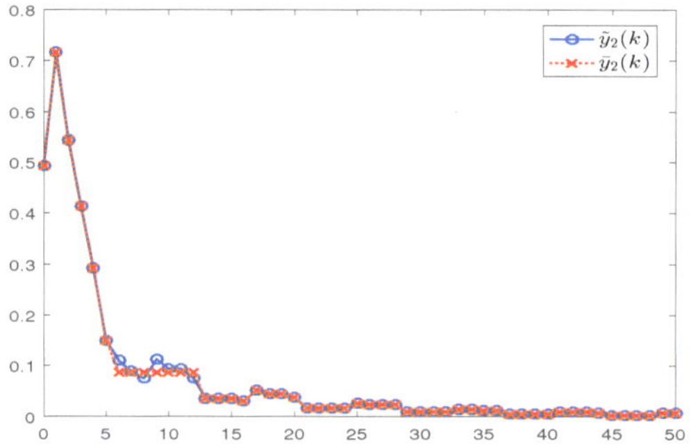

Figure 3. Event-triggered output signal $\tilde{y}_2(k)$ and quantified output signal $\bar{y}_2(k)$.

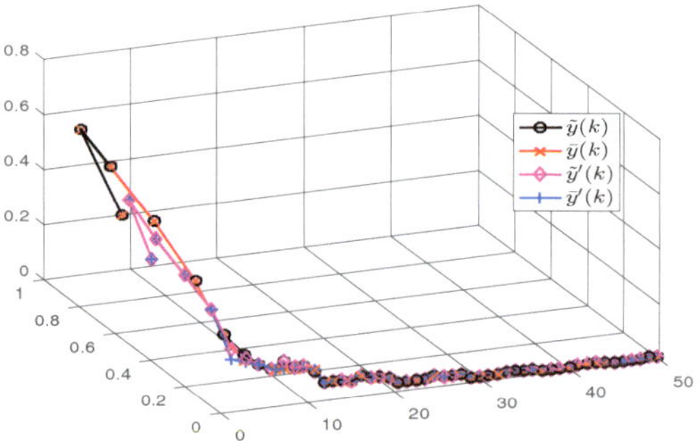

Figure 4. The simulations of $\tilde{y}(k)$ and $\bar{y}(k)$ under different initial conditions.

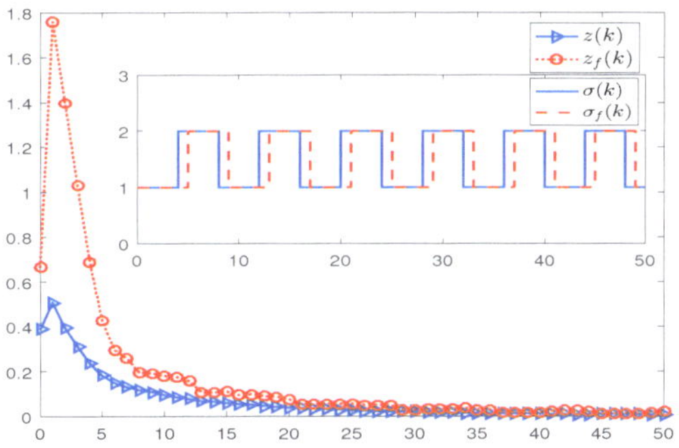

Figure 5. The simulations of $z(k)$ and $z_f(k)$ with an asynchronous switching signal.

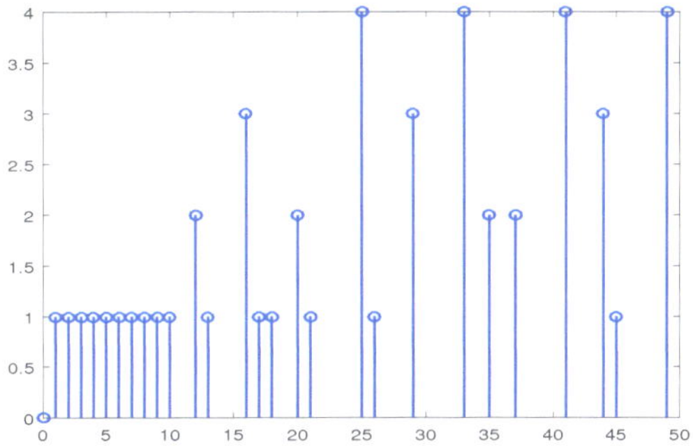

Figure 6. The event-triggering release instants and release intervals.

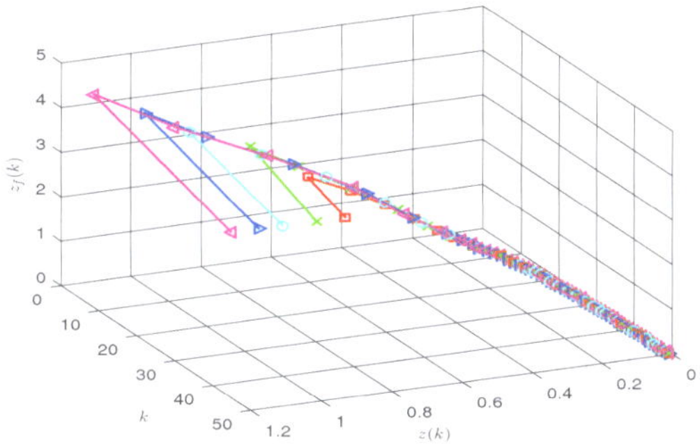

Figure 7. The simulations of $z(k)$ and $z_f(k)$ under different initial conditions.

Example 2. *Consider the system (1) with two subsystems:*

$$A_1 = \begin{pmatrix} 0.0894 & 0.1462 & 0.2763 \\ 0.1945 & 0.1573 & 0.2645 \\ 0.1497 & 0.0154 & 0.1637 \end{pmatrix}, B_1 = \begin{pmatrix} 0.1361 & 0.1104 \\ 0.1476 & 0.0632 \\ 0.0248 & 0.1353 \end{pmatrix}, C_1 = \begin{pmatrix} 0.2 & 0.2 & 0.3 \\ 0.3 & 0.2 & 0.1 \end{pmatrix},$$

$$D_1 = \begin{pmatrix} 0.3 & 0.2 \\ 0.2 & 0.4 \end{pmatrix}, E_1 = \begin{pmatrix} 0.1 & 0.1 & 0.2 \end{pmatrix}, F_1 = \begin{pmatrix} 0.4 & 0.5 \end{pmatrix},$$

$$A_2 = \begin{pmatrix} 0.1523 & 0.0496 & 0.1346 \\ 0.1817 & 0.1977 & 0.0412 \\ 0.1264 & 0.0741 & 0.1255 \end{pmatrix}, B_2 = \begin{pmatrix} 0.1450 & 0.2144 \\ 0.2145 & 0.1562 \\ 0.1855 & 0.1786 \end{pmatrix}, C_2 = \begin{pmatrix} 0.3 & 0.1 & 0.1 \\ 0.1 & 0.3 & 0.2 \end{pmatrix},$$

$$D_2 = \begin{pmatrix} 0.3 & 0.3 \\ 0.2 & 0.1 \end{pmatrix}, E_2 = \begin{pmatrix} 0.2 & 0.1 & 0.2 \end{pmatrix}, F_2 = \begin{pmatrix} 0.3 & 0.3 \end{pmatrix}.$$

Choose the same parameters as in Example 1. Furthermore, under dynamic event-triggering condition (41) and (42), we give $\eta_0 = 0.60, \varrho = 0.50$*. By Theorem 2, the filter gain matrices can be obtained:*

$$A_{f1} = \begin{pmatrix} 0.2388 & 0.2588 & 0.3569 \\ 0.2685 & 0.2962 & 0.3488 \\ 0.2454 & 0.2393 & 0.3122 \end{pmatrix}, B_{f1} = \begin{pmatrix} 0.2952 & 0.0460 \\ 0.2951 & 0.0451 \\ 0.3005 & 0.0550 \end{pmatrix}, E_{f1} = \begin{pmatrix} 0.0039 \\ 0.0034 \\ 0.1127 \end{pmatrix}^\top, F_{f1} = \begin{pmatrix} 1.3585 \\ 0.8493 \end{pmatrix}^\top,$$

$$A_{f2} = \begin{pmatrix} 0.1447 & 0.1751 & 0.2787 \\ 0.1513 & 0.1986 & 0.2497 \\ 0.1386 & 0.1620 & 0.2379 \end{pmatrix}, B_{f2} = \begin{pmatrix} 0.6693 & 0.7391 \\ 0.9234 & 0.6583 \\ 0.8987 & 0.6241 \end{pmatrix}, E_{f2} = \begin{pmatrix} 0.0660 \\ 0.0385 \\ 0.1896 \end{pmatrix}^\top, F_{f2} = \begin{pmatrix} 0.8772 \\ 0.5213 \end{pmatrix}^\top,$$

and the ℓ_1*-gain value is* $\gamma = 1.0796$ *and average dwell time switching satisfies* $\tau_a \geq 4.4728$*. The dynamic event-triggered output signal* $\tilde{y}(k)$ *and the quantified output signal* $\bar{y}(k)$ *are given in Figures 8 and 9. Figure 10 shows the simulation results of the output* $z(k)$ *and the estimated output* $z_f(k)$ *under the asynchronous switching signal. Figure 11 shows the dynamic event-triggering release interval. The simulation results of* $z(k)$ *and* $z_f(k)$ *under different initial conditions are shown in Figure 12.*

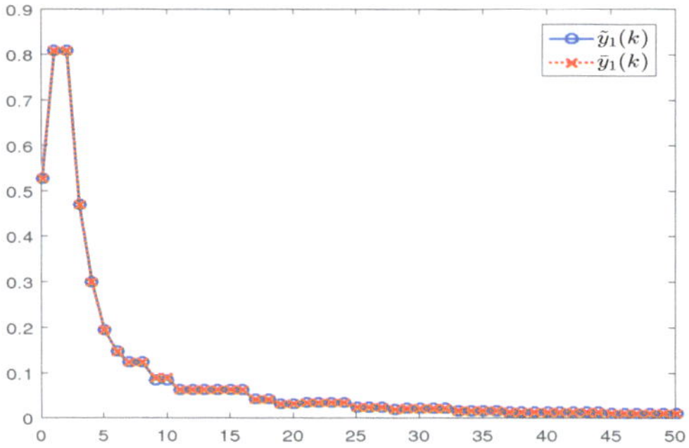

Figure 8. The dynamic event-triggered output signal $\tilde{y}_1(k)$ and quantified output signal $\bar{y}_1(k)$.

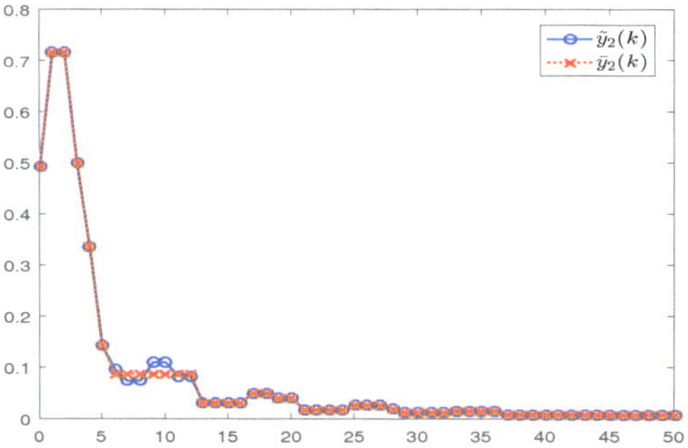

Figure 9. The dynamic event-triggered output signal $\tilde{y}_2(k)$ and quantified output signal $\bar{y}_2(k)$.

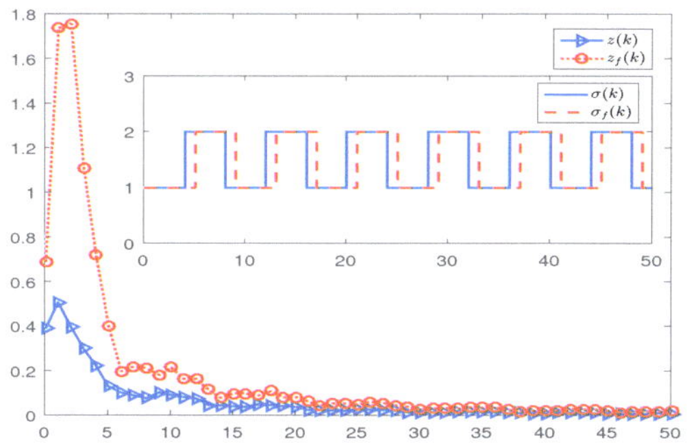

Figure 10. The simulations of $z(k)$ and $z_f(k)$ with asynchronous switching signal.

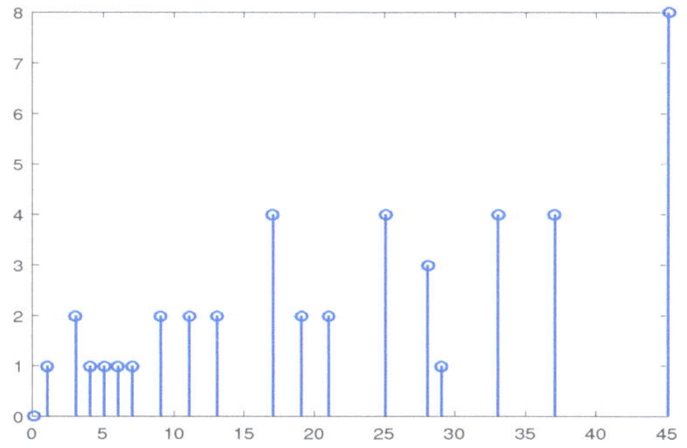

Figure 11. The dynamic event-triggering release instants and release intervals.

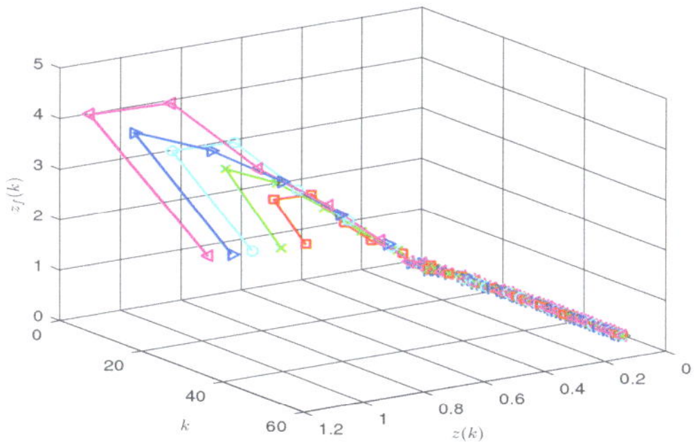

Figure 12. The simulations of $z(k)$ and $z_f(k)$ under different initial conditions.

5. Conclusions

In this paper, we investigated an event-triggered asynchronous filter of nonlinear switched positive systems with output quantization. Based on static and dynamic event-triggering mechanisms, an asynchronous filter was proposed using the matrix decomposition technique. The positivity and L_1-gain stability of the underlying systems were guaranteed by using a linear copositive Lyapunov function and linear programming approach. Then, the issue of output quantization is solved under a quantizer.

Author Contributions: Conceptualization, J.Z. and P.L.; methodology, P.L.; software, P.L.; validation, S.Z. and P.L.; formal analysis, J.Z.; investigation, S.Z., P.L. and J.Z.; resources, S.Z. and P.L.; data curation, P.L.; writing—original draft preparation, S.Z. and P.L.; writing—review and editing, J.Z.; visualization, S.Z. and P.L.; supervision, J.Z.; project administration, P.L. and J.Z. All authors have read and agreed to the published version of the manuscript.

Funding: This work was supported by the Fundamental Research Funds for the Provincial Universities of Zhejiang (Nos. GK219909299001-002, GK219909299001-403, and GK209907299001-007), the National Natural Science Foundation of China (No. 62003119), and the Natural Science Foundation of Zhejiang Province, China (No. LQ21F030014).

Institutional Review Board Statement: Not applicable.

Informed Consent Statement: Not applicable.

Data Availability Statement: Not applicable.

Conflicts of Interest: The authors declare no conflict of interest. The funders had no role in the design of the study; in the collection, analyses, or interpretation of data; in the writing of the manuscript; or in the decision to publish the results.

References

1. Farina, L.; Rinaldi, S. *Positive Linear Systems: Theory and Applications*; John Wiley & Sons: Hoboken, NJ, USA, 2000.
2. Kaczorek, T. *Positive 1D and 2D Systems*; Springer: London, UK, 2001.
3. Hespanha J.P.; Morse A.S. Stability of switched systems with average dwell-time. In Proceedings of the 38th IEEE Conference on Decision and Control, Phoenix, AZ, USA, 7–10 December 1999; Volume 3, pp. 2655–2660.
4. Dinh, T.N.; Marouani, G.; Raïssi, T.; Wang, Z.; Messaoud, H. Optimal interval observers for discrete-time linear switched systems. *Int. J. Control* **2020**, *93*, 2613–2621. [CrossRef]
5. Jadbabaie, A.; Lin, J.; Morse, A. Coordination of groups of mobile autonomous agents using nearest neighbor rules. *IEEE Trans. Autom. Control* **2003**, *48*, 988–1001. [CrossRef]
6. Liberzon, D. *Switching in Systems and Control*; Springer: Berlin/Heidelberg, Germany, 2003.
7. Silva-Navarro, G.; Alvarez-Gallegos, J. On the property sign-stability of equilibria in quasimonotone positive nonlinear systems. In Proceedings of 1994 33rd IEEE Conference on Decision and Control, Lake Buena Vista, FL, USA, 14–16 December 1994; Volume 4, pp. 4043–4048.
8. Ocampo-Martinez, C.; Puig, V.; Cembrano, G.; Quevedo, J. Application of predictive control strategies to the management of complex networks in the urban water cycle. *IEEE Control. Syst. Mag.* **2013**, *33*, 15–41.
9. Fornasini, E.; Valcher, M.E. Stability and stabilizability criteria for discrete-time positive switched systems. *IEEE Trans. Autom. Control* **2011**, *57*, 1208–1221. [CrossRef]
10. Pastravanu, O.C.; Matcovschi, M.H. Max-type copositive Lyapunov functions for switching positive linear systems. *Automatica* **2014**, *50*, 3323–3327. [CrossRef]
11. Xiang, W.; Lam, J.; Shen, J. Stability analysis and L_1-gain characterization for switched positive systems under dwell-time constraint. *Automatica* **2017**, *85*, 1–8. [CrossRef]
12. Zhang, J.; Han, Z.; Zhu, F. L_1-gain analysis and control synthesis of positive switched systems. *Int. J. Syst. Sci.* **2015**, *46*, 2111–2121. [CrossRef]
13. Fainshil, L.; Margaliot, M.; Chigansky, P. On the stability of positive linear switched systems under arbitrary switching laws. *IEEE Trans. Autom. Control* **2009**, *54*, 897–899. [CrossRef]
14. Blanchini, F.; Colaneri, P.; Valcher, M.E. Co-positive Lyapunov functions for the stabilization of positive switched systems. *IEEE Trans. Autom. Control* **2012**, *57*, 3038–3050. [CrossRef]
15. Xiang, M.; Xiang, Z.; Karimi, H.R. Stabilization of positive switched systems with time-varying delays under asynchronous switching. *Int. J. Control. Autom. Syst.* **2014**, *512*, 939–947. [CrossRef]
16. Wang, D.; Wang, Z.; Li, G.; Wang, W. Distributed filtering for switched nonlinear positive systems with missing measurements over sensor networks. *IEEE Sens. J.* **2016**, *16*, 4940–4948. [CrossRef]
17. Zhang, J.; Zhao, X.; Cai, X. Absolute exponential L_1-gain analysis and synthesis of switched nonlinear positive systems with time-varying delay. *Appl. Math. Comput.* **2016**, *284*, 24–36.
18. Zhang, J.; Raïssi, T. Saturation control of switched nonlinear systems. *Nonlinear Anal. Hybrid Syst.* **2019**, *32*, 320–336. [CrossRef]
19. Liu, X.; Yuan, S. Reduced-order fault detection filter design for switched nonlinear systems with time delay. *Nonlinear Dyn.* **2012**, *67*, 601–617. [CrossRef]
20. Zheng, Q.; Zhang, H. H_∞ filtering for a class of nonlinear switched systems with stable and unstable subsystems. *Signal Process.* **2017**, *141*, 240–248. [CrossRef]
21. Aleksandrov, A.; Aleksandrova, E.; Zhabko, A. Stability analysis of some classes of nonlinear switched systems with time delay. *Int. J. Syst. Sci.* **2017**, *48*, 2111–2119. [CrossRef]
22. Baleghi, N.A.; Shafiei, M.H. Stability analysis and stabilization of a class of discrete-time nonlinear switched systems with time-delay and affine parametric uncertainty. *J. Vib. Control* **2019**, *25*, 1326–1340. [CrossRef]
23. Huo, X.; Ma, L.; Zhao, X.; Zong, G. Observer-based fuzzy adaptive stabilization of uncertain switched stochastic nonlinear systems with input quantization. *J. Frankl. Inst.* **2019**, *356* 1789–1809. [CrossRef]
24. Yang, J.; Chen, Y.; Zheng, Z.; Qian, W. Robust adaptive state estimation for uncertain nonlinear switched systems with unknown inputs. *Trans. Inst. Meas. Control.* **2018**, *40*, 1082–1091. [CrossRef]

25. Niu, B.; Zhao, P.; Liu, J.D.; Ma, H.J.; Liu, Y.J. Global adaptive control of switched uncertain nonlinear systems: An improved MDADT method. *Automatica* **2020**, *115*, 108872. [CrossRef]
26. Xiang, W.; Xiao, J. H_∞ filtering for switched nonlinear systems under asynchronous switching. *Int. J. Syst. Sci.* **2011**, *42*, 751–765. [CrossRef]
27. Zheng, Q.; Xu, S.; Zhang, Z. Asynchronous nonfragile H_∞ filtering for discrete-time nonlinear switched systems with quantization. *Nonlinear Anal. Hybrid Syst.* **2020**, *37*, 100911. [CrossRef]
28. Ren, Y.; Er M.J.; Sun, G. Asynchronous ℓ_1 positive filter design for switched positive systems with overlapped detection delay. *IET Control Theory Appl.* **2017**, *11*, 319–328. [CrossRef]
29. Wang, B.; Zhang, H.; Wang, G.; Dang, C. Asynchronous H_∞ filtering for linear switched systems with average dwell time. *Int. J. Syst. Sci.* **2016**, *47*, 2783–2791. [CrossRef]
30. Li, Y.; Du, W.; Xu, X.; Zhang, H.; Xia, J. A novel approach to L_1 filter design for asynchronously switched positive linear systems with dwell time. *Int. J. Robust Nonlinear Control* **2019**, *29*, 5957–5978. [CrossRef]
31. Brockett, R.W.; Liberzon, D. Quantized feedback stabilization of linear systems. *IEEE Trans. Autom. Control* **2000**, *45*, 1279–1289. [CrossRef]
32. Wakaiki, M.; Yamamoto, Y. Stabilization of switched linear systems with quantized output and switching delays. *IEEE Trans. Autom. Control* **2016**, *62*, 2958–2964. [CrossRef]
33. Cheng, J.; Park, J. H.; Cao, J.; Zhang, D. Quantized H_∞ filtering for switched linear parameter-varying systems with sojourn probabilities and unreliable communication channels. *Inf. Sci.* **2018**, *466*, 289–302. [CrossRef]
34. Yang, L.; Guan, C.; Fei, Z. Finite-time asynchronous filtering for switched linear systems with an event-triggered mechanism. *J. Frankl. Inst.* **2019**, *356*, 5503–5520. [CrossRef]
35. Zong, G.; Ren, H.; Karimi, H.R. Event-triggered communication and annular finite-time H_∞ filtering for networked switched systems. *IEEE Trans. Cybern.* **2020**, *51*, 309–317. [CrossRef]
36. Liu, X.; Su, X.; Shi, P.; Nguang, S.K.; Shen, C. Fault detection filtering for nonlinear switched systems via event-triggered communication approach. *Automatica* **2019**, *101*, 365–376. [CrossRef]
37. Wang, Y.L.; Shi, P.; Lim, C.C.; Liu, Y. Event-triggered fault detection filter design for a continuous-time networked control system. *IEEE Trans. Cybern.* **2016**, *46*, 3414–3426. [CrossRef] [PubMed]
38. Zhang, J.; Zheng, G.; Feng, Y.; Chen, Y. Event-triggered state-feedback and dynamic output-feedback control of positive Markovian jump systems with intermittent faults. *IEEE Trans. Autom. Control* **2022**. [CrossRef]
39. Xiao, S.; Zhang, Y.; Xu, Q.; Zhang, B. Event-triggered network-based L_1-gain filtering for positive linear systems. *Int. J. Syst. Sci.* **2017**, *48*, 1281–1290. [CrossRef]
40. Zhang, J.; Raïssi, T.; Shao, Y.; Cai, X. Event-triggered filter design of positive systems with state saturation. *IEEE Syst. J.* **2020**, *15*, 4281–4285. [CrossRef]
41. Shao, Y.; Jiao, C.; Zhang, S.; Zhang, J. Event-triggered filter of switched positive systems with state saturation. In Proceedings of the 2020 7th International Conference on Information, Cybernetics, and Computational Social Systems, Guangzhou, China, 13–15 November 2020; pp. 1–6.
42. Girard, A. Dynamic triggering mechanisms for event-triggered control. *IEEE Trans. Autom. Control* **2014**, *60*, 1992–1997. [CrossRef]
43. Gu, Z.; Tian, E.; Liu, J. Adaptive event-triggered control of a class of nonlinear networked systems. *J. Frankl. Inst.* **2017**, *354*, 3854–3871. [CrossRef]
44. Li, Q.; Shen, B.; Wang, Z.; Sheng, W. Recursive distributed filtering over sensor networks on Gilbert-Elliott channels: A dynamic event-triggered approach. *Automatica* **2020**, *113*, 108681. [CrossRef]
45. Liu, D.; Yang, G.H. Dynamic event-triggered control for linear time-invariant systems with-gain performance. *Int. J. Robust Nonlinear Control* **2019**, *29*, 507–518. [CrossRef]
46. Luo, S.; Deng, F.; Chen, W.H. Dynamic event-triggered control for linear stochastic systems with sporadic measurements and communication delays. *Automatica* **2019**, *107*, 86–94. [CrossRef]
47. Lu, R.; Wu, F.; Xue, A. Networked control with reset quantized state based on Bernoulli processing. *IEEE Trans. Ind. Electron.* **2013**, *61*, 4838–4846. [CrossRef]
48. Dong, S.; Su, H.; Shi, P.; Lu, R.; Wu, Z.G. Filtering for discrete-time switched fuzzy systems with quantization. *IEEE Trans. Fuzzy Syst.* **2016**, *25*, 1616–1628. [CrossRef]
49. Wang, Y.; Han, Q. Network-based modelling and dynamic output feedback control for unmanned marine vehicles in network environments. *Automatica* **2018**, *91*, 43–53. [CrossRef]

MDPI AG
Grosspeteranlage 5
4052 Basel
Switzerland
Tel.: +41 61 683 77 34

Mathematics Editorial Office
E-mail: mathematics@mdpi.com
www.mdpi.com/journal/mathematics

www.ingramcontent.com/pod-product-compliance
Lightning Source LLC
LaVergne TN
LVHW072354090526
838202LV00019B/2541